MICROBIAL BIOGEOCHEMISTRY

Microbial Biogeochemistry

JAMES E. ZAJIC

Faculty of Engineering Science
The University of Western Ontario
London, Canada

1969

ACADEMIC PRESS   New York and London

ACADEMIC PRESS, INC.
111 Fifth Avenue, New York, New York 10003

United Kingdom Edition published by
ACADEMIC PRESS, INC. (LONDON) LTD.
Berkeley Square House, London W.1

LIBRARY OF CONGRESS CATALOG CARD NUMBER: 69-18343

PRINTED IN THE UNITED STATES OF AMERICA

This book is dedicated to
Margaret Sue Clark

PREFACE

I have been interested in soil microbiology for eighteen years. For the last six years, I have studied all publications possible which relate microbiology to geology. Also, during this period, because of my position in research at Kerr-McGee Corporation, I completed biochemical studies on the recovery of copper, uranium, and other metals from low grade ores. This exposure gave me a chance to integrate much of the literature and information of the two important disciplines. Microbiology and geology are combined herein under the heading of "Microbial Biogeochemistry." This term is not new.

The uniformitarian principle proposed by James Hutton of Edinburgh in 1785 is assumed to apply to biogeochemistry as well as to geochemistry to which it was first applied. This principle is based on the premise that rocks formed millions of years ago can be understood and explained in accordance with biochemical and chemical processes now in operation. This concept provides a basis for elucidating the biochemical reactions involving mineral solubilization and deposition.

A general pattern has been utilized in developing the subject matter pertaining to the more important elements. The reader is first introduced to some geochemistry of the minerals involved. With this as a background, biogeochemistry is developed. An attempt is made to present all biochemical reactions involved with the element or mineral under discussion; this is particularly true for the unicellular forms of life. Biogeochemical cycles are presented where evidence is sufficient to support their existence or where bioenergetics support a possible reaction. All microbial reactions discussed are important in weathering and, therefore, are not treated as a special topic.

Some data are given on biogeochemical prospecting, but so little has been done in this area in which microbes are used as an investigative tool that the subject did not warrant a separate chapter. The chapter on sulfur is placed early in the text because the biochemical changes to sulfur are many, and in its different oxidative states sulfur has a profound effect on the solubilization and deposition of other minerals. The same is true for

iron. Since detailed accounts for the cultivation and handling of the microbes are found in the many references, only a few media are described in
detail. The reader will note great similarity between the chapters on bio-
corrosion and bioelectricity. In spite of this, I preferred to treat these
topics separately to emphasize the distinctiveness of the two processes.

The chapter on the biogeochemistry of other elements is short because
suprisingly little is known about the biogeochemistry of strontium, zirconium, hafnium, etc.

It would be impossible to personally acknowledge all those contribut-
ing to this book. I owe a great debt to Mr. D. A. McGee, President and
Chairman of Kerr-McGee Corporation, who stimulated much of my
thinking and introduced me to the field of biogeochemistry.

Appreciation is also extended to Dr. R. Vining, for his contributions
and criticisms on copper, and to Norma Dean Smith, for her efforts on
aluminum. Thanks are due also to Dennis Doughty and Carolyn Kelly who
assisted me with various chapters.

One of the most efficient librarians I have ever met is Lorna C. Wilkie,
now with the U.S. Geological Surveys; her aid is gratefully acknowledged.

London, Ontario   JAMES ZAJIC
May, 1969

CONTENTS

4. Microbial Metabolism of Inorganic Sulfur

5. Sulfate Reduction and Sulfides

6. Biotransformation of Iron

7. Copper Biogeochemistry

8. Microbes and Zinc

19. Biogeochemical Observations on Other Elements

1

ENVIRONMENTAL BIOPHYSICOCHEMISTRY

Microbes are ubiquitous on this planet. Though there are places where microbes do not exist, the chemical and physical conditions there are usually quite abnormal. Pressure is not a problem: barophilic microbes are common. Temperatures exceeding 70°C kill most microbes, but thermophilic bacteria and the spores of some bacteria can tolerate temperatures of 100°C. Low temperatures have both a lethal and preservative action on the bacteria which grow best at moderate temperatures, but there are psychrophilic microbes which grow rapidly at temperatures from 5–20°C under conditions that would kill a mesophile. Halophilic microbes survive in saturated salt solutions or in sugar solutions which exert great osmotic pressure on their cell membranes. The adaptability of microbes gives them the potential to readily develop populations resistant to toxic chemicals. A high natural mutation rate also aids in the selection of mutant populations with highly specialized biochemical properties.

The geologist subdivides our planet into atmosphere, hydrosphere, and lithosphere. To this the biologist adds the biosphere, which, in the oceans, may extend to a depth of 33,000 feet, in the lithosphere to a depth of 6000 feet, and probably reaches several miles into the atmosphere. If anything, these ranges of activity probably represent a conservative estimate.

Dr. Heinz J. Dombrowski of Germany has isolated bacteria from rock salt from the Kali and Zechstein deposits of the Permian period. The bacteria dormant in these samples were estimated to be 180 million years old. Dr. George Claus of the United States reported finding dead microbes possibly microfossils, in the meteorites Orgueil, which fell in 1864, and Ivuna which fell in 1938. Bacteria have also been isolated from Antarctic ice which dated back to the Shackleton Expedition of 1917.

In analyzing the geochemical situations which exist and have existed on our planet, one encounters environments which are not too different from those existing on Mercury, Venus, and Mars. Dr. Salisbury (1962) makes an excellent comparison of the geochemical environments of these planets with earth and concludes that the possibility of a Martian biology is very real.

Since some microbes select and concentrate isotopes, isotopic fractionation studies indicate that microbes have been active on the planet earth since Proterozoic, Paleozoic, and Mesozoic times. The millions of years which have elapsed permit biochemical activities in the biosphere, no matter how slight, to continue to exert a profound geochemical influence on the solubilization and deposition of minerals.

The major rock formations are igneous, sedimentary, and metamorphic. Igneous rocks are formed from molten material in which differential crystallization occurs during cooling. Microbes are not likely to have played a role in the deposition of such rocks. Of the latter two types, even though some heat and pressure may have been involved in their formation, microbes were probably active in carrying out biochemical reactions which influenced mineral deposition or solubilization in ground water passing through the rock. The role of the microbe in these processes forms the basis of this book.

All minerals and rocks are affected by water, which is the prime transporting vehicle for minerals. Water forms part of a triad (Fig. 1.1) in which microbes and minerals are its partners.

In the presence of water, an equilibrium is established between the mineral ions in solution and those in the crystalline state. The microbe acts as a catalyst or functions in other ways to influence this equilibrium. Water-insoluble fractions are referred to as gangue or host rock, and, under some conditions, the gangue rock may be leached out to enrich the mineral content of the remaining rock.

Minerals can be classified in terms of the host or gangue rock. The term syngenesis is used to describe mineral deposits formed concurrently with the host rock, while epigenesis is used to describe minerals deposited subsequent to the deposition of gangue. Minerals deposited independently of the other rocks in the immediate environment are said to have been formed authigenically.

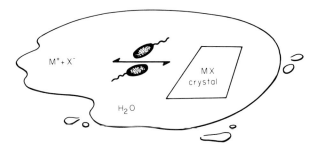

FIG. 1.1. Triad relationship of water, minerals, and microbes.

Marine microbes contribute considerable activity to the biosphere. Since practically all sedimentary deposits are of marine origin, a fossil analysis provides information concerning their deposition. Biologists classify the sea as benthic or pelagic, whereby benthic describes a bottom environment and pelagic an open-water environment. The plant and animal life of the sea are classified as benthic, nektonic, or planktonic or, respectively, bottom dwellers, swimming forms, or floating organisms. This terminology is used more in describing the more complex forms of plant and animal life, although it has some application in describing the ecological habitats of unicellular life.

Microbes can be broadly classified as autotrophic and heterotrophic, depending upon their energy source. Autotrophic bacteria obtain energy by oxidizing minerals or inorganic substances; cellular carbon is obtained from the fixation of carbon dioxide. Heterotrophic bacteria obtain energy from the oxidation of organic compounds. Carbon dioxide is not required by heterotrophs, however, many do fix CO_2. Some microbes are aerobic, utilizing O_2 for growth and as a terminal electron acceptor; others grow anaerobically, utilizing sulfate, nitrate, carbohydrates, etc., as terminal electron acceptors. Photosynthetic microbes (algae and bacteria) exist, obtaining their energy from the sun and synthesizing cellular material from carbon derived from the CO_2 present in the air. Phototrophs function by means of specialized pigments which act as chlorophyll does in plants to utilize the sun's energy. Oxygen is produced as a by-product. The entire microbial world falls under one or more of these classifications. In the bios-complex discussed herein, algae, bacteria, fungi, and protozoa are emphasized. Viruses, higher plants, and animals have been included only where their effect is demonstrable.

Frequent reference is made to the elemental content of the solar atmosphere, the total earth, the earth crust, and seawater. The most complete data available are shown in Table 1.I which is taken from Lowman (1963). For additional information on the abundance of elements see Biogeochemical Periodic Table (pp. 20–21), this volume.

TABLE 1.I

ABUNDANCE[a] OF MAJOR ELEMENTS IN THE SOLAR ATMOSPHERE, THE HYDROSPHERE, AND LITHOSPHERE OF THE EARTH[b]

Element	Solar atmosphere	Total earth	Crust of earth	Seawater[c]
Hydrogen	53.0	Trace	0.14	10.8
Helium	42.0	Trace	0.00000003	0.0000000005
Boron	Trace	Trace	0.0003	0.0005
Carbon	0.012	Trace	0.03	0.003
Nitrogen	0.031	Trace	0.005	0.00005
Oxygen	4.7	28.0	46.6	87.5
Sodium	0.0024	0.14	2.8	1.05
Magnesium	0.043	17.0	2.1	0.13
Aluminum	0.0031	0.4	8.1	0.000001
Silicon	0.029	13.0	27.7	0.0003
Phosphorus	Trace	0.03	0.12	0.000007
Sulfur	0.014	2.7	0.05	0.09
Chlorine	Trace	Trace	0.02	1.90
Potassium	0.00033	0.07	2.6	0.04
Calcium	0.0036	0.61	3.6	0.04
Vanadium	0.000031	Trace	0.02	0.0000002
Manganese	0.00086	0.09	0.10	0.0000002
Iron	0.167	35.0	5.0	0.000001
Cobalt	0.00034	0.20	0.002	0.00000005
Nickel	0.0029	2.7	0.01	0.0000002
Copper	0.000058	Trace	0.01	0.0000003
Zinc	0.00021	Trace	0.01	0.000001
Molybdenum	Trace	Trace	0.0015	0.000001
Iodine	Trace	Trace	0.00003	0.000006

[a] In gm/100 gm.
[b] After Lowman (1963).
[c] Date from J. D. Isaacs (personal communication, 1961).

The term environmental biophysicochemistry as used here relates microbiology to geophysical conditions existing on this planet, and in particular to the extreme physicochemical conditions which have been reported to support some type of life. Primary emphasis has been placed upon microbes because of their environmental and physiological diversity as compared to higher plants and animals.

This subject has been reviewed most recently by Vallentyne (1967). It is of great current interest because of the controversial question as to whether life exists on other planets within the solar system. While much opinion favors a biocentric view, a geocentric position can be supported (Tikov, 1955). In examining a Martian environment, one finds it imposes rather severe physical and chemical conditions for microbes to survive. The average temperature is 50°C below that on earth; a daily fluctuation of temperature of 60°C occurs at the equator. The Martian atmosphere is richer in CO_2 and lower in O_2 than the earth; the pressure is lower, water more scarce, and the level of ultraviolet radiation higher. Analyses of the physicochemical conditions existing on other planets point up similar extremes.

What physicochemical extremes found or reported to have existed within our own planet earth support life? In analyzing this question the important parameters are examined directly. These are temperature, E_h and pH, salinity, pressure, and water.

BIOLOGICAL CONTINUUM

Before discussing these parameters, the concept of a biological continuum must be introduced. In any given parameter or group of parameters selected, microbes will be found to grow within certain minimal and maximal levels. The range over which growth occurs forms a biological continuum. This biological continuum can be compared to a standard curve in which most microbial growth occurs within a restricted range of the optima with limiting growth occurring in the tails representing the lower and upper extremities. It is these extremities with which we will be concerned. If we accept a biological continuum as an operating guide for all ecological habitats controlled by defined selective pressures or environmental variables, space exploration and analysis for life on other planets should be made using the physicochemical limits which exist on such planets as a guide for evaluating the possibility of life. This is far more valid than selecting a norm established by physicochemical conditions existing here on earth.

Temperature

There is a considerable amount of literature on this subject, but most studies are isolated entities and the reports quite old. Many times microorganisms have not been identified, and observations were made in conjunction with some other study. The lowest temperature for cellular growth will probably be found to be less than $-15°C$ and the highest greater than 104°C. These temperature extremes exceed the stability range for pure water under one atmosphere of pressure. Except for atmospheric water, most water in nature is impure and is found under conditions of variable pressure. The stability range in which water exists in a liquid state can be extended by adding solutes or by varying pressures. Ice does not form in seawater with a salinity of 35 parts per thousand until the temperature drops below $-1.9°C$, and water at 105°C will not boil under a hydrostatic pressure of 200 atm. Both Redfort (1932) and ZoBell (1934) have cultivated numerous marine bacteria at a temperature of less than 0°C (Table 1.II). Horowitz-Wlassowa and Grinberg (1933) cultivated five bacterial species at $-5.0°C$; some bacteria multiply in ice cream at $-10°C$ (Weinzirl and

TABLE 1.II

TEMPERATURE EXTREMES SUPPORTING MICROBIAL LIFE

Microbes	Temperature (°C)	Reference
I. Less than 0°C		
Sixty-five marine bacteria	-7.5	Bedford (1933)
Seventy-six marine bacteria	0	ZoBell (1934)
Five bacteria	-5.0	Horowitz-Wlassowa and Grinberg (1933)
Bacteria (in ice cream)	-10.0	Weinzirl and Gerdeman (1929)
Bacteria (in fish)	-11.0	Redfort (1932)
Sporotrichum carnis	-10.0	Haines (1931)
Chaetostylum fresenii	-10.0	Bidault (1921)
Hormodendron clasdosporoides	-10.0	Bidault (1921)
Fungi	-8.0	Tchistiakov and Botcharova (1938)
Pyramimonas (a flagellate)	-7.7	Zernow (1944)
Pyramidomonas and *Dunaliella*	-15.0	Zernow (1944)
Pseudomonads and fungi (osmophilic)	-18 to -20	Borgstrom (1961)
Aspergillus glaucus (in glycerol)	-18	Borgstrom (1961)
Pink yeasts (in oysters)	-18 to -30	McCormack (1950)
Lichens	-20 to -40	Jumelle (1891)
II. Above 60°C		
Bacillus stearothermophilus	80	Baker *et al.* (1955)
Sulfate-reducing thermophiles	65–85	ZoBell (1958)
Sulfate-reducing thermophiles (at 1000 atm)	104	ZoBell (1958)

Gerdeman, 1929) and on refrigerated fish at −11°C (Redfort, 1932). The fungus *Sporotrichum carnis* grows slowly at −10°C (Haines, 1931), and other fungi will grow at −8°C (Tchistiakov and Botcharova, 1938). A flagellate, probably a species of *Pyramidomonas*, was observed by Zernow (1944) swimming in salt water at −7.7°C under an ice cover in Lake Bolpash, Kazakh S.S.R. *Dunaliella* was also observed swimming in soft ice at −15°C.

Molds and pseudomonads grow in concentrated fruit juices and sugar solutions at −18° to −20°C and *Aspergillus glaucus* grows in glycerol at −18°C (Borgstrom, 1961). Pink yeasts grow on oysters between −18° and 30°C (McCormack, 1950).

Older literature indicates that lichens grow and carry out some photosynthetic reactions at −20° to −40°C (Jumelle, 1891). These later findings need correlation. The temperature of 90% of the sea is below 5°C; thus, this environment continually selects microbes which tolerate low temperatures.

Thermophilic microbes have been reported in numerous ecological habitats, such as hot springs, growing at temperatures between 60–85°C (Copeland, 1938; Precht *et al.*, 1955; Allen, 1960). Strains of *Bacillus stearothermophilus* grow well at 80°C. Spore-formers are known to tolerate much higher temperatures. A thermophilic sulfate-reducing bacterium isolated from a subterranean deposit at depths of 6000–12,000 feet has been cultured at 65–85°C (ZoBell, 1958). The *in situ* temperature of this deposit was 60–105°C, and it was under hydrostatic head pressure of 200–400 atm. One culture grew at 1000 atm and produced H_2S at 104°C.

E_h and pH

The environmental limits of E_h and pH for growth and reproduction of microbes have been discussed quite thoroughly by Baas-Becking *et al.* (1956, 1960). The E_h and pH range for the growth of a variety of microbes are summarized in Figure. 1.2. The E_h values do not always represent truly reversible potentials, but they are fairly accurate data. If the E_h and pH characteristic for all these ecological groups are superimposed, a considerable amount of overlapping occurs. On the E_h scale the range extends from 850 to −450 mv, while the pH ranges from 1.0 to 10.2. Vallentyne (1967) emphasizes that these do not represent true extremes because the authors only considered data where paired measurements of E_h and pH were available.

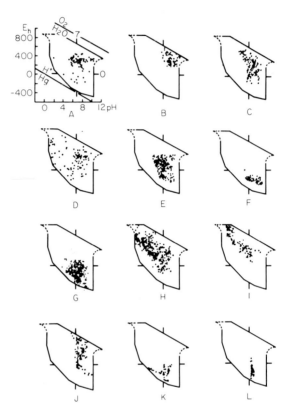

FIG. 1.2. E_h and pH range for microbial growth: (A) green algae and diatoms, (B) *Dunaliella*, (C) *Enteromorpha*, (D) blue-green algae, (E) photosynthetic purple bacteria, (F) photosynthetic green bacteria, (G) sulfate-reducing bacteria, (H) thiobacteria, (I) iron bacteria, (J) denitrifying bacteria, (K) heterotrophic bacteria, and (L) methane bacteria.

It is well known that thiobacilli grow well between pH 1.0 and 3.0. Growth has been reported at pH 0.5 and at values approaching zero. The alga *Cyanidium caldarium*, isolated from a hot spring containing 0.1 N H_2SO_4, has been cultivated in 1.0 N H_2SO_4.

Numerous microbes grow at a pH of 10.0 and more. In one of his early reports, Johnson (1923) demonstrated limited growth of *Penicillium variable* at pH 10.0 to 11.1. *Fusarium bullatum* and *Fusarium oxysporum* were limited by pH values of 9.2 to 11.2. The alkaline lakes of Kenya support populations of algae, rotifers, and copepodes (Jenkin, 1936).

The lakes of Elementeita and Nakaru at pH 10 to 11 contain high concentrations of the blue-green alga *Arthropsira platensis*. Cultures of *Nitrobacter* and *Nitrosomonas* multiply at pH 13.0 but not at 13.4 (Meek and Lipman, 1922). Ludox, a 30% solution of SiO_2 manufactured by DuPont, is stabilized by adding ammonia to give a pH of 13.0. Kingsbury (1954) reported a blue-green alga *Plectonema nostocorum* in solutions of Ludox.

Salinity

Halophilism is a common characteristic of microbes. A biological continuum also exists for this parameter. Heterotrophic bacteria will grow in double distilled water (Kalinenko, 1957) containing only 70 μg of organic matter per liter. At the upper extreme *Aspergillus oryzae* and *Aspergillus terricola* will grow in 4.1 M $MgSO_4$ which is approximately 50% salt (Johnson, 1923). Salt linans and saturated brines are found in so many habitats and used in so many industrial processes that diverse reports have been made on halophilic microbes. The Dead Sea is about 30% salt, and many microbes have been isolated from this ecological habitat (Wilkansky, 1936). Solar evaporation ponds often show fragmented patches and discolorations due to halophilic bacterial and algae growth.

Halophilism is discussed at length in Chapter 17.

Pressure

Although a few individuals such as ZoBell have evaluated and pioneered investigations on barophilic microbes, the primary bio-significance of this parameter still remains to be evaluated. Strughold in 1961 cultivated soil bacteria at a total pressure of 0.1 of the earth's atmosphere, a pressure presumed to be that existing on Mars. At the other extreme ZoBell and Morita (1956) have cultivated barophilic microbes at 1000 atm. Of the many barophilic cultures isolated, most would not grow at a few hundred atmospheres; other cultures were not affected by alternate compression and decompression.

Water

Water is one of the most commonly occurring compounds found in conjunction with life on this planet. Water depletion restricts growth but does not prevent it. Many lichens grow on barren rocks deriving much of their moisture from the atmosphere. Microbes are also found as contaminants in jet fuels which contain very little water. Even in these environments, water is still a vital protoplasmic constituent. If moisture becomes too limited in the environment or if it is removed from a cell, dormancy results. Many microbes survive extreme droughts as do bacterial and fungal spores. *Pleurococcus vulgaris* withstands prolonged periods of drought (Fritch, 1922; Fritch and Haines, 1923; Zeuch, 1934). The fungus *Aspergillus glaucus* grows well on substances where the activity of water (Aw) is as low as 0.65–0.70 (see Chapter 17). The water dependence of microbes on this planet may be directly correlated with its abundance. The types of life on other planets may be determined by a predominance of one or more other liquids which possess some of the qualities comparable to water.

Koyrdum and Bobchenko (1959) believe that many microbes actually reproduce in air, moisture being obtained from the atmosphere. Although the atmosphere is not an optimal medium for microbial growth, it contains most vital elements required for growth, e.g., N_2, CO_2, O_2, H_2O, sulfur as SO_2, dust (minerals). Consult the fine articles by Hutchinson (1954) on the biogeochemistry of the terrestrial atmosphere.

The tolerance of microbes to other chemicals will be emphasized throughout the text. Vallentyne (1967) should be consulted for those specifically interested in biophysicochemistry.

Because radiation tolerance is an important consideration in space travel and planetary life, the subject will be considered briefly. *Sacchoromyces cerevisiae* will tolerate single dosage of gamma rays of 10^6r (Shields *et al.*, 1961). Mainsin *et al.* (1960) using the same microbe found that a daily dosage of 50 mr/day did not prevent cultivation and growth. Maximal tolerances are not known.

REFERENCES

Allen, M. B. (1960). Utilization of thermal energy by living organisms. *Comp. Biochem.* **1**, 487–514.

Baas-Becking, L. G. M., Wood, E. J. F., and Kaplan I. R. (1956). Biological processes in the estuarine environment. *Biochemistry* pp. 96–102.

Baas-Becking, L. G. M., Kaplan, I. R. and Moore, D. (1960). Limits of the pH and oxidation-reduction potentials. *J. Geol.* **68**, 243–284.

Baker, H., Hutner, S. H., and Sobotka, H. (1955). Nutritional factors in thermophily: A comparative study of bacilli and Euglena. *Ann. N.Y. Acad. Sci.* **62**, 349–376.

Bedford, R. H. (1933). Marine bacteria of the Northern Pacific Ocean. The temperature range of growth. *Contrib. Can. Biol. Fisheries* **7**, 433–438.

Bidault, C. (1921). Sur les moisissures de viandes congelées. *Compt. Rend. Soc. Biol.* **85**, 1017–1018.

Borgstrom, G. (1961). Unsolved problems in frozen food microbiology. *Proc. Low Temp. Microbiol. Symp.*, pp. 197–250. Campbell Soup Co., Philadelphia, Pennsylvania, 1961.

Copeland, J. J. (1938). Yellowstone thermal *Myxophyceae*. *Ann. N.Y. Acad. Sci.* **36**, 1–229.

Fritch, F. E. (1922). The moisture relations of terrestrial algae. I. Some general observations and experiments. *Ann. Botany (London)* **36**, 1–20.

Fritch, F. E., and Haines F. M. (1923). The moisture relations of terrestrial algae. The changes during exposure to drought and treatment with hypertonic solutions. *Ann. Botany (London)* **37**, 683–728.

Haines, R. B. (1931). The influence of temperature on the rate of growth of *Sporotrichum carnis*, from $-10°C$ to $+30°C$. *J. Exptl. Biol.* **8**, 379–388.

Horowitz-Wlassowa, L. M., and Grinberg, L. D. (1933). Zur Frage über psychrophile Mikroben. *Zentr. Bakteriol., Parasitenk, Abt. II* **89**, 54–62.

Hutchinson, G. E. (1954). The biogeochemistry of the terrestrial atmosphere. *In* "The Earth as a Planet" (G. P. Kuiper, ed.), pp. 371–433. Univ. of Chicago Press, Chicago, Illinois.

Jenkin, P. N. (1936). Reports on the Percy Sladen expedition to some Rift Valley Lakes in Kenya in 1929. VII. Summary of the ecological results, with special reference to the alkaline lakes. *Ann. Mag. Nat. Hist.* **18**, 133–181.

Johnson, H. W. (1923). Relationship between hydrogen ions, hydroxyl ion and salt concentrations and the growth of seven soil molds. *Iowa State Coll., Agr. Expt. Sta., Res. Bull. No.* **76**, 307–344.

Jumelle, H. (1891). Sur le dégagement d'oxygène par les plantes, aux basses températures. *Compt. Rend. Acad. Sci.* **112**, 1462–1465.

Kalinenko V. O. (1957), Multiplication of heterotrophic bacteria in distilled water. *Mikrobiologiya* **26**, 148–153.

Kingsbury, J. M. (1954). On the isolation, physiology and development of a minute, hardy blue-green alga. Ph.D. Thesis, Harvard University, Cambridge, Massachusetts.

Koyrdum, V. A., and Bobchenko, E. S., (1959). Air as a habitat for microorganisms. *Mikrobiologiya* **26**, 231–235.

Lowman, F. G. (1963). Iron and cobalt ecology. *Proc. 1st Natl. Symp. Fort Collins, Colorado*, 1961, pp. 561–567. Reinhold, New York.

McCormack, G. (1950). "Pink Yeast" isolated from oysters grown at temperatures below freezing. *Comm. Fisheries Rev.* **12**, 28.

Mainsin, J., VanDuyse, E., Dunjic, A., Van Der Merckt, J., and Werbrouck, A. (1960). Acquired radio-resistance, radio-selection, and radio-adaptation. *In* "Intermediate and Low Level Effects of Ionizing Radiation" (A. A. Buzzati-Traverso, ed.), pp. 183–194. Taylor and Francis, London.

Meek, C. S., and Lipman, C. B., (1922). The relation of the reaction and of salt content of the medium to nitrifying bacteria. *J. Gen. Physiol.* **5**, 195–204.

Precht, H., Christophersen, J., and Hensel, H. (1955). "Temperatur und Leben." Springer, Berlin.

Redfort, A. L. (1932). Le sel employé pour combattre la décomposition des fletans. *Bull. Intern. Renseign. Frigorifigues* **4**, 40–43.

Salisbury, F. S. (1962). Martian biology. *Science* **136**, 17–26.

Shields, L. M., Durrell, L. W., and Sparrow, A. H. (1961). Preliminary observations on radiosensitivity of algae and fungi from soils of the Nevada Test Site. *Ecology* **42**, 440–441.

Strughold, H. (1961). Space medicine and astrobiology. *Proc. 11th Intern. Astronaut. Congr. Stockholm*, 1961 Vol. 1, pp. 671–687.

Tchistiakov, F. M., and Botcharova, Z. Z. (1938). Influence of low temperatures on the development of microorganisms. IV. Influence of low temperatures on the development of molds. *Mikrobiologiya* **7**, 838–842.

Tikov, G. A. (1955). Is life possible on other planets? *J. Brit. Astron. Assoc.* **65**, 193–204.

Vallentyne, J. R. (1967). Environmental biophysics and microbial ubiquity. *Ann. N.Y. Acad. Sci.* **108**, 342–352.

Weinzirl, J., and Gerdeman, A. E. (1929). The bacterial count of ice cream held at freezing temperatures. *J. Dairy Sci.* **12**, 182–189.

Wilkansky, B., (1936). Life in the Dead Sea. *Nature* **138**, 467.

Zernow, S. A. (1944). On limits of life at negative temperatures. *Compt. Rend. Acad. Sci. (URSS)* **44**, 76–77.

Zeuch, L. (1934). Untersuchungen Zum Wasserhaushalt von *Pleurococcus vulgaris*. *Planta* **22**, 614–643.

ZoBell, C. E. (1934). Microbiological activities at low temperatures with particular reference to marine bacteria. *Quart. Rev. Biol.* **9**, 460–466.

ZoBell, C. E. (1958). Ecology of sulfate reducing bacteria. *Producers Monthly* **22**, 12–39.

ZoBell, C. E., and Morita, R. Y. (1956). Barophilic bacteria in some deep sea sediments. *J. Bacteriol.* **73**, 563–568.

2

BIOGEOCHEMICAL PERIODIC TABLE

Although the subject of a biogeochemical periodic table is quite new to most individuals, many investigators have centered their research around such a concept. Thatcher (1934) was one of the first investigators to examine periodic behavior in terms of biologically essential elements. He concluded that the essential elements are limited to the first four periods of the Mendeleev-type periodic table. However, such a rule is too limiting because it does not include such important elements as molybdenum and iodine.

In developing this topic, two outstanding recent works on a biogeochemical periodic table should be emphasized. Both papers are by Shaw (1960a,b) who was formerly associated with the Clayton Foundation and the University of Texas. The first paper describes and gives the supporting data for a biogeochemical table, and the second discusses its biological significance. This subject is also treated by Rankama and Sahama (1960). Because of the complexity of the subject the reader is referred to Shaw's original work. His table represents one of the most comprehensive condensations of data completed in the last 50 years. The significance of the table is obvious; however, its use is more difficult.

The elements in the classic periodic table (Table 2.I) are arranged on the basis of electronic distribution. Arrangements normally show atomic number and atomic weight (of the most important species) of each element. The vertical columns are called groups; and all groups have the same

TABLE 2.I

PERIODIC CHART OF THE ELEMENTS

IA	IIA	IIIB	IVB	VB	VIB	VIIB	VIII			IB	IIB	IIIA	IVA	VA	VIA	VIIA	INERT GASES
1 H 1.00797																	2 He 4.0026
3 Li 6.939	4 Be 9.0122											5 B 10.811	6 C 12.01115	7 N 14.0067	8 O 15.9994	9 F 18.9984	10 Ne 20.183
11 Na 22.9898	12 Mg 24.312											13 Al 26.9815	14 Si 28.086	15 P 30.9738	16 S 32.064	17 Cl 35.453	18 Ar 39.948
19 K 39.102	20 Ca 40.08	21 Sc 44.956	22 Ti 47.90	23 V 50.942	24 Cr 51.996	25 Mn 54.9380	26 Fe 55.847	27 Co 58.9332	28 Ni 58.71	29 Cu 63.54	30 Zn 65.37	31 Ga 69.72	32 Ge 72.59	33 As 74.9216	34 Se 78.96	35 Br 79.909	36 Kr 83.80
37 Rb 85.47	38 Sr 87.62	39 Y 88.905	40 Zr 91.22	41 Nb 92.906	42 Mo 95.94	43 Tc (99)	44 Ru 101.07	45 Rh 102.905	46 Pd 106.4	47 Ag 107.870	48 Cd 112.40	49 In 114.82	50 Sn 118.69	51 Sb 121.75	52 Te 127.60	53 I 126.9044	54 Xe 131.30
55 Cs 132.905	56 Ba 137.34	57 'La 138.91	72 Hf 178.49	73 Ta 180.948	74 W 183.85	75 Re 186.2	76 Os 190.2	77 Ir 192.2	78 Pt 195.09	79 Au 196.967	80 Hg 200.59	81 Tl 204.37	82 Pb 207.19	83 Bi 208.980	84 Po (210)	85 At (210)	86 Rn (222)
87 Fr (223)	88 Ra (226)	89 'Ac (227)															

*Lanthanum Series

58 Ce 140.12	59 Pr 140.907	60 Nd 144.24	61 Pm (147)	62 Sm 150.35	63 Eu 151.96	64 Gd 157.25	65 Tb 158.924	66 Dy 162.50	67 Ho 164.930	68 Er 167.26	69 Tm 168.934	70 Yb 173.04	71 Lu 174.97

*Actinium Series

90 Th 232.038	91 Pa (231)	92 U 238.03	93 Np (237)	94 Pu (242)	95 Am (243)	96 Cm (247)	97 Bk (247)	98 Cf (249)	99 Es (254)	100 Fm (253)	101 Md (256)	102 No (253)	103 Lw (257)

( ) Numbers in parentheses are mass numbers of most stable or most common isotope.

Atomic weights corrected to conform to the 1961 values of the Commission on Atomic Weights.

electronic structure in the outer shell. The inert elements have eight electrons in their outer shell yielding a stable condition. The transition elements are arranged in subgroups and all have two electrons in their outer shell with the exception of the subgroup Cu-Ag-Au which has only one. The horizontal rows of elements are called periods. All elements in a given period have the same number of shells of electrons. The lanthanide and actinide series making up the rare earth elements belong, respectively, to periods 6 and 7.

Historically Frey-Wyssling (1935) considered the elements of C, N, P, S, O, H, K, Mg, Ca, Fe, Mn, Cu, Zn, and B as playing an important role in plant nutrition, and from this he drew the *Nährstofflinie* (nutrient line) in the Mendeleev table (Table 2.I). Shaw actually pictured this nutrient line "as two lines extending laterally from carbon which was the common origin. He also describes at least four bioperiodical loops encircling Cu, Zn, Mn, and Fe. One line passed left through B and Al, bisecting Mg and Ca and terminating between Na and K. The second line extended right, bisecting N and P, O, and S and ending with Cl." Bio-loops have been used to link other elements with this nutrient line. Prior to this classification, Fearon (1933) listed 18 elements as being invariably present in living matter and 29 additional elements as variable. The variably present elements were those reported in some species but not in others.

In general the invariable biological elements are those located above the solid line in Table 2.I. The exceptions are Li, Be, and Sc (Webb and Fearon, 1937). The variable bio-elements are those directly below the solid line, and, for our present consideration, all other elements may be regarded as nonbiological elements.

TABLE 2.II

A BIOCHEMICAL AND GEOCHEMICAL ASSIGNMENT FOR NINE SEGMENTS OF IRON

Atomic number

Geochemical character (Atmosphile, Lithophile, Chalcophile, Siderophile)

Abundance in biosphere [1]	[1]26x * SCl *	
Enzymes, Plants, or Animals[2]	[2]EPA (6.0)	Abundance in sea, $-\log A$ (g-atoms/kg)
Element symbol [3]	[3] (.05) Fe (-3.8)	Abundance in crust, $-\log A$ (g-atoms/kg) Cosmic abundance, $-\log A$ (g-atoms/10^4 Si atoms)
Electronegativity[4]	[4]e (com.) 1.6 (Ofn*)	Complexing tendencies: 0 ... bind to ligands through 0
Oxidation state[5]	[5]Z/2/3/	of ... bind through 0 and F
Atomic or ionic radius (CN-6)[6]	[6]r/.74/.64/	n, n* bind through N, n* indicates unusual binding of importance in biology
Ionic potential Z/r[7]	[7]ϕ/2.7/4.7/	s ... bind through S
For element E → E^z, standard electrode potential[8]	[8]E/.44/.33/	ps ... bind through P, S. C or halogen
First and 2nd Ionization potential[9]	[9]I/7.9/16/	

Corollaries from this are:
1. Elements on the nutrient line are essential for growth.
2. Elements nearest the nutrient line are required as micronutrients in certain biological processes.
3. The toxicity of elements increases as the periodicity increases moving away from the nutrient line, e.g., Zn-Cd-Hg.

There are notable exceptions to these corollaries and especially to the third. In Group Ib Cu may be more toxic than Ag and Au, and in Group Vb As is usually more toxic than Sb and Bi.

In a biogeochemical periodic table many characteristics for each individual element must be shown. Shaw (1960a) divided the area around each element in the table into nine segments or sub-areas (Table 2.II). The uppermost sub-area, segment 1, shows the atomic number, the abundance of the element in the biosphere, and the geochemical character. Segment 2 shows whether or not the element is involved with enzymes, plants, or animals; abundance in the sea is expressed as $-\log A$ (g-atoms/kg). The elemental symbol is shown in segment 3. The number to the upper right of the symbol denotes the abundance of the element in the crust and below that, but still in segment 3, is the cosmic abundance. Cosmic abundance is expressed as $-\log A$ in (g-atoms/10^4 Si atoms). In segment 4 the complexing tendency and electronegativity are designated. Segments 5, 6, 7, and 9 show, respectively, the oxidative state, atomic or ionic radius, ionic potential, and the standard electrode potential. Segment 9 shows the first and second ionization potentials. The use of a numbering system for designating the segments is done only in Table 2.II for convenience, and the segments are not numbered in the biogeochemical periodic table.

ELEMENTS IN THE BIOSPHERE

Nearly 200,000 species of plants and 800,000 species of animals exist in the biosphere. These figures should be compared with approximately 2000 minerals and 103 known elements of which only about 60 elements have significance in biogeochemistry. Microbes associated with the plant kingdom utilize these elements in (a) cellular organic material, (b) organic-metal complexes, (c) inorganic skeletal structures or deposits, or (d) inorganic compounds dissolved in body fluids.

If the earth was formed by a process of condensation or accretion of cosmic matter, then cosmic abundance of an element acted as a selective pressure and played an important role in determining the chemical composition of the earth and the type of biological development therein. With geochemical and physical processes effecting many changes a core, a mantle, and a crust were formed. Gases surrounded the planet and ultimately the seas, oceans, and rivers arose. If such a scheme is plausible, then, with the appearance of unicellular life in the seas, it is natural to assume that microbial evolutionary development can be correlated with the abundance of elements in (a) the biosphere, (b) the sea, (c) the crust, and (d) the cosmos. Actually elective cultural conditions existed which acted as selective forces in guiding the types of unicellular forms of life which developed (Kluyver and Van Niel, 1956). These same selective forces are seen in many ecological habitats today.

Because of the lack of quantitation, elemental abundance in the biosphere is expressed using the following arbitrary scale:

XXXX—Constantly present in relatively high concentration (1–60%)

xxxx—Constantly present but less abundant (0.05–1%)

****—Constantly present in traces (less than 0.05%)

::::—Frequently detected in variable low amounts

....—Variable, and still lower in concentration

Note: If no code is given the element has not been detected in microorganisms.

In calculating elemental abundance in either the earth's crust or the ocean, the concentration is expressed in negative logarithms of the sample in g-atoms/kg (Mason, 1958; Vinogradov, 1953). This is designated as a pA value, following the analogy to the pH scale.

Cosmic abundance is also expressed by using the pA* scale, but atoms are expressed in g-atoms/10^4 Si atoms.

$$pA^* = -\log A^*$$

pA* values of less than zero correspond to high abundance levels (pA* = 0 : 1 g-atom/10^4 Si atoms). Increasing positive values of pA* indicate decreasing abundance. If an element is detected in minute amounts in nature, or if it has been synthesized, the symbol "syn" is used.

GEOCHEMICAL CHARACTER OF THE ELEMENTS

Goldschmidt (1954) and many others have found it convenient to classify elements according to their tendency to occur in various geochemical phases. Elements found in the iron phase are called Siderophile (S) elements. Elements occurring in silicate are termed Lithophile (L) and those in sulfide ores are Chalcophile (C). As one would expect, Atmosphile elements (A) exist in the atmosphere as volatile elements or compound gases. The geochemical character of each is designated by the letter S, L, C, or A to the right of the abundance indicator in segment 1 (Table 2.II).

ENZYMES, PLANTS, AND ANIMALS

Data used to represent the existence of the element as in enzyme-element complex or in plant and animal are so diverse and source material so varied that Shaw (1960a) should be consulted for exact references. E is an indication that the element is essential to one or more enzymes. P is used to indicate that the element is required by higher plants; and p is used for lower plants and also to indicate that the requirement in higher or lower plants is variable or that it has not been conclusively established as an essential element. An analogous symbolism for A and a is used to describe the relationship of elements in animals and animal nutrition.

CHEMICAL PROPERTIES OF THE ELEMENTS

The chemical properties for each element are found in segments 4–9 falling below the elemental symbol. In determining the cosmic abundance of elements, properties such as binding energy, half-life, and cross section are of prime importance, since, the biogeochemical fate of an element depends on its extranuclear atomic structure. The geochemical redistribution of the elements after the earth formed depend upon such physio-chemical properties as density, thermodynamic stability, crystal structure, etc. In the hydrosphere and biosphere, solubility is of prime importance. Microbes with concentrating mechanisms may expend energy to increase the availability of certain elements despite a limiting level in the surrounding medium. Thus biochemistry must be added as another criteria in studying elemental movement.

ELECTRONIC CONFIGURATION OF ELEMENTS

The electronic configuration of an element is the basis of classification of elements in the classic periodic table as well as the new long form.

Space limitations prevented Shaw (1960a) from including the detailed configuration for all elements; however, they are shown for the inert gases (Table 2.III). The values for the principal quantum number n are presented directly, under the elemental symbol in segment 4 (Table 2.II). The number of electrons for the s, p, d, and f orbitals are in successive columns adjacent to n values. For detailed tables of electronic configurations, see Moeller (1952). To determine the order in which orbitals are filled, consult the left-hand column in Table 2.III which shows the number of periods. The orbital filling sequence is given beneath the periodic number. With a little practice the approximate electronic configurations and filling sequence can be derived from these data.

ATOMIC NUMBER AND ATOMIC WEIGHT

Atomic number (Z) is shown on the left side of segment 1 (Table 2.II). Atomic weights (A) are presented in a sub-table located in the lower left-hand corner of Table 2.III.

ELECTRONEGATIVITY

Electronegativity has been defined by Pauling (1940) as the power of an atom in a molecule to attract electrons to itself. From this an electronegativity scale based on bond strengths has been devised; Allred and Rothow (1958) have used an electrostatic approach to develop an extensive list of electronegativity values, and these were used in Table 2.III. The electronegativity values are presented in segment 4 just below the letter e. Of the values shown fluorine is the most electronegative (e = 4.1) and cesium the least (e = 0.86). Electronegativity can be used to estimate the ionic character of chemical bonds. The high electonegative elements form anions, while those elements with low electronegative values form cations. As a rough rule if the electronegativity between two elements is equal to or greater than 2.1 units, the bond formed between the atoms will tend to be ionic: 50% or greater. In geochemistry whenever diadochy in a crystal is possible between two elements possessing substantially different electronegativities, the element with the lower electronegativity will be preferentially incorporated because it forms a stronger ionic bond.

TABLE 2.III

THE BIOGEOCHEMICAL PERIODIC TABLE

Legend (key to the entries for each element):

- Atomic number[1]
- Geochemical Character: Atmophile, Lithophile, Chalcophile, Siderophile.[3]
- Abundance in Biosphere[1]
- Needed[2] by: certain or all Enzymes, Plants, or Animals
- Abundance in sea, = log (A in g·atoms/kg). Please note[4].
- = in crust,
- Cosmic Abundance, = log (A in atoms/10^4 Si atoms)
- Element's symbol
- Complexing tendencies:
  - o... bind to ligands through O.
  - of.. bind through O and F.
  - n,n*-bind through N, n* indicates unusual binding of importance in biology.
  - s... bind through S.
  - ps.. bind through P, S, C or halogen.
- Electronegativity
- Oxidation state
- Atomic or ionic radius (CN-6)
- Ionic potential z/r
- First and second ionization potentials
- (See text for details, etc.)
- For element E→E^z, standard electrode potential
- In case of 7b Group, electron affinity appears here. Units[5] and Group O elements[6].

(1) Abundance in biosphere given on basis of following arbitrary scale:
  XXXX - Constantly present in relatively high conc.
  xxxx - Constantly present but less abundant.
  **** - Frequently present in traces.
  :::: - Frequently detected in variable low amounts.
  ..... - Variable, still lower conc.

(2) AP - Well established need by higher and lower forms.
  ap - Less well established, or needed by only one form.

(3) Predominant character placed first.
  Capital letters indicate strong character.
  Lower case indicate less strong.

(4) On this pA scale large negative values correspond to high abundance. Large positive values correspond to low abundance (c.f. pH scale).

(5) Units: e-arbitrary, r-Angstroms z-volts, I-electron volts.

(6) Electronic structures given below symbols.

(7) Relative values indicated by / (This adjusted and should be interpreted as /-.)

Groups: 1a, 2a, 3a, 4a, 5a, 6a, 7a, 8, 1b, 2b, 3b, 4b, 5b, 6b, 7b, 0

Periods (Orbitals filling in each):
- 1 (1s): H, He
- 2 (2s, 2p): Li, Be, B, C, N, O, F, Ne
- 3 (3s, 3p): Na, Mg, Al, Si, P, S, Cl, A
- 4 (4s, 3d, 4p): K, Ca, Sc, Ti, V, Cr, Mn, Fe, Co, Ni, Cu, Zn, Ga, Ge, As, Se, Br, Kr

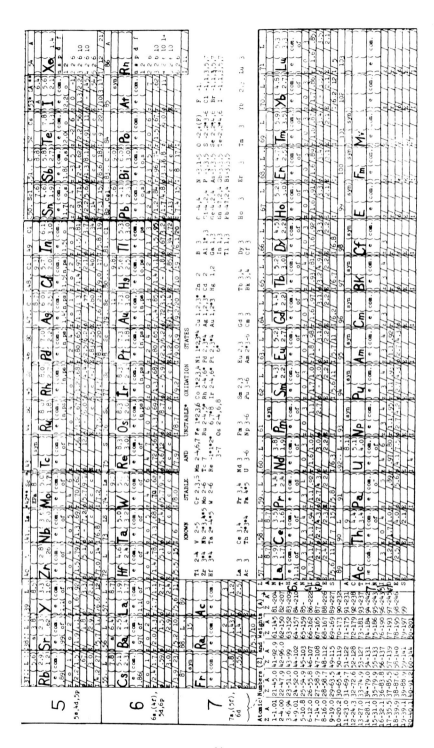

KNOWN STABLE AND UNSTABLE* OXIDATION STATES

21

COORDINATION TENDENCIES

The general subject is discussed more fully later, but because of its importance it is also included in this chapter. Substances composed of a central atom or ion, which is usually a metal, and surrounded by electron donors known as ligands are called coordination compounds. Ligands can be atoms, ions, or molecules. Some ligands are chelators, particularly if the ligand contains two or more electron-releasing centers where the central atom or ion results in a structure containing one or more rings. The bis-ethylenediamine copper-II ion contains two chelate rings. The ligand

$$\begin{bmatrix} H_2C-NH_2 & NH_2-CH_2 \\ & Cu \\ H_2C-NH_2 & NH_2-CH_2 \end{bmatrix}^{2+}$$

formed by ethylenediamine has two electron-releasing centers exemplified by the NH_2-groups attached to the metal ion. A ligand of this type is bidentate or has two couples. Polydentate ligands are well known. Ligands are classified by the types of atoms which they coordinate most firmly. The complexing tendency is shown in Table 2.II in segment 4 under the heading com. Symbols used in the table are as follows:

O: Primarily bind to ligands through oxygen.

of: Most bind through oxygen. Fluorine complexes are also common. Binding tendency with other nonmetals becomes more pronounced with increasing group number.

n: Primarily bind to ligands through nitrogen.

n*:   This symbol is used for iron and magnesium. These biologically important elements usually bind to ligands through oxygen. Porphyrins, phthalocyanins, chlorophyll, hemin, and chelators of this type are discussed in the section under chelation.

s: Bind to ligands through sulfur.

ps: Bind to ligands through phosphorus, sulfur, halogens or carbon. There is a tendency to complex with other nonmetals and lighter halogens which increases with decreasing group and period numbers.

The stability of a complex is directly related to the nature of the ligand and the metal. For the coordination compounds formed by certain cations, a stability sequence can be established that is independent of the nature of the ligand (Mellor and Maley, 1948; Irving and Williams, 1953; Williams, 1953, 1959).

A stability sequence of "Natural Order" with general validity is: $Mn^{2+} < Fe^{2+} < Co^{2+} < Ni^{2+} < Cu^{2+} > Zn^{2+}$. In this sequence, Cu^{2+} is the most stable and Mn^{2+} the least. Basolo and Pearson (1958) extended the above sequence as follows: $Pt > Pd > Hg > (UO_2) > Be > Cu > Ni > Co > Pb > Zn > Cd > Fe > Mn > Ca > Sr > Ba$.

The stability sequence for monovalent cations with one ligand is $Ag > Ti > Li > Na > K > Pb > Cs$; and for tervalent ions: $Fe > Ga > Al > Sc > In > Y > Pr > Ce > La$.

The stability series for monovalent and tervalent ions are applicable with only one oxygen-containing ligand, and supplementary research is required, before applying the series to ligands in general. Ringwood (1955) discusses other geochemical aspects of complex formation with reference to trace element behavior during magmatic crystallization. Note should be taken of the striking parallelism between stability sequences (Shaw, 1960) and the anion affinity sequences of Ahrens (1953).

OXIDATION STATES AND CORRESPONDING RADII

The stable and unstable oxidation states are important both biochemically and geochemically. This importance necessitates inclusion in a biogeochemical periodic table. In Table 2.II the oxidation states Z are presented in segment 5 and the radii r in segment 6. Additional oxidative states are summarized to the right of period 7 in the biogeochemical periodic table. Oxidation states were obtained from numerous source materials, whereas radii were obtained from Sutton (1958), Ahrens (1952), Pauling (1940), Goldschmidt (1954), and Moeller (1952).

The single-bond covalent radii ($Z = 0$) are taken from Pauling (1947). The single-bond radius noted as R (1) is related to the radius of bond order n by:

$$R(1) - R(n) = 0.300 \log n$$

where n is equal to the valence (v) divided by the coordination number. For metallic radii of coordination number 12:

$$R(1) - R(c.n.12) = 0.300 \log v/12$$

In nontransitional elements, the valence is the usual integral valence. For transitional elements nonintegral values for v are used based on magnetic data (Pauling, 1947).

The following criteria were used in selecting radii reported in Shaw's biogeochemical periodic table.

(1) If the element has only one common oxidative state, the ionic radius for that state is given; the single-bond covalent radius is taken from Pauling (1947).

(2) Elements with more than one oxidative state are probably more dependent on covalent bond formation in the biosphere, thus where more than one oxidative state exists the covalent radius and one oxidation state are reported (e.g., carbon and iodine).

(3) Ionic radii are reported for elements which we would expect to function covalently in the biosphere and which possess two or more oxidative states (e.g., N, P, and S). Approximate covalent radii can be determined by interpolation from values reported for neighboring elements. For example, sulfur would be expected to have a covalent radius between that of Si (1.2) and Cl (0.99); the reported value is 1.0.

(4) With elements having several oxidative states such as the transition elements, only the ionic radii for the highest and lowest oxidation states are given. Rough estimates of ionic radii can be made for the intermediate oxidative states by interpolation.

(5) The appearance of A($-$) in segment 7 under ionic potential (ϕ) indicates that other common oxidative states are known.

(6) The oxidation states for the transition elements, lanthanides, and actinides, and groups 3b–7b are shown in the sub-table to the right of period 7 (Table 2.III).

Goldschmidt's empirical rules for predicting the geochemical fate of an element during crystallization from liquids are based upon oxidative states and a knowledge of ionic radii. They provide an important example of the use of z and r in geochemistry.

These rules are summarized for two ions 1 and 2 as follows:

(a) If $Z_1 = Z_2$ and $r_1 = r_2$, the ions will enter a given lattice with equal facility.

(b) If $Z_1 = Z_2$ but r_1 is slightly larger than r_2, then ion 2 enters more readily.

(c) If r_1 is similar to r_2 and Z_1 is greater than Z_2 then ion 1 enters more readily.

On a biochemical basis oxidation states and ionic radii are important:

(1) Elements with several oxidative states are metabolized more easily than elements with only one, and certain microbes may even derive energy from oxidizing such elements (e.g., S and Fe).

(2) Ionic radii can be used to predict the permeability of ions of similar radii across cell membranes.

One can expect many additional and more detailed correlations between oxidative state and ionic radii and their relation to biological phenomena.

IONIC POTENTIAL

The ionic potential ϕ is defined as the oxidative state divided by the ionic radius Z/r. Since most of these values can be calculated from data presented in segments 5 and 6 of Table 2.II, ϕ values could be omitted. However, since many correlations have been made with ionic potential they have been included by Shaw (Table 2.III). Ionic potentials have been correlated with:

1. Periodic table, Cartledge (1928a,b, 1930)
2. Acid-base behavior of ions, Cartledge (1928b)
3. Thermochemical data, Cartledge (1951)
4. Hydration, Mason (1958)
5. Coordination chemistry, Bailar (1956)
6. Solubilization in terrestrial plants, Hutchinson (1943)
7. Affinity of cations for ion-exchange resins, Kunin (1958).

The correlation of ionic potential with hydration and coordination chemistry has obvious biochemical geochemical significance. Comparative studies on ionic potential and solubilization should be made with microbes; they already have been with terrestrial plants (Hutchinson, 1943). Other items stressed by Shaw (1960a) and Kunin (1958) are:

1. Cations with similar ϕ are frequently geochemically associated
2. Elements with a $\phi = 4-5$ are electron-releasing base formers that exist as soluble cations at ordinary pH.
3. Elements with $\phi = 4-5$ or $10-12$ are amphoteric and tend to form insoluble oxides or hydroxides.
4. Elements with $\phi = 10-12$ are electron acceptors and acid formers that tend to form soluble complex anions.
5. In general, coordinating ability increases with an increasing ϕ of the central cation.
6. Cations of high ϕ are more polarizing and more highly hydrated than cations of low ϕ.
7. The affinity of cations for ion-exchange resins is inversely related to ϕ. High unconventional negative values of ϕ correspond to high conventional positive values. Because of space limitations the Z, r,

and ϕ values for oxygen and the halogens are omitted from the biogeochemical periodic table. They are, respectively O: 6, 0.10, 60; F: 7 (hypothetical), 0.08, 87; Cl: 7, 0.27, 26; Br: 7, 0.39, 18; I: 7, 0.50, and 14 (see Ahrens, 1952).

IONIZATION POTENTIAL

Ionization potentials (I) are important in all phases of inorganic chemistry, coordination chemistry, biology, and geochemistry. General conclusion are:

1. For cations of the same Z, an increased I parallels an increase in anion affinity and complex stability.
2. Cations with similar ϕ's and similar I's are geochemically quite coherent.
3. Elements with a low I form soluble cations.
4. Elements with a very high I form soluble anions or are inert. Ionization potentials for oxidation states greater than 2 are presented in Table 2.IV (Ahrens, 1953).

TABLE 2.IV

IONIZATION POTENTIALS FOR ELEMENTS
WITH OXIDATION STATES (Z) GREATER THAN 2[a]

		Oxidation state		
Z = 3	Z = 4	Z = 5	Z = 6	Z = 7
B-38	C-64	N-97	O-137	F-184
Al-28	Si-45	P-65	S-88	Cl-114
Sc-25	Ti-43	V-65	Cr-96	Mn-122
V-26	V-48	V-65	Se-82	Br-104
Cr-32	Ge-46	As-63	Mo-70	Tc-95
Ga-31	Zr-34	Nb-52	Te-72	I-90
Y-20	Sn-41	Sb-56	W-61	Re-79
In-28	Ce-37	Ta-45		
La-19	Hf-31	Bi-56		
Tl-30	Pb-39			

[a] After Ahrens (1953).

STANDARD ELECTRODE POTENTIALS

The biological implication of standard electrode potential (E) has been discussed in the last two chapters. Reference should also be made to application of E in biology by George and Griffith (1959). The application of standard electrode potentials to geochemistry is covered by Mason (1958) and to chemistry by Latimer (1952).

APPLICATION OF A BIOGEOCHEMICAL PERIODIC TABLE

The application of a biogeochemical periodic table are many. Shaw (1960b) has summarized the more significant features. The nutrient line (or *Nährstofflinie*) is quite obvious. The use of bio-loops to include additional elements above and below the nutrient line does not appear necessary. In most instances there appears to be at least two elements in each group of biochemical significance. Steinberg (1938) has made several correlations between biological essentiality and atomic structure. He concluded that there should be three essential elements in each periodic group such as Mg, Ca, and Zn in Group II.

Further analysis shows that elements in the biogeochemical periodic table can be segregated into three natural groups: (I) essential alkali and alkaline earth metals; (II) essential transitional metals, and (III) essential nonmetals (Table 2.V). Division I containing Na, K, H, Mg, Ca, Sr, and Ba, consists of low electropositive elements existing as cations in many water-soluble compounds. Magnesium is the most electronegative element of this division and forms the most stable complexes.

The Division II elements have intermediate ϕ and e values. They form water-insoluble compounds. Division II has partially filled d orbitals which may explain the catalytic activity of these elements. Their value as trace elements are associated with a pronounced chalcophile and siderophile character and a low hydrophilicity. Division III is characterized by high electronegativity and ϕ values so that these elements form anions in many water-soluble substances.

In comparative biochemistry, the biological importance of an element can be related to both biospheric abundance and nutritional essentiality. Shaw (1960b) has presented excellent correlations between oceanic abundance and essentiality. As a working rule he proposed that, within a periodic

TABLE 2.V

COMPARATIVE PHYSICOCHEMICAL PROPERTIES OF
THREE BIO-DIVISIONS OF ELEMENTS[a]

Division I		Division II		Division III	
Group 1a	Na (pA)	Period 4	V (pa)	Group 3b	B (P)
	K (EPA)		Mn (EPA)	Group 4b	C (EPA)
	H (EPA)		Fe (EPA)		Si (?pa)
			Co (EPA)		
Group 2a	Mg (EPA)		Zn (EPA)	Group 5b	N (EPA)
	Ca (EPA)				P (EPA)
	Sr (a)	Period 5	Mo (EPA)		
	Ba (a)			Group 6b	O (EPA)
					S (EPA)
					Se (pa)
				Group 7b	F (a)
					Cl (pA)
					Br (a)
					I (A)
					H (EPA)

Parameter	Division I	Division II	Division III
e:	$0.91 \leq e \leq 1.2$	$1.4 \leq e \leq 1.8$	$2.0 \leq e \leq 4.1$
ϕ:	$0.77 \leq \phi \leq 3.0$	$1.0 \leq \phi \leq 15$	$14 \leq \phi \leq 87$[b]
	$0.77 \leq \phi \leq 2.0$	$2.3 \leq \phi \leq 15$	
	(Without Mg)	Cu omitted	
I:	$4.3 \leq I \leq 7.6$	$6.7 \leq I \leq 9.4$	$8.3 \leq I \leq 17$
	$4.3 \leq I \leq 6.1$		
	(Without Mg)		
Com: Complex stability sequence:	of, n[b]	of, n, n[b], ps	o
	Ba < Sr < Ca < Mg	< Mn < Fe < Co < Cu < Zn	Most frequently forms ligands

[a] After Shaw (1960b).
[b] Hypothetical.

group, biological importance tends to parallel oceanic abundance. In
Group Ia, hydrogen, sodium, and potassium are biologically most important and are found in greatest abundance in seawater. In Group IIa, magnesium and calcium have the highest oceanic abundance, and both play

vital roles in animal and plant chemistry. Similar analogies can be drawn with Division II and III elements. One notable exception to this rule may be the positions of Mo and Cr.

Division II contains the essential "trace" transitional metals. In addition it possesses the largest of the two major empty spaces in the periodic table. The largest empty space is made up of elements 43–46 and 75–78. Osmium is a typical element of this group. It has the highest known density of any element, which is why it is used in electron micrographic studies. Osmium also has strong siderophile tendencies, is found in low abundance, possesses extremely low hydrophilicity, has negative E-values and general chemical nobility, and has a high complex stability. These characteristics are common to all the elements of this empty space.

A second smaller empty space is found for elements 21, 22, 29, and 30 which are, respectively, scandium, titanium, yttrium, and zirconium. Future studies may demonstrate that one or all of these elements are essential to biological systems. However, at present, considerable information is required to relate their biochemical significance to the biosphere.

In Division III there is still a strong correlation between oceanic abundance and microbiological utilization. An exception must be made for aluminum which is more abundant than boron in both the cosmos and the earth's crust. Aluminum is not essential to life; however, it appears to be involved in some biochemical processes, whereas boron, a strong hydrophile, is needed by many plants. In group 4b, carbon, which is concentrated in oceans, is more important than silicon which is concentrated in the earth's crust. In this instance, the atmospheric availability of carbon in CO_2 must be weighed in evaluating the greater biological requirement for carbon rather than silica.

UNDISCOVERED ESSENTIAL AND TRACE ELEMENTS

The nutrient line concept indicates that the elements essential for biological life should be extended to include all elements in the first four periods. Elements which need biochemical supporting data for this justification are Li, Be, Si, Ti, Ga, and Ge. Thus the nutrient line must be broadened.

Elements in period 5 should show variable essentiality for life. Oceanic and crustal abundance can be used in estimating the potential involvement of these elements in biochemical systems.

REFERENCES

Ahrens, L. H. (1952). The use of ionization potentials. I. Ionic radii of the elements. *Geochim Cosmochim. Acta* **2**, 155–169.

Ahrens, L. H. (1953). The use of ionization potentials. II. Anion affinity and geochemistry. *Geochim. Cosmochim. Acta* **3**, 1–29.

Allred, A. L., and Rothow, E. G. (1958). A scale of electronegativity based on electrostatic force. *J. Inorg. Nucl. Chem.* **5**, 264–268.

Bailar, J. C. (1956). "The Chemistry of Coordination Compounds." Reinhold, New York.

Basolo, F., and Pearson, R. G. (1958). "Mechanisms of Inorganic Reactions." Wiley, New York.

Cartledge, G. H. (1928a). Studies on the periodic system. I. The ionic potential as a periodic function. *J. Am. Chem. Soc.* **50**, 2855–2863.

Cartledge, G. H. (1928b). Studies on the periodic system. II. The ionic potential and related properties. *J. Am. Chem. Soc.* **50**, 2863–2872.

Cartledge, G. H. (1930). Studies on the periodic system. III. The relation between ionizing potentials and ionic potentials. *J. Am. Chem. Soc.* **52**, 3076–3083.

Cartledge, G. H. (1951), The correlation of thermochemical data by the ionic potential. *J. Phys. Chem.* **55**, 248–256.

Fearon, W. R. (1933). A classification of the biological elements, with a note on the biochemistry of beryllium. *Sci. Proc. Roy. Dublin Soc.* **20**, 531.

Frey-Wyssling, A. (1935). Die unentbehrclihen Elemente der Pflanzennahrung. *Naturwissenschaften* **23**, 767–769.

George, P., and Griffith, J. S. (1959). Electron transfer and enzyme catalysis. *In* "The Enzymes" (P. D. Boyer, H. Lardy, and K. Myrbäck, eds.), 2nd ed., Vol. I, Chapter 8, pp. 347–389. Academic Press, New York.

Goldschmidt, V. M. (1954). "Geochemistry." Oxford Univ. Press, London and New York.

Hutchinson, G. E. (1943). The biogeochemistry of aluminum and of certain related elements. *Quart. Rev. Biol.* **18**, 1–363.

Irving, H., and Williams, R. J. P. (1953). Stability of the complexes of the divalent transition metal ions. *J. Chem. Soc.*, pp. 3192–3210.

Kluyver, A. J., and Van Niel, C. B. (1956). "The Microbes' Contribution to Biology." Harvard Univ. Press, Cambridge, Massachusetts.

Kunin, R. (1958). "Ion Exchange Resins," 2nd ed., pp. 29–31. Wiley, New York.

Latimer, W. M. (1952). "Oxidation potentials," 2nd ed. Prentice-Hall, Englewood Cliffs, New Jersey.

Mason, B. (1958), "Principles of Geochemistry," 2nd ed. Wiley, New York.

Mellor, D. P., and Maley, L. (1948). Order of stability of metal complexes. *Nature* **161**, 436–437.

Moeller, T. (1952). "Inorganic Chemistry, An Advanced Text Book." Wiley, New York.

Pauling, L. (1940). "The Nature of the Chemical Bond." Cornell Univ Press, Ithaca, New York.

Pauling, L. (1947). Atomic radii and interatomic distances in metals. *J. Chem. Soc.* **69**, 542–553.

Rankama, K., and Sahama, T. G. (1960). "Geochemistry," Chapter 8. Univ. of Chicago Press, Chicago, Illinois.

Ringwood, A. E. (1955). The principles governing trace element behavior during magmatic crystallization. II. The role of complex ion formation. *Geochim. Cosmochim. Acta* **7**, 242–254.

Shaw, W. H. R. (1960a). Studies in biogeochemistry. I. A biogeochemical periodic table. The data. *Geochim. Cosmochim. Acta* **19**, 196–207.

Shaw, W. H. R. (1960b). Studies in biogeochemistry. II. Discussion and references. *Geochim. Cosmochim Acta* **19**, 207–215.

Steinberg, R. A. (1938). Correlations between biological essentiality and atomic structure of the chemical elements. *J. Agri. Res.* **57**, 851–858.

Sutton, L. E. ed. (1958). "Tables of Interatomic Distances and Configuration in Molecules and Ions," Spec. Publ. No. 11. Chem. Soc., London.

Thatcher, R. W. (1934). A proposed classification of the chemical elements with respect to their function in plant nutrition. *Science* **79**, 463–466.

Vinogradov, A. P. (1953). "The Elementary Chemical Composition of Marine Organisms," Sears Found. Marine Res. Memoir II. Yale Univ. Press, New Haven, Connecticut.

Webb, D. A., and Fearon, W. R. (1937). Studies on the ultimate composition of biological materials. Pt. I. Aims, scope, and methods. *Sci. Proc. Roy. Dublin Soc.* **21**, 487.

Williams, R. J. P. (1953). Metal ions in biological systems. *Biol. Rev.* **28**, 381–415.

Williams, R. J. P. (1959). Coordination, chelation and catalysis. *In* "The Enzymes" (P. D. Boyer, H. Lardy, and K. Myrbäck, eds.), 2nd ed., Chapter 9, pp. 391–441. Academic Press, New York.

3

LABORATORY AND *IN SITU* LEACHING TECHNIQUES

Both the microbiologist and geologist are concerned with how microbes can be used to leach elements from ores. The technology is quite simple, although close analysis shows that the reactions, the engineering, and geology can be complicated. Since descriptive material describing the operations used are not readily available to those interested in the subject, the main techniques are summarized here. The chief requirements for success are an understanding of microbiology, geology, hydrology, and some basic engineering. Only a general outline is given.

AIRLIFT PERCOLATOR

The most important laboratory technique developed to date for the microbial leaching of minerals has been the airlift percolator. Its design and simplicity of operation make it an attractive tool (Fig. 3.1). It is similar to a soil perfusion apparatus (Audus, 1946; Lees and Quastel, 1944), and consists of a metal or glass cylinder, filled from one-half to two-thirds capacity with crushed ore. A good laboratory unit will often possess about a 1000-ml total volume. Ores are ground to a certain mesh size (-20 to -400). The effect of particle size on leaching is often one of the primary parameters investigated. If ores are prone to plug while leaching, sand may be added to improve the porosity and to give better flow rates. After the ore is added, all external orifices should be plugged with cotton. Small units of this type can be steam-sterilized in an autoclave at 120°C at 15 psi for 1–3 hours. This procedure may have to be repeated for ores which are difficult to sterilize. It is better not to add the leaching solutions until after the ore in the airlift percolator has been steam-sterilized. After cooling, the leaching solution, which has been sterilized separately, is added. The microbe used for leaching is introduced into the system by standard aseptic inoculation procedures. Since it is usually desirable to determine the leaching effect of one particular microbe, the use and knowledge of aseptic techniques are absolutely essential. The leachate perfuses through the ore by gravity flow and collects below the perforated disk. Pressure can be used, but it is not required in early studies. Leach liquor is recycled by sterile air which is filtered through glass wool prior to injection into the airlift. Leaching reactions of this type are normally conducted at room temperature (25°C). Samples can be withdrawn as frequently as necessary from the reservoir underneath the perforated disk for the determination of the rate of leaching by chemical analysis.

Batteries of these columns are easily handled (Fig. 3.2). At times it may be necessary to use open unsterile systems for field work. Natural conditions must be used for controlling microbial growth under such circumstances. Airlift percolators have been described by Malouf and Prater (1961), Razzell (1962), and Sutton and Corrick (1963).

ECOLOGICAL FLASK

A second system used for leaching is a shake flask unit of the type used for studying antibiotic and vitamin fermentations. In an open flask system 1–30 gm ore is added to a 500-ml Erlenmeyer flask, the flask is stoppered with cotton, and the system is steam-sterilized in an autoclave using standard sterilizing procedures. The leaching solution is sterilized separately, and 100 ml of the cool leaching solution is added to each flask aseptically. The system is mixed by shaking and then inoculated with one or more known microbes. Autotrophic microbes are the most commonly used. The flasks are incubated, usually at 25°C, using either "still" or "shaker" culture systems. Still cultures are simply placed on the incubator shelf, whereas, common reciprocal or rotary shakers are used for aerating and agitating shake flasks. Agitation usually increases the rate of leaching, particularly in aerobic systems. First, it increases oxygen transfer to the

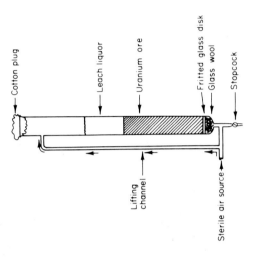

FIG. 3.1b

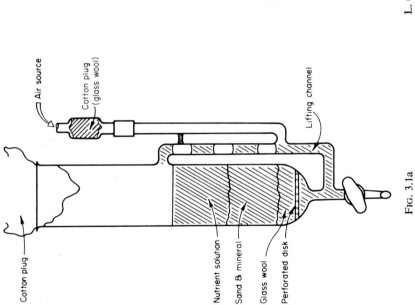

FIG. 3.1a

FIG. 3.1. Airlift percolator. Reproduced with permission from L. C. Bryner, Brigham Young University, Provo, Utah.

34

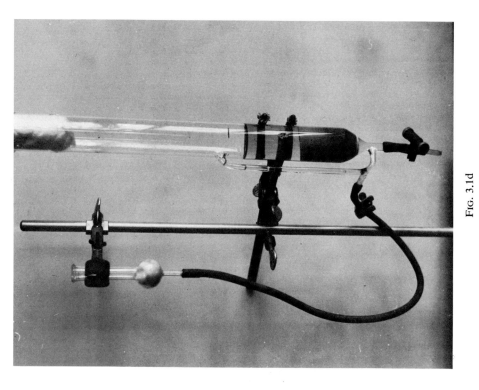

FIG. 3.1d

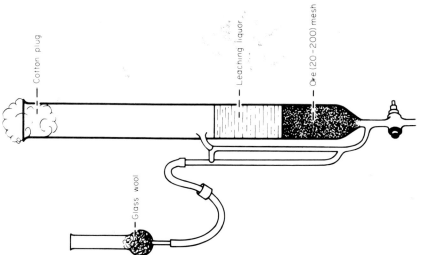

Cotton plug

Leaching liquor

Ore (20 - 200) mesh

Glass wool

FIG. 3.1c

FIG. 3.2. Multiple-column leaching apparatus. Reproduced with permission from V. F. Harrison, Canadian Department of Energy, Mines and Resources, Ottawa, Canada.

liquid phase as it is depleted, and, second, it gives a uniform mixing condition.

A variant of the shake flask technique is the closed ecological flask (Fig. 3.3). This system is identical to the shake flask unit in all respects, except a 500-ml vacuum flask is used. It is stoppered and equipped with a stainless steel tube gassing port. A rubber septum is inserted into the

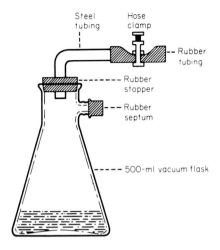

FIG. 3.3. Closed ecological flask.

lateral glass suction port. Sterilization, addition of leaching solution, and inoculation steps are similar to those used for standard aerobic shake flask systems. After inoculation the system can be coupled to a vacuum pump, evacuated, and sterile gases of known composition metered into the system. The system is closed off by a screw clamp. The gaseous phase is analyzed by inserting a needle through the rubber system and collecting a sample with a syringe. Gases are separated and analyzed in a gas chromatograph. Leaching solutions are removed by simply tilting the flask until it spills over into the rubber septum. Changes in O_2 and CO_2 concentration in leaching can easily be evaluated in a closed ecological flask.

PRESSURE LEACHING

Leaching in the field is often conducted through drilling holes. One such technique is known as a hole-to-mine leaching. Large hydrostatic pressures are reached which usually aid in leaching, especially in moving leaching solutions through an ore body. If the ore body is impervious, fracting techniques are employed to open up the formation.

Pressure leaching can be simulated in the laboratory. Ore blocks of 50—400 lb are placed in large vats (Fig. 3.4). Stainless steel or polyvinylchloride tubing is set into the ore block and carefully sealed with a good mastic. The leaching hole should always extend 6–10 inches below the lowest point of the tube. Leaching solution is added to cover the ore block

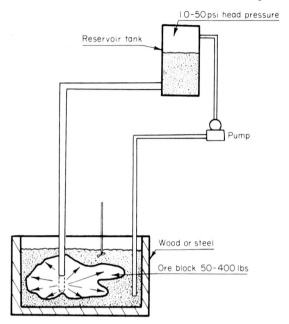

FIG. 3.4. Pressure leaching with blocks of ore.

and pumped to the reservoir tank where it is maintained under a constant head pressure, e.g., 1.0–50 psi. The head pressure used depends on the permeability of the ore block and the general physical characteristics of the ore. Channeling is common. Microbes which grow in these systems are not harmed by changes in pressure when recycled through the system. Specific penetration of rocks by microbes can be seen in such a system. Studies of this type are few; in one study, Myers and McCready (1966) used *Serratia marcescens* containing radioactive phosphorus (^{32}P) and determined the penetration of several rocks. A 14.2-in. core of Berea sandstone and 3–in. cores of Mississippian limestone, early Devonian limestone, and late Mesozoic sandstone were completely penetrated at atmospheric pressure by this microbe.

The movement of leaching solution through ores can be fast but normally it is so slow that its movement is laminar. Laminar flow is smooth, continuous, and traceable. The laws of ground-water movement were discovered and formulated in 1850 by Henry Darcy. Darcy's law relates the velocity of the ground water to the coefficient of permeability and the hydraulic gradient.

$$V = PI$$

where V = velocity of ground water; P = coefficient of permeability (usually calculated in units volume per unit time through a unit cross section); and I = hydraulic gradient or the slope of the water table.

Since the geologist is more concerned with the quantity rather than the velocity of the water, the quantity of water Q moving through a cross-sectional area A is used to replace the velocity:

$$V = \frac{Q}{A}$$

where $Q = PIA$ or quantity of water (gallons per day), and A = cross-sectional area through which water moves.

The movement of some aquifers varies with the permeability of the formation. Velocities vary with permeability from a few feet to several feet per day.

IN SITU AND ALTERNATE LEACHING TECHNIQUES

The technology for leaching minerals is old. Engineers have not described the techniques used in great detail because much of it is conducted empirically. The reasons for the use of *in situ* leaching procedures are quite basic. Only high-grade ores can be mined by conventional processes which means that high concentrations of minerals are left in the halos which surround a mined-out area. In addition mined ores are often hauled to the surface before it is found that they are too low in metal values to be worth hauling to the plant for recovery. *In situ* leaching is used to concentrate mineral values in liquors which can either be piped and pumped or hauled to the plant for recovery. With the improvement made in liquid-liquid solvent extraction technology the last ten years, it has become possible to process tremendous volumes of aqueous leaching solutions containing low amounts of metals.

The technology used in *in situ* leaching consists of (1) heap leaching, (2) hillside heap leaching, (3) hole-to-hole leaching, (4) hole-to-mine leaching, and (5) routine spray leaching of stopes and mined-out areas.

Heap Leaching

Heap leaching technology is quite simple. It is used for leaching low-grade ores which are mined and moved to the surface, but which cannot be hauled to the plant because of prohibitive haulage costs. Although heap leaching technology varies from mine to mine, the general procedures are quite similar.

Landscaping the site used to promote drainage to control runoff is required. An asphalt pad (Fig. 3.5) up to 4 in. thick is laid to prevent leaching solutions from soaking into the ground after percolating through the heap. Pads may also be constructed of cement, rubber, or plastic liners. Pads are first covered with a 6–8 inches layer of fine ore and topped with a 5-ft layer of regular "run of mine ore" which provides a cushion for large boulders. This cushion protects the pad from damage by cracking and reduces unnecessary loss of pregnant solutions.

Bulldozers are usually used to spread the ore and water pumped from the mine to irrigate the heaps. Leaching temperatures should not exceed 37°C; although they often do, it does not appear to severely affect microbiological activity.

Heap leaching has only been successful thus far in processes where sulfuric acid is the prime leaching agent. If the leaching solution is not acid, sulfuric acid is added or the system is inoculated with the proper autotrophic bacteria. In grading and landscaping, the heap furrows are left to increase aeration and to improve the control of movement of the leach water. Dikes must be firm, and ore is used to subdivide the surface into smaller leaching ponds. This reduces loss of leaching solutions when

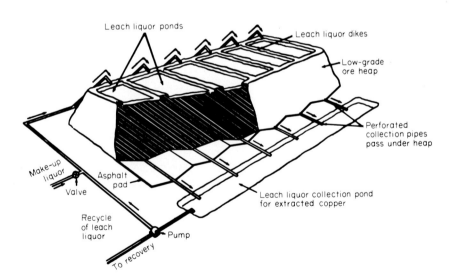

FIG. 3.5. Heap leaching on a level area.

dikes break. Surfaces of the heaps are irrigated or flooded rather than sprinkled. Sprinkling gives better aeration of the leach liquid; however, irrigation gives a more even flow of leach liquid through the bed. Pregnant leaching solutions flow by gravity. When the solution reaches the pad it follows the contour to a basin which is dozed out at the base of the heap. Solutions are collected and recycled or pumped to the plant for recovery.

Heap leaching is used by Western Nuclear Corp. in the Gas Hills of Wyoming for leaching low-grade uranium ores containing 0.05–0.10% uranium (Mashbir, 1964). Ore size is usually 90% minus 1 in. Heaps are built to a depth of 17 ft and contain 25,000 tons of ore. The ratio of solution to ore for leaching is 2.4:1.0. Average flow rate of solution through ore is 0.40 gal/hr/ft². Approximately 250,000 gallons of leaching solution is added per paddie. Ponds or sumps receiving discharge from paddies are lined with 8-mil polyethylene or rubber liners. Sumps are roughly 60 × 40 × 5 ft deep; each one holds about 100,000 gallons. Up to 90% recovery of uranium has been obtained.

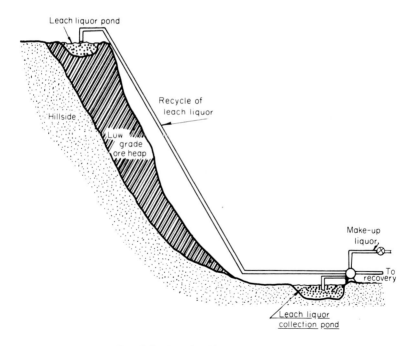

Fig. 3.6. Heap leaching on a hillside.

Hillside Heap Leaching

To avoid hauling, low-grade ores may also be bulldozed over a hillside onto areas which contain impervious surface rocks to minimize solution losses. Sites should be selected where drainage can be controlled (Fig. 3.6). The procedures for leaching such areas are similar to those used for heap leaching. Tremendous tonnages of low-grade copper ores are leached by this procedure.

Hole-to-Hole Leaching

In ore bodies pervious to leaching solutions good recovery of solutions can be expected, and hole-to-hole leaching has potential. The entire geology of the area must be known, because unusual faulting may cause misdirection of liquors and complete loss of solutions.

One system used is the 5 spot program (Fig. 3.7) in which five holes are drilled. Holes are drilled at four corner points and one hole is located centrally. The distance between the holes is completely dependent upon the porosity of the ore body and other geological factors. The central hole is drilled several hundred feet below the peripheral holes. Thus all leach liquors pumped to the peripheral holes will tend to gravitate to the center hole. Liquors from the center hole may be recycled through the peripheral holes and back through the formation or pumped or hauled to the plant for recovery.

Hole-to-Mine Leaching

Once an area within the mine has been completely mined and abandoned, halos of low-grade ore surrounding tunnels, stopes, rises, and pillars may be leached by hole-to-mine leaching techniques. The engineering required in underground mining normally provides the basic needs for this operation. The main shaft is almost always a main drainage reservoir. Tunnels are always run upgrade so that condensate and drainage water gravitates to main shaft. In many mines water must be continuously pumped from the shaft, and this water is often quite rich in the element desired. With proper hole-to-mine leaching the level always increases.

A highly diagrammatic sketch of hole-to-mine leaching is shown in Fig. 3.8. A cheap water supply is required. Often raffinate liquor or tailings pond waste water from the mill is used. When economics permit, the leach liquor can be supplemented by adding chemicals to a make-up tank located at the surface. Leach liquor is pumped to the hole which is kept about

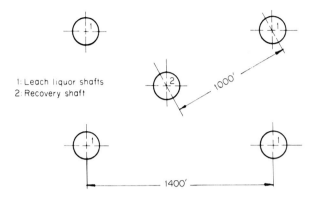

1: Leach liquor shafts
2: Recovery shaft

1000'

1400'

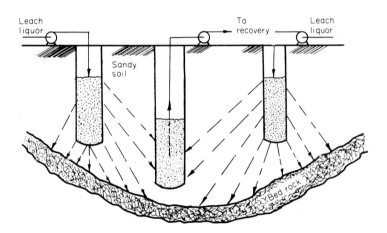

Leach liquor

To recovery

Leach liquor

Sandy soil

Bed rock

FIG. 3.7. Hole-to-hole leaching of low-grade ores.

two-thirds full of liquor. Leaching liquor permeates the area and ultimately drains to the tunnels and main shaft.

Semi-pregnant liquor from the shaft can be recycled or pumped to the plant for recovery.

An ancillary technique associated with hole-to-mine leaching is the manual routine spraying of stopes with leach liquor. This has been used

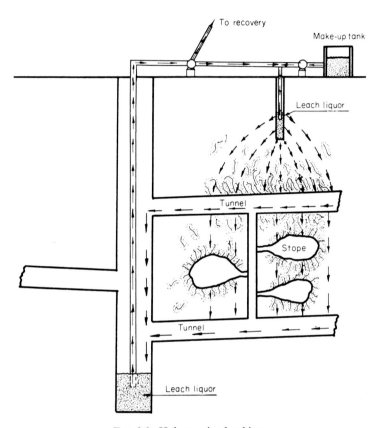

FIG. 3.8. Hole-to-mine leaching.

particularly in recovering uranium values from underground mined-out areas. Stanrock Uranium Corp. mines up to 1200 lb of U_3O_8 per month using this technique (MacGregor, 1966).

After the rate of leaching drops off leaching should be discontinued, and the area should be dried for one to several months to permit additional oxidation of the ore. Renewed leaching usually begins at a much higher rate.

The leaching of elements by any of these procedures is limited by many factors:

1. Availability of cheap water.
2. Ores which contain elements which can be oxidized by microbes to form oxidized metals, e.g., Fe^{3+}, Mn^{2+}, which in turn oxidize

or solubilize the desired metal or mineral sulfides such as FeS_2 which can be oxidized by microbes to yield ferric sulfate and sulfuric acid.

3. Growth of a suitable microbe to catalyze the desired reactions.
4. The availability of other cheap chemical additives which aid in leaching.
5. Cheap solvent extraction of the desired element from dilute aqueous solutions.

REFERENCES

Audus, L. J. (1946). A new soil perfusion apparatus. *Nature* **158**, 419.

Lees, H., and Quastel, J. H. (1944). A new technique for the study of soil sterilization. *Chem. & Ind. (London)*, **26**, 238.

MacGregor, R. A. (1966). Recovery of U_3O_8 by underground leaching. *Can. Mining Met. Bull.* **59**, pp. 583–587.

Malouf, E. E., and Prater, J. D. (1961). Role of bacteria in the alteration of sulfide minerals. *J. Metals* (May) pp. 353–356.

Mashbir, D. S. (1964). Heap leaching low grade uranium ore. *Mining Congr. J.* (Dec.) pp. 50–54.

Myers, G. E., and McCready, R. G. L. (1966). Bacteria can penetrate rock. *Can. J. Microbiol.* **12**, 477–484.

Razzell, W. E. (1962). Bacterial leaching of metallic sulphides. *Trans. Can. Inst. Mining Met.* **65**, 135–136.

Sutton, J. A., and Corrick, J. D. (1963). Microbial leaching of copper minerals. *Mining Eng.* **15**, 37–40.

4

MICROBIAL METABOLISM
OF INORGANIC SULFUR

Sulfur is a structural unit of many biochemical intermediates supporting life processes. It is a constituent of cysteine, cystine, glutathione,
adenosylmethionine, methionine, biotin, thiamine, and diphosphothiamine.
The metabolic functions of these compounds are quite complex. The present discussion is directed to the supply of sulfur in nature, the role of
microbes in the sulfur cycle, and, especially, the release of sulfur from
mineral sulfides.

Microbes decompose sulfur and many sulfide minerals. The sulfide
moiety is easily attacked and possesses a number of oxidative states. This
property alone makes it attractive in biochemical oxidations. The important
mineral sulfides are shown in Table 4.I. The related but rarer selenides,
tellurides, arsenides, and antimonides could also be classed in this group.
Important minerals in the latter categories are calaverite ($AuTe_2$), sylvanite
[$(Au, Ag)Te_2$], and skutterudite [$(Co, Ni, Fe)As_3$].

There are many other metal sulfides of varying degrees of importance.
Sulfide does not have to move far in ground water before it complexes
with some metal, e.g., La_2S_3, MgS, MnS, SnS, Sb_2S_3, FeS_n, WS_2, ZnS,
CaS, PbS, CuS, Cu_2S, HgS, RuS_2, and Ag_2S.

TABLE 4.I

IMPORTANT MINERAL SULFIDES

Mineral	Formula	Mineral	Formula
Argentite	Ag_2S	Covellite	CuS
Chalcocite	Cu_2S	Cinnabar	HgS
Bornite	Cu_5FeS_4	Realgar	AsS
Galena	PbS	Orpiment	As_2S_3
Sphalerite	ZnS	Stibnite	Sb_2S_3
Chalcopyrite	$CuFeS_2$	Bismuthinite	Bi_2S_3
Stannite	Cu_2FeSnS_4	Pyrite	FeS_2
Greenockite	CdS	Cobaltite	$(Co, Fe)AsS$
Pyrrhotite	$Fe_{1-x}S$	Marcasite	FeS_2
Niccolite	$NiAs$	Arsenopyrite	$FeAsS$
Millerite	NiS	Molybdenite	MoS_2
Pentlandite	$(Fe, Ni)_9S_8$		

OXIDATIONS OF INORGANIC SULFUR

The formation of sulfate as $CaSO_4$ (gypsum) or other salts by oxidation of reduced sulfur compounds is quite common in nature. Sulfate is also formed from organic sulfates by the enzyme sulfatase. Stable intermediates synthesized from reduced sulfur compounds are sulfite, thiosulfate, and elemental sulfur. Tetrathionate and other polythionates are also observed (Vishniac and Santer, 1957), Although heterotrophic microbes oxidize sulfur compounds, the thiobacilli and photosynthetic microbes possess special enzymes which catalyze sulfur oxidation. These latter microbes have been the most thoroughly investigated.

The photosynthetic anaerobic bacteria of the families Chlorobacteriaceae and Thiorhodaceae oxidize reduced inorganic sulfur compounds to sulfur or sulfate. At present the photosynthetic bacteria are not known to obtain energy from the oxidation of reduced organic sulfur compounds; however, oxidation of inorganic sulfur compounds appears to proceed as in other microbes.

Elemental sulfur is formed as a transient intermediate during the oxidation of thiosulfate (Lasada et al., 1961). This reaction is directly associated with photosynthesis. Polythionates are utilized and are probably formed during the oxidation of sulfide.

Reaction	ΔF (kcal)
SH^- $\longrightarrow$ S	-40
(Sulfide) (Sulfur)	
S $\longrightarrow$ $S_2O_3^{2-}$	-15
(Thiosulfate)	
$2\,S_2O_3^{2-}$ $\longrightarrow$ $S_4O_6^{2-} + 2^{e-}$	-18
$\longrightarrow$ (Tetrathionate)	
$S_4O_6^{2-}$ $\longrightarrow$ SO_4^{2-}	-100
(Sulfate)	
$2\,S_4O_6^{2-}$ $\longrightarrow$ $S_5O_6^{2-} + S_3O_6^{2-}$	4
$S_3O_6^{2-} + 3\,O_2 + 4\,e^-$ $\longrightarrow$ $3\,SO_4^{2-}$	-355

Thiobacillus

The genus was first described because of the unique ability of its members to obtain all their energy from the oxidation of inorganic sulfur compounds and all their carbon from CO_2.

Most thiobacilli are capable of oxidizing sulfide, elemental sulfur, thiosulfate, and tetrathionate and, in some instances, sulfite to sulfate. In addition, each species possesses unique characteristics which distinguishes it metabolically from other closely related members.

Thiobacillus thioparus

Thiobacillus thioparus oxidizes thiosulfate, tetrationate, and trithionate to sulfate. It is a strict autotroph which does not oxidize ferrous salts.

$$S_2O_3^{2-} \longrightarrow S_4O_6^{2-}$$
(Thiosulfate)   (Tetrathionate)
$$2\,S_4O_6 \longrightarrow S_5O_6^{2-} + S_3O_6^{2-}$$
(Trithionate)
$$S_3O_6 \longrightarrow SO_4^{2-}$$
(Sulfate)

Thiobacillus thioparus does not fix CO_2 unless thiosulfate is present. Low concentrations of 2,4-dinitrophenol of 10^{-4} M give 40–60% inhibition of CO_2 fixation.

Sulfide can be substituted for thiosulfate (Kelly and Syrett, 1963).

The sulfur present in calcareous rocks and limestone ores is oxidized when stored under humid conditions. Cultures of *T. thiooxidans* and *T. thioparus* oxidize native sulfur according to the reaction:

$$S + 1.5\,O_2 + CaCO_3 \rightharpoondown CaSO_4 + CO_2$$

During oxidation the sulfur content decreases, heat is produced, and gypsum is synthesized (Karavaiko *et al.*, 1962).

Cell-free extracts of *T. thioparus* were used to study the oxidation of ^{18}O-labeled compounds of thiosulfate to sulfate (Peck and Stulberg, 1962; Peck and Fisher, 1962). The initial reaction is a reductive cleavage of thiosulfate to sulfite and sulfide (Skarzynski *et al.*, 1957; Kelly and Syrett, 1964).

$$S_2O_3^{2-} \rightarrow SO_3^{2-} + S^{2-}$$

During the oxidation the ^{18}O is transferred from orthophosphate to adenosine monophosphate (AMP) and from AMP to sulfate as follows:

$$2\,SO_3^{2-} + 2\,AMP \rightarrow 2\,\text{Adenosine 5'-phosphosulfate (APS)} + 4e^-$$

$$2\,APS + P_i \rightarrow 2\,\text{Adenosine diphosphate (ADP)} + 2\,SO_4^{-2}$$

In the oxidation of $S_2O_3^{2-}$ to SO_4^{2-} by *T. thioparus*, the ^{18}O of $P^{18}O_4$ is incorporated into SO_4^{2-} (Santer, 1959). This suggested the formation of a S—O—P linkage which was later identified as adenosine 5'-phosphosulfate. The enzyme catalyzing this reaction is APS-reductase. In the fourth reaction catalyzed by ADP-sulfurylase, there is a cleavage of the P—O bond of APS while the S—O bond of the sulfate produced remains intact. Gluthathione is required in the oxidative step.

Tetrathionate is oxidized by *T. thioparus* after the lag phase of culture growth. Tetrathionate oxidation is increased by adding trace amounts of thiosulfate. Tetrathionate was metabolized anaerobically with the formation of thiosulfate (Trudinger, 1964a).

In the oxidation of ^{35}S-labeled $S_2O_3^{2-}$, ^{35}S was found in cystine, cysteine, methionine, cysteic acid, cysteinesulfonic acid, and a compound thought to be α-thioctic acid (Ostrowski *et al.*, 1959). In the oxidation of thiosulfate, small amounts of trithionate, pentathionate, and sulfur are produced and the pH drops from 7 to 3 (Pankhurst, 1964).

Trudinger (1964b,c), using a culture identified as *Thiobacillus*, observed that tetrathionate is hydrolyzed anaerobically to $S_2O_3^{2-}$, SO_4^{2-}, and small amounts of trithionate and pentathionate:

$$S_4O_6^{2-} + 2\,OH^- \leftrightarrow SO_3^{2-} + S_2O_3^{2-} + H_2SO_2$$

Trithionate oxidation increases with the addition of $S_2O_3^{2-}$. The oxidation proceeds rapidly in air in the presence of a dense bacterial population. A gas phase of 100% oxygen inhibits trithionate oxidation.

Under anaerobic conditions, trithionate is metabolized to SO_4^{2-} and $S_2O_3^{2-}$:

$$S_3O_6^{2-} + H_2O \leftrightarrow S_2O_3^{2-} + SO_4^{2-} + 2\,H^+$$

Several enzymes of the tricarboxylic acid cycle are present in whole cells and cell-free extracts of *T. thioparus*, e.g., aconitase, isocitric dehydrogenase, malic dehydrogenase, and succinic dehydrogenase (Cooper, 1964).

A new cytochrome, cytochrome *s*, was obtained from *T. thioparus*. It is similar to cytochrome *c* and cytochrome *f*; however, it differs from these by its oxidation-reduction potential: E_0 of -0.14 to -0.15 at pH 7.0 (Skarzynski *et al.*, 1956).

Thiobacillus concretivorus

Thiobacillus concretivorus was found by Parker (1945a,b) to be responsible for the corrosion of pipe. C. B. Taylor and Hutchinson (1947) isolated and cultivated this bacterium *in vitro* in a broth medium supplemented with Portland cement (Ciment fondu). *Thiobacillus concretivorus* differed from *T. thiooxidans* by oxidizing thiosulfate to free sulfur and sulfuric acid; *T. thiooxidans* produced only sulfuric acid.

Thiobacillus concretivorus is similar to *T. thiooxidans* and *T. thioparus* in the production of sulfuric acid from elemental sulfur (Parker and Prisk, 1953):

$$S_{\text{(Elemental)}} + 1.5\,O_2 + H_2O \leftrightarrow H_2SO_4$$

Thiobacillus concretivorus also oxidizes hydrosulfide to sulfuric acid:

$$HS_2 + 2\,O_2 \leftrightarrow H_2SO_4$$

No isotopic fractionation occurred when *T. concretivorus* was cultivated on labeled elemental sulfur. Upon addition of ^{32}S-labeled H_2S, an enrichment of ^{32}S was observed in sulfur and sulfate (Kaplan and Rafter, 1958; Kaplan and Rittenberg, 1962). Sulfur was also incorporated into cells of *Chromatium* and *Saccharomyces cerevisiae* without a change in valence.

Ammonium sulfate and ammonium phosphate induce microbial growth and corrosion of limestone monuments (Kauffmann and Taussant, 1954). Copper sulfate inhibited corrosion. Although nitrifying bacteria were present, *T. concretivorus* was the prime offender.

Concrete degradation or corrosion is greatly increased by both sulfate-reducing and sulfur-oxidizing microbes (Blondiau, 1960). Some super-sulfated cement improves the stability. Resistance of cement to microbial degradation depends upon: (a) hardness, (b) porosity and permeability, and (c) chemical nature of aggregate.

Cement should be low in CaO to minimize reaction with sulfate waters. Low CaO cements form Ca-sulfaluminate which strengthens concrete by crystallizing in the pores. If the CaO is high swelling, which physically aids corrosion, occurs. A low sulfide content in cement is desirable because *T. concretivorus* and other autotrophs oxidize sulfides to sulfuric acid. Sulfates of any alkaline earth form $CaSO_4$ and alkalies which promote hydrolysis of 2 $CaO \cdot Al_2O_3$. The most resistant cements are $CaO \cdot SiO_2$ which are low in CaO. Supersulfated cement requires less water for preparation but contains more water of hydration.The hardened cement is less porous and possesses compressive strengths up to 5768 psi.

The question still arises as to whether sulfur is formed as an intermediate in H_2S or mineral sulfide oxidation. In the oxidation of labeled H_2S by *T. concretivorus*, Kaplan and Rafter (1958) observed an enrichment of ^{32}S over ^{34}S and pointed out that sulfur was formed as an intermediate. This indicates that the sulfur was not formed from chemical oxidation. ^{32}S enrichment was greater in sulfate than in sulfur.

Thiobacillus thiooxidans

Thiobacillus thiooxidans was originally found in soils containing free sulfur and rock phosphate (Waksman and Joffe, 1922). It is a strict aerobe. Normally, extracellular deposition of either sulfur or polythionates does not occur.

The free energy available in sulfur and thiosulfate oxidation are adequate to support excellent growth.

Sulfur oxidation

$$S + 1.5\,O_2 + H_2O \rightarrow H_2SO_4$$
$$-\Delta F = 141,800 \text{ cal}$$

Thiosulfate oxidation

$$Na_2S_2O_3 + 2\,O_2 + H_2O \rightarrow Na_2SO_4 + H_2SO_4$$
$$-\Delta F = 211,100 \text{ cal}$$

Sulfuric acid is formed and the pH often drops to 0.4. Cells will survive pH 0.4, but cultures should not be maintained at such low pH levels, because growth is erratic on transfer to fresh media. Glucose is broken down but not metabolized. Ammonium salts are preferred sources of nitrogen.

The ratio of sulfur oxidized to carbon assimilated was determined by Waksman and Starkey (1923):

$$\frac{\text{Sulfur oxidized}}{\text{Carbon assimilated}} = 31.82$$

The free energy efficiency is:

1 gm carbon requires 9846 cal for assimilation
32 gm sulfur liberates 141,800 cal
31.82 gm sulfur liberates 140,900 cal

$$\frac{9846}{140,900} \times 100 = 6.2\%$$

Both *T. thiooxidans* and *T. concretivorus* are cultivated on medium at pH 3 containing 0.25 gm powder sulfur; 0.03 gm $(NH_4)_2SO_4$; 3.50 gm KH_2PO_4; 0.50 gm $MgSO_4 \cdot 7H_2O$; 0.25 gm $CaCl_2$; 0.01 gm $Fe_2(SO_4)_3 \cdot 9H_2O$; 1000 ml water. Nitrate may be substituted for ammonium for the growth of *T. thiooxidans*.

Sulfur oxidation by *T. thiooxidans* is highest at pH 3.0 although some activity is observed at pH 7.3 (Imai *et al.*, 1962). Sulfur undergoes an enzymic condensation reaction with SO_3^{2-} to form thiosulfate.

$$S + SO_3^{2-} \overset{\text{(Cells, Extracts)}}{\longleftrightarrow} S_2O_3^{2-}$$

Thiosulfate is only utilized at a pH below 4.5. Tetrathionate is formed as an intermediate.

$$2 S_2O_3^{2-} \longleftrightarrow S_4O_6^{2-}$$

Thiosulfate acts as an electron donor for the oxidation of sodium-α-glycerol phosphate. Trithionate undergoes a reductive cleavage to thiosulfate and sulfite at pH 7.0:

$$S_3O_6^{2-} \longleftrightarrow S_2O_3^{2-} + SO_3^{2-}$$

Thiosulfate is oxidized to tetrathionate and sulfate (Parker and Prisk, 1953);

$$3 S_2O_3^{2-} + 2.5 O_2 \longleftrightarrow S_4O_6^{2-} + 2 SO_4^{2-}$$

Tetrathionate is subsequently oxidized to sulfate and sulfuric acid:

$$S_4O_6^{2-} + O_2 \longleftrightarrow SO_4^{2-} + H_2SO_4$$

The actual molar ratio of sulfate and sulfuric acid formed by *T. thiooxidans* should be determined.

Sulfides are oxidized by intact cells to thiosulfate, sulfate, and polythionates (Suzuki and Werkman, 1959).

$$\text{Sulfide} \xrightarrow[\text{(Cells)}]{} S_2O_3^{2-} + SO_4^{2-} + \text{Polythionates}$$

$$\text{Sulfide} \xrightarrow[\substack{\text{(Cell} \\ \text{extracts)}}]{} S_2O_3^{2-} + \underset{\text{(Elemental)}}{S}$$

$$\underset{\text{(Elemental)}}{S} + \text{Glutathione} \leftrightarrow S_2O_3 + \text{Polythionates}$$

Cell-free extracts oxidize sulfide to thiosulfate and elemental sulfur. Elemental sulfur is oxidized by cell-free extracts to thiosulfate and polythionates only if glutathione is present.

It is doubtful that mineral sulfides are actively oxidized by *T. thiooxidans* (Bryner and Jameson, 1958). Pyrite may be one exception.

Phosphate at a concentration of 0.3% has been observed to retard bacterial oxidation (*T. thiooxidans*) of pyrite (Quispel *et al.*, 1952). This oxidation is also inhibited at pH 7.0 or over.

The oxidation of pyrite occurs at the surface of the crystal where ferrous ions are oxidized to Fe_2O_3. The sulfur from the surface layer is released or reacts to form polysulfides with pyrite molecules found in the deeper layers. The excess sulfide is oxidized by bacteria to produce sulfates.

3-Phosphoglyceric acid is one of the primary products formed in the early stages of the fixation of CO_2 and oxidation of sulfur by *T. thiooxidans* (Iwatsuka *et al.*, 1962). Cells exposed to sulfur or sulfide under aerobic conditions retain the ability to fix CO_2 under anaerobic conditions. Post-oxidative fixation of CO_2 requires a reducing substance such as S^{2-} plus energy-rich compounds which form during the pre-oxidation phase (Iwatsuka, 1962). Eighty percent of the ^{14}C of $NaHCO_3$ was found initially in phosphoglyceric acid and 8.0% in aspartic acid (Suzuki and Werkman, 1958). Preferential labeling was found in the carboxyl groups of phosphoglyceric acid and serine, the β-carboxyl groups of aspartic acid, and the α-carboxyl group of glutamic acid. An enzyme in *T. thiooxidans* catalyzes an irreversible carboxylation of phosphenolpyruvate to oxalacetate (Suzuki and Werkman, 1957). Thus, two carboxylating systems are functional in *T. thiooxidans*: the carboxylation of ribose diphosphate to phosphoglyceric acid and carboxylation of phosphenolpyruvate to oxalacetate.

Although no cytochromes were observed in earlier analyses of *T. thiooxidans* the microbe does contain cytochrome *c* (London, 1963), cytochrome 550 (Cook and Umbreit, 1963), and coenzyme Q_8. The formation of cytochrome s may depend upon the medium used for growth (Szczepkowski and Skarzynski, 1952).

The surfactant or wetting agent produced by *T. thiooxidans* which aids in sulfur oxidation is phosphatidylinositol (Schaeffer and Umbreit, 1963). Synthetic wetting agents also aid in elemental sulfur oxidation; sodium octylsulfate, polyoxyethylene sorbitan monooleate, α-naphthalene sulfonate, alkylbenzene sodium sulfonate, and triethanolamine dodecyl sulfate have been used successfully (Starkey *et al.*, 1956).

Thiobacillus ferrooxidans

Thiobacillus ferrooxidans, T. thiooxidans, and *Ferrobacillus ferrooxidans* have all been isolated from acid coal mine waters. Powell and Parr (1919) stated that sulfide oxidation in coal appeared to be hastened by bacteria or some catalytic agent. Later Colmer and Hinkel (1947), Temple and Delchamps (1953), and Temple and Koehler (1954) showed the relation of *T. thiooxidans* and *T. ferrooxidans* to sulfuric acid formation. The pH range in acid mine waters where these microbes normally subsist is usually 2.5–3.0.

Thiobacillus ferrooxidans is easily isolated from acid water draining from from bituminous coal mines. Isolation is completed on a medium at pH 3.5 containing (Leathen *et al.*, 1951): 1.0% $FeSO_4$ (sterilize by filtration and add separately); 0.015% $(NH_4)_2SO_4$; 0.005% KCl; 0.005% K_2HPO_4; 0.050% $MgSO_4 \cdot 7H_2O$; 0.001% $Ca(NO_3)_2$; 100 ml H_2O.

The optimal pH range for growth is 2–4 and should not exceed 4.5. The E_h range during growth extends from 16 to 30. The population of cells is increased by the addition of manganese sulfate, sodium sulfate, and greater quantities of ferrous iron. Pure cultures have been obtained by growth on solid media. Growth was best in either sterile mine water containing 3% agar or silica gel plates impregnated with the medium described by Leathen. Colonies formed on solid media are quite small. Cells are monotrichous and a chromatinlike substance is concentrated centrally in each cell. Colmer *et al.* (1949) observed growth on liquid and solid media supplemented with hyposulfite. Morphologically *T. ferrooxidans* is similar to other thiobacteria.

Although skepticism still exists concerning the role of microbes in iron oxidation, developments in the last decade demonstrate unequivocally the vital role of microbes in this process. This became possible when cultures of *T. ferrooxidans* oxidized iron under acid conditions where strictly chemical oxidation was limiting. For example, Leathen *et al.* (1956), using ferrous iron in acid mine water, obtained 200 mg/liter of ferric iron in

3 days using *T. ferrooxidans* as a biocatalyst. Chemical oxidation of a sterile solution by oxygen required more than 2 years.

During the oxidation of 120 gm of ferrous sulfate, 16.06 mg of carbon was fixed (Temple and Colmer, 1951). Assuming that the oxidation of iron yields 11.3 kcal/g-atom and the fixation of 1 g-atom of carbon requires 115 kcal, the efficiency of the reaction was 3.2%. This is quite low. Higher efficiencies of 20.5 and 30.0% have been reported (Silverman and Lundgren, 1959a,b; Lyalikova, 1958). Low efficiencies may be correlated with the larger expenditure of energy required to carry out metabolic reactions under highly acid conditions.

The rate of oxidation of ferrous salts is increased by addition of orthophosphate to aged phosphate-depleted cells of *T. ferrooxidans* (Beck and Shafia, 1964). The oxygen uptake was proportional to the phosphate added. Approximately 1.7 micromoles of CO_2 was fixed per 100 micromoles of O_2 absorbed. This increased to 8.0 micromoles of CO_2 fixed per 100 micromoles O_2 utilized upon oxidation of sulfur,

The question arises as to the importance of *T. ferrooxidans* in geochemistry and to whether sulfur or sulfides are oxidized. Hyposulfite is oxidized, while molecular sulfur is oxidized to sulfuric acid. Since these are not primary characteristics of the *T. ferrooxidans*, it is possible for this microbe to oxidize and leach iron from certain deposits leaving other sulfide ore deposits in place.

The presence of organic matter influences the metabolic activity and growth of *T. ferrooxidans*. Agar is believed by some to inhibit growth. Addition of organic acids like formate, acetate, etc., inhibit iron oxidation, whereas common carbohydrates have no effect. In oxidizing ferrous sulfate to ferric sulfate, 2 moles of sulfuric acid are required:

$$4 FeSO_4 + 2 H_2SO_4 + O_2 \rightarrow 2 Fe_2(SO_4)_3 + 2 H_2O$$

However, ferric sulfate is easily hydrolyzed with the formation of 2 moles of ferric hydroxide and 6 moles sulfuric acid. Thus, there is a net gain of 4 moles of sulfuric acid:

$$2 Fe_2(SO_4)_3 + 12 H_2O \rightarrow 2 Fe_2(OH)_6 + 6 H_2SO_4$$

Such reactions make it possible for *T. ferrooxidans* to induce acid conditions selective for its own growth.

Sulfide ores containing different metals are oxidized by *T. ferrooxidans* which is also one of the most active autotrophs catalyzing the oxidation of pyrite:

$$FeS_2 + H_2O + 3.5 O_2 \rightarrow FeSO_4 + H_2SO_4$$

Lyalikova (1961) used *T. ferrooxidans* to oxidize the iron sulfides of pyrite and marcasite which form inclusions in coal. About three-fourths of the sulfuric acid formed in the drainage of coal mines was synthesized by microbes. The biochemical oxidation of sulfides is three to ten times faster than chemical oxidation. Metals existing as sulfides are solubilized by this process. In addition to acting directly on the sulfides of metals, *T. ferrooxidans* oxidizes ferrous sulfate to ferric sulfate.

$$4 \, FeSO_4 + 2 \, H_2SO_4 + O_2 \rightarrow 2 \, Fe_2(SO_4)_3 + 2 \, H_2O$$

or:

$$12 \, FeSO_4 + 3 \, O_2 + 6 \, H_2O \rightarrow 4 \, Fe_2(SO_4)_3 + 4 \, Fe(OH)_3$$

The ferric sulfate will in turn oxidize metals sulfides.

$$FeS_2 + Fe_2(SO_4)_3 \leftrightarrow 3 \, FeSO_4 + 2 \, S$$
$$CuFeS_2 + 2 \, Fe_2(SO_4)_3 + H_2O + 3 \, O_2 \leftrightarrow CuSO_4 + 5 \, FeSO_4 + 2 \, H_2SO_4$$
$$2 \, ZnS + 2 \, Fe_2(SO_4)_3 + 2 \, H_2O + 3 \, O_2 \leftrightarrow 2 \, ZnSO_4 + 4 \, FeSO_4 + 2 \, H_2SO_4$$

The biochemical oxidation of chalcocite, bornite, and tetrahedrite has been observed by Bryner *et al.* (1954).

Chalcocite
$$4 \, Cu_2S + O_2 \rightarrow 4 \, CuS + 2 \, Cu_2O$$
$$Cu_2S + 2 \, O_2 \rightarrow CuSO_4 + Cu$$
$$4 \, CuS + 9 \, O_2 \rightarrow 4 \, CuSO_4 + 2 \, Cu_2O$$
$$4 \, Cu_2S + 9 \, O_2 \rightarrow 4 \, CuSO_4 + 2 \, Cu_2O$$
Bornite
$$2 \, Cu_5FeS_4 + 18 \, O_2 + 3 \, H_2O \rightarrow 8 \, CuSO_4 + Cu_2O + 2 \, Fe(OH)_3$$
Tetrahedrite
$$(6 \, Cu_2S \cdot Sb_2S_3) + 27 \, O_2 \rightarrow 12 \, CuSO_4 + 2 \, SbO_3$$

Sulfide is also leached from chalcopyrite by *T. ferrooxidans* (V. I. Ivanov *et al.*, 1961). Copper yields from chalcopyrite of one gm/per liter in 28 days have been observed (Razzell and Trussell, 1963). In these tests sodium chloride inhibited iron oxidation but not oxidation of sulfides.

If sulfur is limiting, not all the copper goes into solution; thus some native copper is left. In many instances, ferric sulfate is also formed biogenically and serves to oxidize chalcocite and many other minerals (see p. 71).

In addition to the copper minerals, *T. ferrooxidans* oxidizes the sulfides present in sphalerite (Malouf and Prater, 1961), millerite (Razzell and Trussell, 1963), and molybdenite (Bryner and Jameson, 1958).

Sphalerite
$$ZnS + 4 \, H_2O \underset{(O_2)}{\leftrightarrow} Zn^{2+} + SO_4^{2-} + 8 \, H^+ + 2e^-$$

Millerite

$$NiS + 4 H_2O \rightsquigarrow Ni^{2+} + SO_4^{2-} + SH^+ + 2e^-$$
$$(O_2)$$

Molybdenite

$$MoS_2 + O_2 + H_2O \rightsquigarrow H_2MoO_4, Mo^{5+} + H_2SO_4$$

or

$$MoS_2 + 9 O_2 + 6 H_2O \rightsquigarrow 2 H_2MoO_4 + 4 H_2SO_4$$

Covellite, arsenopyrite, pyrrhotite, and galena are all oxidized in a similar manner (V. I. Ivanov *et al.*, 1961).

Zimmerly *et al.* (1958) developed strains of *T. ferrooxidans* which would actively oxidize $FeSO_4$ to $Fe_2(SO_4)_3$ in solutions containing 17,000 ppm Zn, 12,000 ppm Cu, and 6290 ppm Al. The pH range was 1.5–7.5. Ammonium ions up to a concentration of 200 ppm increased bacterial activity. The copper concentration was increased in certain leaching solutions to 11.74 gm/liter. Leaching studies were completed upon copper-bearing pyrite ores, molybdenite concentrates from copper sulfide ores, low-grade chromite ore (high in iron), and titaniferous ore (high in iron and zinc sulfide) (Marmatite).

Sulfur-containing rocks in the Guardak, Nor-Su, and Vodin deposits contain high concentrations of cells of *T. thiooxidans* and *T. ferrooxidans*. *Thiobacillus ferrooxidans* was the most frequently observed microbe in the Akhtal deposits of polymetallic ores. Pyrite and chalcopyrite were oxidized with the formation of sulfuric acid. The lead deposits of galenite and fluorite in Takob did not contain cultures of *T. ferrooxidans* (M. V. Ivanov *et al.*, 1958).

Thiobacillus denitrificans

One nonphotosynthetic thiobacterium species possesses the ability to grow autotrophically under anaerobic conditions. This facultative anaerobe is known as *Thiobacillus denitrificans*. Nitrate is used in place of oxygen as an oxidizing agent. Under anaerobic conditions, *T. denitrificans* oxidizes inorganic sulfur to sulfate while nitrate is reduced to free nitrogen (Baalsrud and Baalsrud, 1954).

Denitrification:
$$6 NO_3^- + 2 H_2O \rightarrow 4 OH^- + 3 N_2 + 16 [O]$$
Sulfur oxidation:
$$5 S + 6 K^+ + 4 OH^- + 16 [O] \rightarrow K_2SO_4 + 4 KHSO_4$$
Overall reaction:
$$5 S + 6 KNO_3 + 2 H_2O \rightarrow K_2SO_4 + 4 KHSO_4 + 3 N_2$$
Thiosulfate oxidation:
$$2 Na_2S_2O_3 + 2 NaNO_2 + 2 H_2O \rightarrow Na_2S_4O_6 + 2 NO + 4 NaOH$$

Hydroxyl ions produced during denitrification tend to neutralize the hydrogen ions released in sulfur oxidation. Thus, the pH never changes as rapidly as it does with the aerobic thiobacilli. *Thiobacillus denitrificans* also grows aerobically and, under aerobic conditions, nitrate is not reduced. For anaerobic growth, the molar ratio of nitrate to thiosulfate should not exceed 1.25 as NO_2 accumulates and poisons the cells. The preferred molar ratio of nitrate/thiosulfate is 0.4. Nitrite inhibits sulfur oxidation by the oxidation of thiosulfate. In the oxidation of thiosulfate, nitrite is reduced to nitric oxide, nitrous oxide, and free N_2.

Carbon dioxide fixation by *T. denitrificans* is similar to that observed in *T. thiooxidans* cultures which is analogous to that presented by Calvin for photosynthetic organisms (Aubert *et al.*, 1957). *Thiobaccillus denitrificans* synthesizes cytochrome *c* (Aubert *et al.*, 1958).

The free energy available from the anaerobic oxidation of sulfur (Baas-Becking and Parks, 1927) and thiosulfate (Beijerinck, 1921) is given in the literature.

Sulfur
$$6\ KNO_3 + 5\ S + 2\ CaCO_3 \rightarrow 3\ K_2SO_4 + 2\ CaSO_4 + 2\ CO_2 + 3\ N_2$$
$$-\Delta F^{\circ}_{25^{\circ}C} = 660\ \text{kcal}$$

Thiosulfate
$$5\ Na_2S_2O_3 + 8\ KNO_3 + 2\ NaHCO_3 \rightarrow 6\ Na_2SO_4 + 4\ K_2SO_4 + 4\ N_2 + 2\ CO_2 + H_2O$$
$$-\Delta F^{\circ}_{25^{\circ}C} = 893\ \text{kcal}$$

Thiobacillus denitrificans fixes 1.0 gm of carbon/100 gm of sodium thiosulfate oxidized. The free energy efficiency of the organisms is calculated as follows:

$$780\ \text{gm}\ Na_2S_2O_3\ \text{liberates}\ 893,000\ \text{cal}$$
$$1\ \text{gm}\ Na_2S_2O_3\ \text{liberates}\ 1130\ \text{cal}$$
$$6\ CO_2 + 6\ H_2O \rightarrow C_6H_{12}O_6 + 6\ O_2$$
$$(3 \times 10^{-4}\ \text{atm})\qquad (0.2\ \text{atm})$$
$$-\Delta F^{\circ} = 708,900\ \text{cal}$$

Under these conditions, 1.0 gm of carbon requires:

$$\frac{708,900}{72} = 9846\ \text{cal}$$

for fixation. This energy is supplied by oxidation of 100 gm of sodium thiosulfate which liberates 1130×100 cal. The free energy efficiency is:

$$\frac{9846}{1130 \times 100} \times 100 = 8.7\%$$

Thiobacillus thiocyanoxidans

There are many microbes which obtain their energies by oxidizing compounds toxic to other forms of life. *Thiobacillus thiocyanoxidans* falls into this microbial category. Energy is obtained by oxidation of thiocyanate to sulfate, carbonate, and ammonia. A culture of *T. thiocyanoxidans* was originally isolated by Happold and Key (1934) from gas work effluent on a simple medium of thiocyanate and phosphate. Growth on organic media is not observed unless thiocyanate is present. Organic acids inhibit growth. Sulfur and thiosulfate may be substituted for thiocyanate. The oxidation of thiocyanate probably proceeds as follows:

$$NH_4CNS + 2 O_2 + 2 H_2O \rightarrow (NH_4)SO_4 + CO_2$$
$$-\Delta F^\circ = 220 \text{ kcal}$$

Youatt's investigations (1954) indicate that a stepwise hydrolysis of thiocyanate occurs; cyanate is formed first:

$$CNS^- + H_2O \rightarrow HCNO + SH^-$$
(Thiocyanate)　　　(Cyanate)

This is followed by hydrolysis of the cyanate to ammonia and carbon dioxide.

$$HCNO + H_2O \rightarrow NH_3 + CO_2$$

The sulfide synthesized in the first step is oxidized to sulfate via the mechanism characteristic of other thiobacilli. The second reaction is quite unique because both the carbon and nitrogen required for growth are produced. Concentrations of sulfide added as substrate should not exceed 16 M. Sulfide formed by hydrolysis of thioacetamide and thioacetate is oxidized to sulfate. The oxidation of tetrathionate is questionable.

Polythionates are synthesized from thiosulfate by *T thiocyanoxidans*. The production of Folin-Ciocalteu reaction substances parallels the accumulation of polythionate (Pratt, 1958). During thiosulfate oxidation, tetrathionate, trithionate, pentathionate, and elemental sulfur are produced (Pankhurst, 1964). Polarographic techniques and Starkey's alkali method for thionate analysis are required to quantitate the sulfur intermediates formed.

The British are using *T. thiocyanoxidans* to remove thiocyanate from either spent-gas works or coke-oven liquors. The bacteria are coated on the internal surfaces of tanks, on baffles, etc. Thiocyanate destruction after 17 day's operation was 1.21 gm of CNS^-/liter/day with an efficiency of 93.3%. Increasing the internal surface area 70% increased the CNS^- destruction rate to 1.65 gm of CNS^-/liter/day (Badger and Graham, 1961).

Thiobacillus novellus

Thiobacillus novellus is a facultative autotroph which grows well upon both autotrophic and heterotrophic media (Santer *et al.*, 1959). On organic media, thiosulfate is oxidized as a sole source of energy, and carbon dioxide provides the carbon. Growth also occurs on organic mineral salts medium containing either glutamate-citrate or glutamate as the primary source of energy and carbon. It is interesting to note that cells grown upon a heterotrophic medium are more resistant to radiation damage than those grown upon an autotrophic medium (Cheosakul, 1962).

Sulfite accumulates as an intermediate during thiosulfate oxidation by resting cells of *T. novellus*. When the ratio of cells to thiosulfate is increased, the oxidation proceeds all the way to sulfate (Deley and Van Pouche, 1961).

Polythionates or sulfur are not formed as intermediates in the oxidation of $Na_2S_2O_3$.

The ratio of sulfur oxidized to carbon assimilated is 56 (Starkey, 1934a,b):

$$Na_2S_2O_3 + H_2O + 2\,O_2 \rightarrow Na_2SO_4 + H_2SO_4$$
$$-\Delta F° = 211,100 \text{ cal}$$

The energy efficiency for carbon fixation is 5.3%. *Thiobacillus novellus* grows best at an alkaline pH of 8.0–9.0. Its tolerance in acid conditions is quite limited, and a pH of 5.0–5.5 restricts growth.

Other Thiobacilli

There are other less well-known species of thiobacillus such as *Thiobacillus coproliticus* which has been isolated from dung and oxidizes sulfur to obtain energy and *Thiobacillus neopolitanus* which has been isolated from concrete undergoing corrosion and oxidizes both sulfur and hydrogen sulfide to sulfuric acid.

To further confuse this genus, another facultative autotroph, *Thiobacillus trautweinii*, possessing characteristics similar to *T. novellus* and *T. denitrificans* has been described.

The close interrelationship of the common iron and sulfur autotrophs are shown in Fig. 4.1. *Thiobacillus thiooxidans* and *T. concretivorus* grow over a broad pH range extending from pH 0.5–6.0, *T. ferrooxidans* grows less well below pH 2.0, and *T. thioparus*, *T. novellus*, and *T. denitrificans* grow over pH ranges from 4.5–8.5. These last two autotrophs differ markedly from the first. *Thiobacillus denitrificans* is a facultative anaerobe, and *T. novellus* and *T. thioparus* are facultative autotrophs. Apparently the autotrophic thiobacteria that have adopted or retained their ability

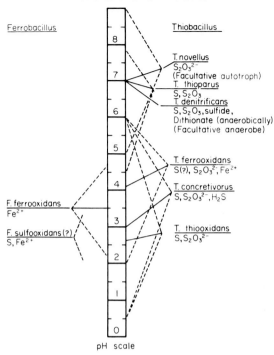

FIG. 4.1. pH range for sulfur autotrophs.

to grow at or near neutrality possess the enzymatic apparatus necessary for competition as heterotrophs. The *Ferrobacillus* cultures grow over a narrow pH range, probably from 2.0–4.7, although *F. sulfooxidans* may possibly grow under pH levels as low as 0.5.

Photosynthetic Microbes

Many of the photosynthetic microbes oxidize sulfur. They are quite ubiquitous and both the purple and the green types are easily isolated from muds. Oxygen is not produced as an end product as in green plants, but sulfide is required and is oxidized to sulfur by the green bacteria.

$$2\ H_2S + CO_2 \xrightarrow[\text{(Energy)}]{\text{(Light)}} (CH_2O) + H_2O + \quad 2\ S$$
$$\text{(Elemental)}$$

The photosynthetic purple bacteria oxidize sulfide directly to sulfate.

$$H_2S + 2\ H_2O + 2\ CO_2 \xrightarrow[\text{(Energy)}]{\text{(Light)}} 2\ (CH_2O) + H_2SO_4$$

Thus, sufficient oxidant capacity is generated photosynthetically to supply the energy to oxidize sulfide to either sulfur or sulfate. The growth requirements of this ecological group are (1) light; (2) minerals (carbonate, phosphate, chloride, iron, and sulfide); (3) alkaline pH; and (4) strict anaerobiosis.

The green sulfur bacteria comprise six genera falling in the family Chlorobacteriaceae. Of this family the genus *Chlorobium* has received the most study (Larsen, 1954).

Two species are well known: *C. limicola* which uses only sulfide as an electron donor and *C. thiosulfatophilum* which utilizes thiosulfate as an electron donor. In the presence of light *Chlorobium thiosulfatophilum* converts sulfides to elemental sulfur (Fedorov and Maksimov, 1965). Addition of either S^{2-} or $S_2O_3^{2-}$ gives rise to an increase of SO_4^{2-} in the culture medium. One percent sodium chloride insures good growth, and iron added in small quantities will give a good yield of chlorophyll. The iron is not complexed with the chlorophyll itself which is a magnesium porphyrin complex.

In *C. thiosulfatophilum*, sulfide is oxidized to sulfate. Sulfur is synthesized as an intermediate.

$$H_2S + CO_2 \xrightarrow[\text{(Energy)}]{\text{(Light)}} (CH_2O) + \quad 2\,S \quad + H_2O$$
$$\text{(Elemental)}$$

$$2\,S + 3\,CO_2 + 5\,H_2O \xrightarrow[\text{(Energy)}]{\text{(Light)}} 3\,(CH_2O) + 2\,H_2SO_4$$

Both thiosulfate and tetrathionate are oxidized by *C. thiosulfatophilum*. Thiosulfate oxidation occurs with the formation of sulfate:

$$Na_2S_2O_3 + 2\,CO_2 + 3\,H_2O \xrightarrow[\text{(Energy)}]{\text{(Light)}} 2\,(CH_2O) + Na_2SO_4 + H_2SO_4$$

Thus, the photosynthetic oxidation of sulfur compounds yields acid conditions which eventually provides suitable conditions for the growth of the chemosynthetic sulfur-oxidizing autotrophs. *Chlorobium* cultures can also use H_2 as well as sulfide as an electron donor.

$$2\,H_2 + CO_3 \xrightarrow[\text{(Energy)}]{\text{(Light)}} (CH_2O) + H_2O$$

Higher light intensities are required for optimal growth of photosynthetic bacteria in the presence of reducing agents such as Na_2S, $Na_2S_2O_3$, and sodium thioglycollate (Table 4.II) (Shaposhnikov *et al.*, 1963). In *Chloropseudomonas ethylicum*, addition of a reducing agent may increase the light intensity requirements from 34×10^3 to 211×10^3 erg/cm^2.

TABLE 4.II

LIGHT INTENSITY REQUIRED FOR GROWTH
OF *Chloropseudomonas ethylicum* IN THE
PRESENCE OF REDUCING AGENTS[a]

Reducing agents	Light intensity ($ergs/cm^2$)
Control	34×10^3
Na_2S	211×10^3
$Na_2S_2O_3$	211×10^3
Sodium thioglycollate	151×10^3
NaOAc	151×10^3
Ethyl alcohol	151×10^3

[a] From (Shaposhnikov *et al.*, 1963).

The formation of the pigment bacterioviridin increases with the increased requirement for higher light intensity.

The purple sulfur bacteria of the family Thiorhodaceae are more divergent in their metabolic capacities. They not only use sulfur compounds and hydrogen as electron donors, but they may also substitute certain organic compounds (A) for this purpose:

$$2 H_2A + CO_2 \xrightarrow[\text{(Energy)}]{\text{(Light)}} (CH_2O) + H_2O + 2 A$$

Since the A produced can be carbon dioxide, there has to be more of it produced than is needed for cellular synthesis before it can be observed.

Cultures of *Chromatium* of the family Thiorhodaceae, grow best at pH 8.6 with 0.25% $Na_2S \cdot 9 H_2O$ added as substrate. Magnesium (Mg^{2+}) is required (Naguib, 1959). The photochemical oxidation of H_2S to S and further to SO_4^{2-} appears to be readily reversible.

The turnover of sulfur compounds in *Chromatium okenii* at light saturation is controlled by a linear rate of fixation of carbon dioxide which amounts to 0.069 micromoles of CO_2/minute/mg cell protein (Trueper, 1964). Sulfate is produced with an increase in intracellular sulfur. The rate of oxidation of intracellular sulfur is highest after complete consumption of sulfide. The time during which sulfide is oxidized to intracellular sulfur amounts to one-third to one-fourth the time required for quantitative oxidation of sulfide to sulfate.

Chromatium okenii oxidizes a number of sulfur compounds: sulfide, thiosulfate, tetrathionate, sulfite, dithionite, thioglycollate, sulfoxylate, thioacetamide, dithiooxamide, 2-mercaptoethanol, and potassium tri-thiocarbonate (Trueper and Schlegel, 1964).

There is another group of sulfur bacteria which have been thoroughly studied. These are the nonsulfur purple bacteria belonging to the family Athiorhodaceae. *Rhodomicrobium vannielii* is the best known. These bacteria require growth factors (B vitamins) and are unusual in that they are able to grow anaerobically under photosynthetic conditions and aerobically in the dark while using the same substrates.

J. J. Taylor (1958) made a study of the ability of 22 pure cultures of photosynthetic bacteria to utilize sulfur compounds as hydrogen donors during photosynthesis. These cultures were isolated from ponds and marshes located on Pike Island (Ontario) and the Bass Islands (Western Lake Erie). Sulfur, thiosulfate, dithionite, sulfhydrate, and thioglycollate supported growth. Compounds not supporting growth were: sulfite, bisulfite, meta-bisulfite, pyrosulfate, thiocyanate, ethyl xanthate, methyl bisulfide, thioacetate, thioacetamide, thiourea, thiouracil, thiomalate, cysteine, and cystine.

Iron Bacteria

Sphaerotilus natans when exposed to H_2S will deposit globules of sulfur intracellularly:

$$2\,H_2S + O_2 \rightarrow \quad 2\,S \quad + 2\,H_2O$$
$$\text{(Intracellular)}$$

Thiothrix carries out the same reaction (Skerman *et al.*, 1957).

Fungi

The SH moiety in cysteine is oxidized to inorganic sulfate by *Aspergillus niger*, *A. oryzae*, *A. ruber*, and *A. repens*. Inorganic sulfide was not formed from cysteic acid, taurine, dithiodilactic acid, dithiodiglycol, and thioglycollic acid (Obata and Ishikawa, 1954).

Sulfate is stored as choline sulfate by many fungi (Spencer and Harada, 1960). Choline sulfate synthesis requires ATP and Mg^{2+} and involves the intermediate formation of adenosine 3′-phosphate 5′-sulfatophosphate (APS). The sulfate is transferred from APS to choline by a transferase or choline sulfokinase. The pathway of reutilization of choline sulfate is believed to involve transfer of sulfate to adenosine 3′,5′-diphosphate to form APS.

Algae

A question may be raised as to the ability of microbes such as algae to accumulate either sulfides or sulfates. In tests on six species of algae, 8–16 days were required before equilibrium with SO_4^{2-} occurred. Accumulation of S^{2-} was rapid with saturation occurring within three hours (Polikarpov, 1960). The internal cellular concentration of sulfate never increases above the external concentration; however, *Ceramium rubrum* will accumulate S^{2-} to a concentration twice that of the external level.

Heterotrophic oxidation of thiosulfate also occurs. *Pseudomonas aeruginosa, P. fluorescens,* and *Achromobacter stutzeri* oxidize thiosulfate to tetrathionate while increasing the alkalinity of the medium.

$$2 Na_2S_2O_3 + H_2O + 0.5 O_2 \rightarrow Na_2S_4O_6 + 2 NaOH$$
$$- \Delta F° + 21,000 \text{ cal}$$

How much of this energy goes to support heterotrophic growth is not known.

ELIMINATION OF SULFUR FROM COAL

Acid coal waters contain *Thiobacillus ferrooxidans* and *T. thiooxidans.* It has been suggested that *Thiobacillus ferrooxidans* catalyzes the rapid oxidation of pyrite and marcasite present in coal (27–30 days). The smaller the particle size of ore, the greater the rate of oxidation (Zarubina *et al.,* 1959).

Sulfuric acid is a common constituent of bituminous coal mine effluents. These effluents are also high in iron sulfate. Coal contains pyrite (FeS_2), marcasites, and other sulfuritic compounds, e.g., sulfur balls. In the absence of moisture, pyrite is oxidized as follows:

$$FeS_2 + 3 O_2 \rightarrow FeSO_4 + SO_2$$

In the presence of moisture, sulfuric acid is formed:

$$2 FeS_2 + 7 O_2 + 2 H_2O \rightarrow FeSO_4 + 2 H_2SO_4$$

The ferrous sulfate is oxidized further by oxygen with the reaction greatly facilitated by the presence of bacteria. The ferric sulfate formed is rapidly hydrolyzed to basic ferric sulfates:

$$4 FeSO_4 + 2 H_2SO_4 + O_2 \rightarrow 2 Fe_2(SO_4)_3 + 2 H_2O$$

The degree of hydrolysis depends on the acidity, temperature, and relative ionic concentrations. Hydrolysis can continue:

$$Fe_2(SO_4)_3 + 2\ H_2O \rightarrow 2\ Fe(OH)SO_4 + H_2SO_4$$

Hydrolysis may proceed until all the iron is in the form of ferric hydroxide. When no acid is present, ferrous sulfate is oxidized to basic ferric sulfate:

$$4\ FeSO_4 + O_2 + 2\ H_2O \rightarrow 4\ Fe(OH)SO_4$$

Basic ferric sulfate and ferric hydroxide give the characteristic gold and yellow color to deposits found in streams, etc.

Ferric sulfate per se is an excellent oxidant and, in combination with pyrite, ferrous sulfate and sulfide are formed:

$$Fe(SO_4)_3 + FeS_2 \rightarrow 3\ FeSO_4 + 2\ S^{2-}$$

The sulfide may also be oxidized chemically with ferric sulfate:

$$2\ S^{2-} + 6\ Fe_2(SO_4)_3 + 8\ H_2O \rightarrow 12\ FeSO_4 + 8\ H_2SO_4$$

Any elemental sulfur liberated in the penultimate reaction is likely to be oxidized to sulfuric acid through the catalytic action of *Thiobacillus thiooxidans* (Temple and Delchamps, 1953).

Leathen *et al.* (1953a,b) observed acid formation by *Thiobacillus thiooxidans* from elemental sulfur but not from marcasite or "sulfur balls."

Thiobacillus ferrooxidans produces sulfuric acid and sulfate from "sulfur ball" material and marcasite but not from pyrite. This microbe does oxidize ferrous iron in acid solution (Leathen *et al.*, 1953a,b,).

Ferrobacillus ferrooxidans and *Thiobacillus ferrooxidans* accelerate the solution of iron from pyrite and copper pyrites through the formation of ferric sulfate. Copper was leached from chalcocite, covellite, and bornite. *Thiobacillus concretivorus* was not capable of oxidizing the sulfide present in these minerals (Sutton and Corrick, 1963).

Ferrobacillus ferrooxidans increases the oxidation of pyrite 8–13-fold in coal mine waters. *Thiobacillus thiooxidans* does not function in pyrite oxidation. Sulfur is solubilized and removed from coal by this process (Rogoff *et al.*, 1960).

In addition to the autotrophic microorganisms cited, there are many acid-tolerant microflora in mine waters (Joseph, 1953). Bacteria are represented by the species *Bacillus, Micrococcus, Sarcina, Escherichia, Aerobacter, Crenothrix, Microsporium; Actinomycetes* were not found.

Fungi in the following genera were isolated: *Aspergillus, Trichoderma, Helminthosporium, Alternaria, Pencillium, Trichothecium,* and *Cladosporium.*

Of the diatoms, *Navicula viridis* was most abundant and may be used as an index of acidity. Protozoa were represented by *Euglena viridis, E. mutabilis,* ameba, and paramecium.

The knowledge gained in recent years concerning the biochemical reactions involved in acid mine drainage has not helped alleviate this problem of acid production. In a study of reaction mechanisms involved in acid production in nature, greater effort should be directed to the biochemistry.

More quantitative and kinetic data on the relative rates of chemical and biochemical actions are required. The microbes and enzymes catalyzing each reaction must be more effectively studied. Differences in pyrite and marcasite oxidation catalyzed by *Thiobacillus* sp. must be resolved (Hanna *et al.,* 1963). Table 4.III shows metals and minerals utilized by *T. thio-*

TABLE 4.III

ACTIVITY[a] OF MICROBES FROM MINE WATERS[b]

Material utilized	*Thiobacillus thiooxidans*	*Ferrobacillus ferrooxidans*	*Thiobacillus ferrooxidans*
Thiosulfate	+	0	+[c]
Sulfur	+	+	+
Iron	0	+	+
Sulfur balls	V	+	+
Pyrite	0	0	+
Marcasite	+	+	—
Chalcopyrite	0	—	+
Molybdenite	0	—	+

[a] V = variably active; + = active; 0 = not active.
[b] From Hanna *et al.* (1963).
[c] Activity later lost.

oxidans. F. ferrooxidans, and *T. ferrooxidans.* Even though there are distinct differences one can find merit in combining *F. ferrooxidans* and *T. ferrooxidans* cultures.

SULFUR METABOLISM IN SOIL

Thiosulfate treatment of soils increases manganese uptake by plants. This is particularly true of beets. Sulfur initiates the same reaction. This may be partially due to the thiosulfate synthesized. Thiosulfate affected root growth adversely in peas, cress, and cabbage (Quastel *et al.*, 1948).

In soil thiosulfate is converted to either sulfate and tetrathionate or sulfate and sulfur. The former two products are more commonly observed. Relatively high concentrations of phosphates tend to favor sulfur and sulfate synthesis.

Tetrathionate is converted to sulfate in soil and is a common intermediate in the oxidation of thiosulfate to sulfate. Amino acids and certain organic acids such as succinic acid increase the rate of oxidation of thiosulfate to tetrathionate; however, glucose and other carbohydrates favor the reduction of tetrathionate to thiosulfate. Thiosulfate, trithionate, and tetrathionate are degraded in soil; dithionate is not.

DESULFURIZING PETROLEUM

In the same way that the sulfur-oxidizing bacteria are active in the weathering of sulfide minerals, other microbes are active in desulfurizing petroleum (Kirshenbaum, 1961). Sulfate generated by this biochemical conversion may be removed with an electrolyte having an affinity for sulfate or with cultures of *Desulfovibrio* reducing the sulfate to H_2S (Harrison, 1963).

The production of sulfuric acid by sulfur-oxidizing bacteria both conditions and improves the filtration characteristics of sewage sludge. The use of elemental sulfur gave better results than $Na_2S_2O_3$ (Johnston and Ives, 1960).

BIOGEOCHEMICAL SULFUR CYCLE

Elemental sulfur is oxidized by thiobacteria through several intermediates to sulfate (Fig. 4.2). The important inorganic sulfur intermediates formed are sulfide, thiosulfate, tetrathionate, polythionates, trithionate, sulfite, and sulfate. These intermediates may exist in ground water or may

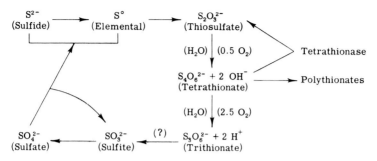

FIG. 4.2. Biogeochemical sulfur cycle.

be complexed with organic compounds. Many heterotrophic microbes produce the enzyme tetrathionase which catalyzes the reduction of tetrathionate to thiosulfate (Pollock and Knox, 1943), and of course, the sulfate reducers are active in reducing sulfate to sulfur and sulfide. While many of the thiobacteria oxidize trithionate, sulfite has not been established as an intermediate in sulfate formation. Sulfur is formed by biochemical processes both by the reduction of sulfates and the oxidation of sulfides. It should be noted (Table 4.IV) that the $^{32}S/^{34}S$ ratio is

TABLE 4.IV

DISTRIBUTION OF SULFUR[34] IN NATURE[a]

Location	Origin	$^{32}S/^{34}S$
Dorchester, Ontario	Organic (sulfureted water)	22.71
Gulf of Mexico	Organic	22.37
Gulf of Mexico	Organic	22.29
Tillsonburg, Ontario	Organic (sulfureted water)	22.29
Sudbury, Ontario	Organic (Copper Cliff Mine)	22.23
Italy	Volcanic	22.07
Sicily	Volcanic	22.03
Mt. Vesuvius, Italy	Volcanic	21.81

[a] From Macnamara and Thode (1951).

higher in deposits of organic origin than in volcanic deposits (Macnamara and Thode, 1951). The $^{32}S/^{34}S$ ratio of organically derived ore from Dorchester, Ontario was 22.71, whereas, this ratio decreased to 21.81 in volcanic minerals from Mt. Vesuvius, Italy.

HABITAT OF SULFUR OXIDIZERS

The sulfur oxidizers are found wherever sulfur or mineral sulfides occur. The more common habitats are summarized in Table 4.V. *Thiobacillus thioparus* and *T. thiooxidans* are commonly found in sulfur ores.

TABLE 4.V

HABITATS OF *Thiobacillus*

Habitat	Microbe	Product	Reference
Tarnabzhey S deposit	*T. thioparus*	—	Rogovskaya and
	T. denitrificans	—	Lazareva (1961)
Rozdol ore S	*T. thioparus*	SO_4^{2-}	Sokolova and Karavaiko
			(1963)
Rozdol ore, 25% S	*T. thioparus*	SO_4^{2-}	Sokolova and Karavaiko
			(1962)
Rozdol ore (S mesh limestone)	*T. thiooxidans*	$CaSO_4$	Karavaiko (1959)
Sulfur ore (limestone)	*T. thiooxidans*	—	Karavaiko (1959)
	T. thioparus	—	Karavaiko *et al.* (1962)
Oil brine (Kalinovka)	*T. thioparus*	S	Sokolova (1960)
Sergiev Springs (Na₂S)	*T. thioparus*	S	Ivanov (1957)
Venera "S" Spring (Italy)	*T. neopolitanus*	SO_4^{2-}	Cocuzza and Nicoletti
	T. concretivorus	SO_4^{2-}	(1961)
Elbe River (S_2O_3)	*T. thiooxidans*	—	Mutze and Engel (1960)
	T. denitrificans	—	
Coal water (S)	*T. thiooxidans*	—	Marchlewitz and
	T. ferrooxidans	—	Schwartz (1961)
Coal (pyrite)	*Ferrobacillus ferrooxidans*	—	Rogoff *et al.* (1960)
	T. thiooxidans	—	
Mine waste (S^{2-})	*T. thiooxidans*	SO_4^{2-}	Malouf and Prater
			(1961)
Peat soil (Kerala, India)	Sulfur bacteria		Subramoney (1960)
Marsh (Kzyl-Orda region)	*T. thioparus*	—	Ilyaletdinov and
	T. denitrificans	SO_4^{2-}	Kanatchinova (1964)
Marine Algonkian sediments	Sulfur bacteria	—	Kantor (1954); Fusan and Kantor (1953)
Black Sea (S)	Sulfur bacteria	SO_4^{2-}	Kriss (1954)
Concrete (S, H_2S)	*T. concretivorus*	H_2SO_4	Forrester (1959)
Concrete (H_2S)	*T. thioparus*	H_2SO_4	Gilchrist (1953)
	T. thiooxidans	—	
Budding stone	Thiobacteria	H_2SO_4	Pochon and Tchan (1948)

The former has been isolated from oil brine (Sokolova, 1960) and springs high in sulfide (M. V. Ivanov, 1957). Mutze and Engel (1960) isolated thiosulfate-oxidizing cultures of *T. thiooxidans* and *T. denitrificans* from the Elbe River. Coal supports a high population of sulfide oxidizers

TABLE 4.VI

MINERAL SULFIDES OXIDIZED AND LEACHED BY MICROBES

Mineral sulfide	Microbe	Location	Reference
Covellite (CuS)	*T. ferrooxidans*	Canada	Duncan and Trussell (1964)
Chalcocite (Cu$_2$S)	*T. ferrooxidans*	Canada	Duncan and Trussell (1964)
			Bryner and Jameson (1958)
Bornite (Cu$_5$FeS$_4$)	*T. ferrooxidans*	—	Duncan and Trussell (1964)
Chalcopyrite (CuFeS$_2$)	*T. ferrooxidans*	—	V. I. Ivanov *et al.* (1961)
			Duncan and Trussell (1964)
Cu and Co ores (S)	*T. thiooxidans*	—	Sutton and Corrick (1961)
CuS	*T. thioparus*	—	Sokolova (1960)
CuS (coal)	*T. ferrooxidans*	—	Kuznetsov (1963)
Marcasite (FeS$_2$)	*T. ferrooxidans*	Russia	Lyalikova (1961)
Pyrite (FeS$_2$)	*T. ferrooxidans*	Russia	Lyalikova (1961)
			V. I. Ivanov *et al.* (1961)
			Bryner and Jameson (1958)
Millerite (NiS)	*T. ferrooxidans*	—	Duncan and Trussell (1964)
Pyrrhotite (FeS)	*Arthrobacter*	—	Ehrlich (1963)
Sphalerite (ZnS)	*Arthrobacter*	—	Ehrlich (1963)
			V.I. Ivanov *et al.* (1961)
Galena (PbS)	*Arthrobacter*	—	Ehrlich (1963)
Realgar (AgS)	*Arthrobacter*	—	Ehrlich (1963)
	Hymphomicrobium	—	Ehrlich (1963)
Cu$_2$S	*T. ferrooxidans*	—	Ehrlich (1963)
Molybenite (MoS$_2$)	*T. ferrooxidans*	—	Bryner and Jameson (1958)
S (elemental)	*T. thiooxidans*	—	Parker and Prisk (1953)
	T. concretivorus	—	
	T. thioparus	—	
H$_2$S	*T. concretivorus*	—	Parker and Prisk (1953)

(Rogoff et al., 1960). Also peat soils (Subramoney, 1960) and marsh lands contain a high number of sulfur bacteria (Ilyaletdinov and Kanatchinova, 1964). The Black Sea is a prime habitat for sulfur bacteria (Kriss, 1954). Concrete containing even small quantities of either S^{2-} or H_2S supports a definite population of thiobacteria.

MINERAL SULFIDES OXIDIZED BY MICROBES

The mineral sulfides may be oxidized by thiobacteria which attack either or both the metallic and sulfide components of the mineral (Table 4.VI). Except for the iron-containing mineral sulfides, the sulfide is the preferred locus of attack. When iron is involved, ferrous forms may be oxidized to the ferric ion which may either precipitate as an hydroxide if the pH is sufficiently high or possibly may oxidize other metals and become reduced again to supply additional substrate for microbial growth. Of the copper minerals covellite, chalcocite, bornite (Duncan and Trussel, 1964), and chalcopyrite are oxidized by T. ferrooxidans (Duncan and Trussell, 1964; V. I. Ivanov et al., 1961). Marcasite and pyrite are also oxidized by T. ferrooxidans (Lyalikova, 1961). The mineral millerite (Duncan and Trussel, 1964) containing nickel and molybenite (Bryner and Jameson, 1958) are decomposed by T. ferrooxidans. Ehrlich (1963) found certain Arthrobacter active in oxidizing pyrrhotite, sphalerite, galena, and realgar. In addition, a culture of Hyphomicrobium was capable of decomposing realgar.

REFERENCES

Aubert, J. P., Milhaud, G., and Millet, J. (1957) .The path of carbon dioxide assimilation in chemo-autotrophic bacteria. Ann. Inst. Pasteur 92, 515–524.
Aubert, J. P., Milhaud, G., Moniel, C., and Millet, J. (1958). The existence, isolation, and physio-chemical properties of cytochrome c of Thiobacillus denitrificans. Compt. Rend. 246, 1616–1619.
Baalsrud, K., and Baalsrud, K. S. (1954). Thiobacillus dentitrificans. Arch. Mikrobiol. 20, 34–62.
Baas-Becking, L. G. M., and Parks, G. S. (1927). Energy relations in the metabolism of autotrophic bacteria. Physiol. Rev. 7, 85–106.
Badger, E. H. M., and Graham, P. W. (1961). Removal of thiocyanate. British Patent 884, 825.
Beck, J. V., and Shafia, F. M. (1964). Effect of phosphate ion and 2,4-dinitrophenol on the activity of intact cells of Thiobacillus ferrooxidans. J. Bacteriol. 88, 850–857.

Beijerinck, M. W. (1921). *Azotobacter chroococcum* as an indicator of the fertility of soil. *Koninkl. Ned. Akad. Wettenschap., Verslag Gewone Vergader, Afdel. Nat.* **30**, 431–438.

Blondiau, L. (1960). Suitability of supersulfated cement for the construction of sewers. *Rev. Mater. Construct. Trav. Publ.* C. **535**, 91–98.

Bryner, L. C., and Jameson, A. K. (1958). Microorganisms in leaching sulfide minerals. *Appl. Microbiol.* **6**, 281–287.

Bryner, L. C., Beck, J. V., Davis, D. B., and Wilson, D. G. (1954). Microorganisms in leaching sulfide minerals. *Ind. Eng. Chem.* **46**, 2587–2592.

Cheosakul, U. (1962). Comparison of the radiation sensitivity of cells of *Thiobacillus novellus* when grown and recovered on autotrophic and saprophytic media. *J. Natl. Res. Council Thailand*, **3**, 1–37.

Cocuzza, G., and Nicoletti, G. (1961). Metabolic characteristics of certain chemosynthesizing species of autotrophic thiobacteria isolated. *Atti. Soc. Peloritana Sci. Fis. Mat. Nat.* **7**, 769–775.

Colmer, A. R., and Hinkle, M. E. (1947). The role of microorganisms in acid mine drainage: A preliminary report. *Science* **106**, 253–256.

Colmer, A. R., Temple, K., and Hinkle, M. (1949). An iron-oxidizing bacterium from the drainage of some bituminous coal mines. *J. Bacteriol.* **59**, 317–328.

Cook, T. M., and Umbreit, W. W. (1963), The occurrence of cytochrome and coenzyme Q in *Thiobacillus thiooxidans*. *Biochemistry* **2**, 194–196.

Cooper. R. C. (1964). Evidence for the presence of certain tricarboxylic acid cycle enzymes in *Thiobacillus thioparus*. *J. Bacteriol.* **88**, 624–629.

Deley, J., and Van Pouche, M. (1961). Formation of sulfite during the oxidation of thiosulfate by *Thiobacillus novellus*. *Biochim. Biophys. Acta* **50**, 371–373.

Duncan, D. W., and Trussell, P. C. (1964). Advances in the microbiological leaching of sulfide ores. *Can. Met. Quart.* **3**, 43–55.

Ehrlich, H. L. (1963). Microbial association with some mineral sulfides. *Biochem. Sulfur Isotopes, Proc. Symp., Yale Univ.*, 1693, pp. 153–158. Yale Univ. Press, New Haven, Connecticut.

Fedorov, V. D., and Maksimov, V. N. (1965), Metabolism of sulfur compounds in cultures of photosynthesizing green sulfur bacteria *Chlorobium thiosulfatophilum*. *Dokl. Akad. Nauk SSSR* **160**, 1185–1186.

Forrester, J. A. (1959). Destruction of concrete caused by sulfur bacteria in a purification plant. *Surveyor* **118**, 881–884.

Fusan, O., and Kantor, J. (1953), Mineralgraphic observation of the "Alzbeta" sulfide deposits in Bystry Potok. *Geol. Sbornik (Bratislava)* **4**, 623–684.

Gilchrist, F. M. C. (1953). Microbiological studies of the corrosion of concrete sewers by sulfuric acid producing bacteria. *S. African Ind. Chemist* **7**, 214–215.

Hanna, G. P., Jr., Lucas, J. R., Randles, E. I., Smith, E. E., and Brant, R. A. (1963). Acid mine drainage research potentialities. *J. Water Pollution Control Federation* **35**, 275–296.

Happold, F. C., and Key, A. (1934). Purification of gas work liquors in the bacterial flora of sewage. *J. Hyg.* **32**, 573–580.

Harrison, W. M. (1963). Bacterial treatment of media containing hydrocarbons and sulfides. U.S. Patent 3,105,014.

Ilyaletdinov, A. N., and Kanatchinova, M. K. (1964). Microbiological transformation of sulfur compounds in periodically flooded soils of Kzyl-Orda region. *Mikrobiologiya* **33**, 118–125.

Imai, K., Okleuzumi, M., and Katagiri, H. (1962), Pathways of sulfur oxidation by *Thiobacillus thiooxidans. Koso Kagaku Shimpoziumu* **17**, 132–143.

Ivanov, M. V. (1957). Role of microorganisms in depositing sulfur in hydrogen sulfide springs of Sergleo mineral waters. *Mikrobiologiya* **26**, 338–345.

Ivanov, M. V., Lyalikova, N. N., and Kuznetsov, S. I. (1958). The role of sulfur bacteria in the decay of mountain rocks and sulfide ores. *Izv. Akad. Nauk SSSR, Ser. Biol.* **2**, 183–192.

Ivanov, V. I., Nagirnyak, F. I., and Stepanov, B. A. (1961). Bacterial oxidation of sulfide ores. I. The role of *Thiobacillus ferrooxidans* in the oxidation of Chalcopyrite and Sphalerite. *Mikrobiologiya* **30**, 688–692.

Iwatsuka, H. (1962). Metabolism of sulfur-oxidizing bacterium. III. Post-oxidative fixation of CO_2. *Plant Cell Physiol. (Tokyo)* **30**, 167–173.

Iwatsuka, H., Kuno, M., and Maruyama, M. (1962), Metabolism of sulfur-oxidizing bacterium. II. Systems of CO_2-fixation in *Thiobacillus thiooxidans. Plant Cell Physiol. (Tokyo)* **3**, 157–166.

Johnston, W. E., and Ives, K. J. (1960). Sewage-sludge conditioning with sulfur-oxidizing bacteria. *J. Biochem. Microbiol. Technol. Eng.* **2**, 401–410.

Joseph, J. M. (1953). Microbiological study of acid mine waters: Preliminary report. *Ohio J. Sci.* **53**, 125–127.

Kantor, J. (1954). The problem of the so-called mineralized sulfur bacteria and their stratigraphic distribution. *Geol. Prace* **2**, 29–41.

Kaplan, I. R., and Rafter, T. A. (1958). Fractionation of stable isotopes of sulfur by Thiobacilli. *Science* **127**, 517–518.

Kaplan, I. R., and Rittenberg, S. C. (1962). Fractionation of isotopes in relation to the problem of elemental sulfur transport of microorganisms. *Nature* **194**, 1098–1099.

Karavaiko, G. I. (1959). Biological influence in oxidation of sulfur compounds of Rozdolski deposits. *Mikrobiologiya* **28**, 846–849.

Karavaiko, G. I., Ivanov, M. V., and Srebrodolskii, B. I. (1962). Oxidation of stored sulfur ore. *Sov. Geol.* **5**, 133–139.

Kauffmann, J., and Taussant, P. (1954). Corrosion of stone. *Corrosion Anti-Corrosion* **2**, 240–244.

Kelly, D. P., and Syrett, P. J. (1963). Effect of 2,4-dinitrophenol on carbon dioxide fixation by *Thiobacillus. Nature* **197**, 1087–1089.

Kelly D. P., and Syrett, P. J. (1964). Inhibition of formation of adenosine triphosphate (ATP) in *Thiobacillus thioparus* by 2,4-dinitrophenol. *Nature* **202**, 597–598.

Kirshenbaum, I. (1961). Bacteriological desulfurization of petroleum. U.S. Patent 2,975,103.

Kriss, A. E. (1954). The role of microorganisms in the biological productivity of the Black Sea. *Uspekhi Sovremennoi Biol.* **38**, 86–110.

Kuznetsov, S. I. (1963). Biochemistry of sulfur. *Izv. Akad. Nauk. SSSR, Ser. Biol.* **28**, 668–680.

Larsen, H. (1954), "Autotrophic Microorganisms." Cambridge Univ. Press, London and New York.

Lasada, M., Nozaki, M., and Arnon, D. I. (1961). Photoproduction of molecular hydrogen from thiosulfate by *Chromatium* cells. *In* "Light and Life" (W. D. McElroy and B. Glass, eds.), pp. 570–575. Johns Hopkins Press, Baltimore, Maryland.

Leathen, W. W., McIntyre, L. D., and Braley, S. A. (1951). A medium for the study of the bacterial oxidation of ferrous iron. *Science* **114**, 280.

Leathen, W. W., Braley, S. A., and McIntyre, L. D. (1953a). The role of bacteria in the formation of acid from certain sulfuric constituents associated with bituminous coal. I. *Thiobacillus thiooxidans*. *Appl. Microbiol.* **1**, 61–64.

Leathen, W. W., Braley, S. A., and McIntyre, L. D. (1953b). The role of bacteria in the formation of acid from certain sulfuritic constituents associated with bituminous coal. II. Ferrous iron oxidizing bacteria. *Appl. Microbiol.* **1**, 65–68.

Leathen, W. W., Kinsel, N. A., and Braley, S. A. (1956). *Ferrobacillus ferrooxidans:* A chemosynthetic autotrophic bacterium. *J. Bacteriol.* **72**, 700.

London, J. (1963). Cytochrome in *Thiobacillus thiooxidans*. *Science* **140**, 409–410.

Lyalikova, N. N. (1958), Study of chemosynthetic processes in *Thiobacillus ferrooxidans*. *Mikrobiologiya* **27**, 656–659.

Lyalikova, N. N. (1961). Bacteria in the oxidation of sulfide ores. *Inst. Mikrobiol., Akad. Nauk SSSR* **9**, 134–143.

Macnamara, J., and Thode, H. G. (1951). The distribution of sulfur^{34} in native sulfur deposits. *Research (London)* **4**, 582–583.

Malouf, E. E., and Prater, J. D. (1961). Role of bacteria in the alteration of sulfide minerals. *J. Metals* **13**, 353–356.

Marchlewitz, B., and Schwartz, W. (1961). Microbe association of acid mine waters. *Z. Allgem. Mikrobiol.* **1**, 100–114.

Mutze, B., and Engel, H. (1960). Bacterial oxidation of sulfur in the River Elbe. *Arch. Mikrobiol.* **35**, 303–309.

Naguib, N. (1959). The microbiological activities of the purple sulfur bacteria. Mass assimulation. *Zentr. Bacteriol. Parasitenk., Abt. II* **112**, 649–665.

Obata, Y., and Ishikawa, Y. (1954). Biochemical studies of sulfur containing amino acids. Sulfur production from cysteine by *Aspergillus* and its mechanisms. *Nippon Nogeikagaku Kaishi* **28**, 70–73.

Ostrowski, W., Skarzynski, B., and Szczepkowski, T. W. (1959). Identification of organic sulfur compounds formed during thiosulfate oxidation by *Thiobacillus thioparus*. *Proc. 2nd U.N. Intern. Conf. Peaceful Use: At. Energy, Geneva, 1958* Vol. 24, pp. 43–46. United Nations, New York.

Pankhurst, E. S. (1964) Polarographic evidence of the production of polythionates during the bacterial oxidation of thiosulfate. *J. Gen. Microbiol.* **34**, 427–439.

Parker, C. D. (1945a). Corrosion of concrete. I. Isolation of a species of bacterium associated with the corrosion of concrete exposed to atmospheres containing hydrogen sulfide. *Australian J. Exptl. Biol. Med. Sci.* **23**, 81–90.

Parker, C. D. (1945b). Corrosion of concrete. II. Formation of *Thiobacillus concretivorus* in the corrosion of concrete exposed to atmospheres containing hydrogen sulfide. *Australian J. Exptl. Biol. Med. Sci.* **23**, 91–98.

Parker, C. D., and Prisk, J. (1953). Oxidation of inorganic compounds of sulfur by various sulfur bacteria. *J. Gen. Microbiol.* **8**, 344–364.

Peck, H. D., Jr., and Fisher, E., Jr. (1962). Oxidation of thiosulfate and phosphorylation in extracts of *Thiobacillus thioparus*. *J. Biol. Chem.* **237**, 190–197.

Peck, H. D., Jr., and Stulberg, M. P. (1962). O^{18} studies on the mechanism of sulfate formation and phosphorylation in extracts of *Thiobacillus thioparus*. *J. Biol. Chem.* **237**, 1648–1652.

Pochon, J., and Tchan, Y-T. (1948). Role of sulfur bacteria in the disintegration of building stone. *Compt. Rend.* **226**, 2188–2189.

Polikarpov. G. G. (1960). The role of marine benthos in the migration of sulfates and sulfides. *Nauchn. Dokl. Vysshei Shkoly, Biol. Nauki* pp. 103–106.

Pollock, M. R., and Knox, R. (1943). Bacterial reduction of tetrathionate. *Biochem. J.* **37**, 476–481.

Powell, A. R., and Parr, S. W. (1919). Forms in which sulfur occurs in coal. *Bull. Am. Inst. Mining Met. Eng.* pp. 2041–2049.

Pratt, D. P. (1958). Detection of polythionates on cultures of *Thiobacilli* by means of the Folin-Ciocalteu reagent. *Nature* **181**, 1075.

Quastel, J. H., Hewitt, E. J., and Nicholas, D. J. D. (1948). Control of manganese efficiency in soils. *J. Agr. Sci.* **38**, 315–322.

Quispel. A., Harmsen, G. W., and Otzen, D. (1952). The chemical and bacteriological oxidation of pyrite in soil. *Plant Soil* **4**, 43–55.

Razzell, W. E., and Trussell, P. C. (1963). Isolation and properties of an iron-oxidizing *Thiobacillus*. *J. Bacteriol.* **85**, 595–603.

Rogoff, M. H., Silverman, M. P., and Wender, I. (1960). Elimination of sulfur from coal by microbiological action. *Am. Chem. Soc., Div. Gas Fuel Chem.* **2**, 25–36 (Preprints).

Rogovskaya, T. I., and Lazareva, M. F. (1961). Intensification of biochemical purification of industrial sewage. III. Microbiological characteristics of the biofilm purifying H_2S containing sewage. *Mikriobiologiya* **30**, 699–702.

Santer, M. (1959). The role O^{18} phosphate in thiosulfate oxidation by *Thiobacillus thioparus*. *Biochem. Biophys. Res. Commun.* **1**, 9–12.

Santer, M., Bryer, J., and Santer, U. (1959). *Thiobacillus novellus*. I. Growth or organic acid inorganic media. *J. Bacteriol.* **78**, 197–202.

Schaeffer, W. I., and Umbreit, W. W. (1963). Phosphatidylinositol as a wetting agent in sulfur oxidation by *Thiobacillus thiooxidans*. *J. Bacteriol.* **85**, 492–493.

Shaposhnikov, V. N., Balitskaya, R. M., and Kondrateva, E. N. (1963). Effect of some reducing agents on development of green sulfur bacteria and the synthesis by them of bacterioviridin under various light intensities. *Dokl. Akad. Nauk SSSR* **151**, 708–711.

Silverman, M. P., and Lundgren, D. G. (1959a). Studies on the chemoautotrophic iron bacterium *Ferrobacillus ferrooxidans*. I. An improved medium and a harvesting procedure for securing high cell yields. *J. Bacteriol.* **77**, 642–647.

Silverman, M. P., and Lundgren, D. G. (1959b). Studies on the chemoautotrophic iron bacterium *Ferrobacillus ferrooxidans*. II. Manometric studies. *J. Bacteriol.* **78**, 326–331.

Skarzynski, B., Klimek, B., and Szczepkowski, T. W. (1956). Cytochrome in *Thiobacillus thioparus*. *Bull. Acad. Polon. Sci., Class (II)* **4**, 299–304.

Skarzynski, B., Ostrowski, W., and Krawizyk, A. (1957). Sulfur metabolism in *Thiobacillus thioparus*. *Bull. Acad. Polon. Sci., Ser. Sci. Biol.* **5**, 194–264.

Skerman, V. D. B., Dementjeva, G., and Carey, B. J. (1957). Intracellular deposition of sulfur by *Sphaerotilus natans*. *J. Bacteriol.* **73**, 504–512.

Sokolova, G. A. (1960). Microbiological production of sulfur from sulfidic water strata. *Mikrobiologiya* **29**, 888–893.

Sokolova, G. A., and Karavaiko, G. I. (1962). Biogenic oxidation of sulfur of the Rozdol ore under laboratory conditions. *Mikrobiologiya* **31**, 984–989.

Sokolova, G. A., and Karavaiko, G. I. (1963). Biogenic oxidation of sulfur in Rozdol ore in laboratory conditions. *Mikrobiologiya* **31**, 797–800.

Spencer, B., and Harada, T. (1960). Role of choline sulfate in the sulfur metabolism of fungi. *Biochem. J.* **77**, 305–315.

Starkey, R. L. (1934a). Products of the oxidation of thiosulfate by bacteria in mineral media. *J. Gen. Physiol.* **18**, 325–349.

Starkey, R. L. (1934b). The cultivations of organisms concerned in the oxidation of thiosulfate in mineral media. *J. Bacteriol.* **27**, 52–53.

Starkey, R. L., Jones G. E., and Frederick, L. R. (1956). Effects of medium agitation and wetting agents on oxidation of sulfur by *Thiobacillus thiooxidans. J. Gen. Microbiol.* **15**, 329–334.

Subramoney, N. (1960). Sulfur bacterial cycle. Probable mechanism of toxicity in acid soils of Kerola. *Sci. Cult. (Calcutta)* **25**, 637–638.

Sutton, J. A., and Corrick, J. D. (1961). Bacteria in mining and metallurgy. Leaching selected ores and mineral. Experiments with *Thiobacillus thiooxidans. U.S., Bur. Mines, Rept. Invest.* 5839, 1–16.

Sutton, J. A., and Corrick, J. D. (1963). Leaching copper minerals by means of bacteria. *Mining Eng.* **15**, 37–40.

Suzuki, I., and Werkman, C. H. (1957). Phosphoenol pyruvate carboxylase in extracts of *Thiobacillus thiooxidans*, a chemoautotrophic bacterium. *Arch. Biochem. Biophys.* **72** 514–515.

Suzuki, I., and Werkman, C. H. (1958). Chemoautotrophic fixation of carbon dioxide by *Thiobacillus thiooxidans. Iowa State Coll. J. Sci.* **32**, 425–483.

Suzuki, I., and Werkman, C. H. (1959). Glutathione and sulfur oxidation by *Thiobacillus thiooxidans. Proc. Natl. Acad. Sci. U.S.* **45**, 234–244.

Szczepkowski, T. W., and Skarzynski, B. (1952). The biochemistry of autotrophic sulfur bacteria. I. Cytochromes and hemoprotein enzymes in *Thiobacillus thioparus* and *Thiobacillus thiooxidans. Acta Microbiol. Polon.* **1**, 93–106.

Taylor, C. B., and Hutchinson, G. H. (1947). Corrosion of concrete caused by sulfur-oxidizing bacteria. *J. Soc. Chem. Ind. (London)* **66**, 54–57.

Taylor, J. J. (1958). Thiorhodaceae. I. The effect of sodium thioglycolate on the photosynthetic and dark metabolism of purple sulfur bacteria. *Can. J. microbiol.* **4**, 425–435.

Temple, K. L., and Colmer, A. R. (1951). The autotrophic oxidation of iron by a new bacterium, *Thiobacillus ferrooxidans. J. Bacteriol.* **62**, 605–611.

Temple, K. L., and Delchamps, E. W. (1953). Autotrophic bacteria and the formation of acid in bituminous coal mines. *Appl. Microbiol.* **1**, 255–258,

Temple, K. L., and Koehler, W. A. (1954). Drainage from bituminous coal mines. *West. Va. Univ., Eng. Expt. Stat., Res. Bull.* **25**.

Trudinger, P. A. (1964a). Products of anaerobic metabolism of tetrathionate by *Thiobacillus X. Australian J. Biol. Sci.* **17**, 446–498.

Trudinger, P. A. (1964b). Effects of thiosulfate and oxygen concentration on tetrathionate oxidation by *Thiobacillus X and Thiobacillus thioparus. Biochem. J.* **90**, 640–646.

Trudinger, P. A. (1964c). Metabolism of trithionate by *Thiobacillus X. Australian J. Biol. Sci.* **17**, 459–468.

Trueper, H. G. (1964). Sulphur metabolism in Thiorhodaceae. II. Stoichiometric relation of CO_2 fixation to oxidation of hyrogen sulfide and intracellular sulfur in *Chromtium okenii. Antonie van Leeuwenhoek, J. Microbiol. Serol.* **30**, 385–394.

Trueper, H. G., and Schlegal, H. G. (1964). Sulfur metabolism in Thiorhodaceae. I. Quantitative measurements on growing cells of *Chromatium okenii, Antonie van Leeuwenhoek, J. Microbiol. Serol.* **30**, 225–238.

Vishniac, W., and Santer, M. (1957). The *Thiobacilli. Bacteriol. Rev.* **21**, 195–213.

Waksman, S. A., and Joffe, J. S. (1922). Microorganisms concerned in the oxidation of sulfur in the soil. II. *Thiobacillus thiooxidans*, a new sulfur-oxidizing organism. *J. Bacteriol.* **7**, 239–256.

Waksman, S. A., and Starkey, R. L. (1923). The growth and respiration of sulfur-oxidizing bacteria. *J. Gen. Physiol.* **5**, 285–310.

Youatt, J. B. (1954). The metabolism of *Thiobacillus thiocyanoxidans*. *J. Gen. Microbiol.* **11**, 139–149.

Zarubina, Z. M., Lyalikova, N. N., and Shmuk, Ye. I. (1959). Investigation of microbiological oxidation of coal pyrite. *Izv. Akad. Nauk SSSR., Otd. Tekhn. Nauk., Met. i. Toplivo* **1**, 117–119.

Zimmerly, S. R., Wilson, D. G., and Prater, J. D. (1958). Cyclic leaching process employing iron-oxidizing bacteria. U.S. Patent 2,829,964.

5

SULFATE REDUCTION
AND SULFIDES

The involvement of sulfate reduction in the syngenesis of sulfur and metal sulfide ores makes this bioreaction the most important in mineral deposition (Temple, 1964). Sulfate-reducing microbes are ubiquitous (Postgate, 1959), but certain conditions are required for inducing their growth or control. Sulfate reducers are active in the corrosion of iron and aluminum alloys, desulfurization of oil, deposition of mineral sulfides, etc.

ENVIRONMENT

For efficient reduction of sulfate certain ecological conditions are necessary. An anaerobic environment with hydrogen or an adequate supply of organic matter are required. Sulfate reduction has a standard free energy change of $+ 14$ kcal. Normally these microbes tolerate high acidities, halophilic conditions, high concentrations of sulfides, and toxic levels of metals. Almost any form of organic matter will support growth of one or more of the microbes falling in this group, e.g., cellulose, lignin, sewage, etc. These microbes are good scavengers in organic waste products from natural or industrial sources.

Sulfide as used here connotes the most reduced form of sulfur. The particular form of sulfide which predominates depends upon the equilibrium of hydrogen sulfide, hydrosulfide, and sulfide ions, as well as the total sulfide concentration, pH, temperature, and pressure.

$$H_2S \rightleftharpoons HS^- + H^+ \rightleftharpoons S^{2-} + H^+$$

Biochemical sulfate reduction is of two types: (1) dissimilatory and (2) assimilatory. It is the dissimilatory type reduction in which sulfate is reduced in excess of that required for cellular growth. Dissimulatory sulfate reduction is of primary concern in biogeochemistry.

SULFATE-REDUCERS

The important sulfate-reducing bacteria are: *Desulfovibrio desulfuricans, Desulfovibrio aestuarii, Desulfovibrio rubentschikii, Desulfovibrio orientis,* and *Clostridium nigrificans.*

Desulfovibrio desulfuricans

Desulfovibrio desulfuricans grows readily upon an anaerobic liquid medium at pH 6.8 containing 0.3% yeast extract, 0.04% NH_4Cl, 0.025% K_2HPO_4, 0.02% $FeSO_4$, and adequate sulfate. The gaseous phase should be hydrogen. To obtain a suitable E_h, ascorbic acid (0.01%) should be added as a reducing agent.

When yeast extract is eliminated, no hydrogen consumption occurs (Goldner *et al.*, 1963). Gas chromatographic studies on these systems have shown that gaseous ammonia, hydrogen sulfide, methane, and carbon dioxide do not accumulate.

Inorganic Nutrient Requirements

Ammonium is the most readily assimilable nitrogen source for *Desulfovibrio desulfuricans* although ZoBell (1958) has used certain amino acids as a nitrogen source. Nitrate is also assimilable and is reduced to ammonia (Baumann and Denk, 1952). Most of the cellular carbon is derived from yeast extract, but bicarbonate will provide 3–20% of the carbon required (Postgate, 1960). The sulfate may be derived from almost any mineral containing this radical. Sodium sulfate, sodium thiosulfate, taurine, and choline sulfate are reduced, and hydrosulfide is formed (Murakami, 1952).

Marine isolates of *Desulfovibrio desulfuricans* will tolerate H_2S up to 1000–2500 ppm (Miller, 1950a,b). High concentrations of H_2S apparently do not affect hydrogenase activity. Absorbents such as activated charcoal (10 gm/100 ml medium) increases the rate of hydrogen utilization (Goldner *et al.*, 1963) while in natural environments absorbents such as clay, mud, sediments, or ferrous sulfide may act in a manner similar to activated charcoal.

The optimal molar ratio of $H_2:SO_4$ for good sulfate reduction is $4:1$. This ratio is influenced by the basal medium, especially NH_4Cl, yeast extract, and sodium bicarbonate. Phosphate has only a slight effect on the ratio of hydrogen to sulfate required.

Although *Desulfovibrio desulfuricans* is an anaerobe, it contains cytochrome c_3. Strict autotrophy is not a genus characteristic since all the carbon utilized is not derived from CO_2. Energy is obtained by the oxidation of molecular hydrogen, and hydrogenase is active in most cultures. The characteristic of hydrogen utilization may be a consequence of an endogenous dissimilatory reaction involving the phospho-clastic breakdown of pyruvate in which hydrogen oxidation is coupled to sulfate reduction (Peck, 1962):

$$SO_4^{2-} + 8\,H \rightarrow 4\,H_2O + S^{2-}$$

Since they are incapable of using molecular O_2, these microbes substitute the oxygen atoms present in the sulfate ion (SO_4^{2-}) and oxidize either organic compounds by removing hydrogen of elemental H_2.

The concentration of hydrogenase and cytochrome c_3 present in the cell is controlled by the amount of iron in the medium. Lactate and glucose are not used for growth in the absence of either iron or sulfate (Rossmoore *et al.*, 1964).

Desulfovibrio desulfuricans isolated from an ammonia plant operation reduced sulfate in the presence of lactate (Katagiri *et al.*, 1947) as follows:

$$4\,CH_3CH(OH)COONa + 2\,MgSO_4 + H_2 \rightarrow 3\,CH_3COONa + 2\,MgCO_3$$
$$+ NaHCO_3 + 8\,CO_2 + 8\,H + 2\,H_2S$$

$$2\,CH_3CH(OH)COONa + 3\,MgSO_4 \rightarrow 3\,MgCO_3 + Na_2CO_3 + 2\,CO_2$$
$$+ 2\,H_2O + 3\,H_2S$$

An acidity of pH 4.3 or less controls the rate of growth.

Cell-free extracts of *Desulfovibrio desulfuricans* reduce sulfate to sulfide in the presence of adenosine triphosphate (ATP) and hydrogen (Peck, 1962). The sulfate is in the form of adenosine 5'-phosphosulfate (APS). The APS is formed by ATP-sulfurylase and is subsequently reduced to sulfite and adenosine monophosphate by APS-reductase. Microbes which only reduce enough sulfate for self-assimilation utilize a different sulfate-reducing pathway in which 3'-phosphoadenosine 5'-phosphosulfate is the first intermediate formed.

Arylsulfatases which are concerned with the hydrolysis of aromatic sulfates have also been involved in the formation of active forms of sulfate. A conventional arylsulfatase was not found in *Desulfovibrio desulfuricans*.

The deoxyribonucleic acid content of the sulfate-reducing bacteria differ. At least four subgroups characterized by significantly different ratios of adenine plus thymine to guanine plus cytosine have been described by Sigal *et al.* (1963). The nonsporulating polar flagellated forms containing cytochrome c_3 and desulfoviridin are divided into two groups. Freshwater nonhalophilic strains exhibit a base ratio of 0.54–0.59 and halophilic or halotolerant strains a ratio of 0.74–0.77. The sporulating peritrichous strains without cytochrome and *Clostridium nigrificans* and *Desulfovibrio orientis* are separate and distinct from the above and each other possessing, respectively, base ratios of 1.20 and 1.43.

Desulfovibrio orientis

Desulfovibrio orientis, a new sulfate-reducing form, has been isolated from a soil near Singapore (Bawn *et al.*, 1958). Sulfate reduction in a lactate medium proceeds in a manner similar to that seen with other sulfate-reducing bacteria, yielding a metal sulfide and acetate.

$$Na_2SO_4 + 2\ NaOOCCHOHCH_3 \leftrightarrow Na_2S + 2\ H_2O + 2\ NaOOCCH_3$$

Either sulfite or thiosulfate may be substituted for sulfate but, in contrast to *Desulfovibrio desulfuricans*, this microbe does not reduce tetrathionate. Acetate, butyrate, and propionate cannot be substituted for lactate and this distinguishes the microbe from *D. rubentschickii*.

Desulfovibrio orientis forms spores and closely resembles *Clostridium nigrificans*, though it will not normally grow at 43°C or higher. Further differentiation between these two cultures is needed.

Clostridium nigrificans

Another active sulfate-reducing microbe is the thermophilic anaerobic bacterium *Clostridium nigrificans*. Hydrogenase-producing strains are quite common and are implicated in the corrosion of iron. Both freshwater and marine forms are encountered.

Freshwater cultures may be grown upon 0.1% NH_4Cl, Na_2SO_4, 0.01% $CaCl_2 \cdot 6\ H_2O$, 0.02% $MgSO_4$, 0.35% sodium lactate, 0.1% yeast extract, 0.05% K_2HPO_4, 0.01% glucose, 0.40% $FeSO_4$, and 0.01% ascorbic acid medium at pH 7.3. Thermophilic strains grow best at 55°C but certain isolates will grow well at 30°C (Sealey, 1951) making it hard to distinguish them from *Desulfovibrio orientis*.

Other Sulfate-Reducing Microorganisms

Many other microbes reduce sulfate by dissimilatory mechanisms. *Bacillus megatherium* produces H_2S from $(NH_4)_2SO_4$ over a pH range 5–8 (Bromfield, 1953).

Desulfovibrio hydrocarbonoclastica, isolated from alkaline ground water, produces H_2S. The water contained a low level of bi- or trivalent cations and a high level of bicarbonate. A bituminous oillike substance was formed when formate and bicarbonate were added as a sole source of carbon (Hvid-Hansen, 1951).

In dissimilatory sulfate reduction by *Aspergillus niger*, the formation of choline sulfate and "sulfur-proteins" occurs in the upper and lower levels of mycelial mats (Yanagita and Kogane, 1963). In the mycelia of *Penicillium chyrsogenum*, sulfate accumulated in 3'-phosphoadenosine 5'-phosphosulfate and choline-*O*-sulfate (Segel and Johnson, 1963).

Ascosporic members of *Aspergillus nidulans* assimilate sulfur from sulfate, sulfite, hyposulfite, thiosulfate, metabisulfite, methionine, cystine, and thiourea (Agnihotri, 1963).

Algae actively reduce sulfate. The sulfate reduced by *Chlorella pyrenoidosa* and even Euglena species appear in cysteine, glutathione, and homocysteine (Schiff, 1964). Free hydrogen sulfide production was not reported.

ORGANIC SUBSTRATES

The sulfate reducers utilize sulfate as an oxidant for numerous organic substrates in addition to hydrogen. Oxygen itself is one of the most potent inhibitors of sulfate reduction. The organic materials providing energy for sulfate reduction have not been enumerated in detail. Organic acids and carbohydrates are utilized rapidly. Grease, keratens, organic sludges (Pipes, 1960), spent distillery liquor (Ghose and Basu, 1961; Ghose *et al.*, 1964), and waste from viscose rayon plant (Mongait *et al.*, 1962) are all utilized.

The chemical oxygen demand (COD) of raw distillery liquor can be reduced from 158,000 ppm to 34,400 ppm in 24 hours by continuous anaerobic digestion with *Desulfovibrio desulfuricans*, *Desulfovibrio rubentschikii*, and *Clostridium nigrificans*. Approximately 35% of the organic matter is destroyed in 48 hours. The gases evolved comprised 40% CH_4, 30% CO_2, 6–8% H_2S, 1–2% H_2; the balance was nitrogen (Barta, 1962).

HYDROGEN SUBSTRATES

Desulfovibrio desulfuricans and other sulfate reducers possess the enzyme hydrogenase. Hydrogen is rapidly consumed in the presence of sulfate. Oxygen reversibly inhibits this activity. When both hydrogen and pyruvic acid are present, they are metabolized simultaneously (Senez, 1955). Hydrogen is oxidized in excess of that required for the hydrogen sulfide synthesis from sulfate (Sorokin, 1954). The energy produced is used for assimilation of carbon dioxide and synthesis of cellular organic matter. The enzyme hydrogenase is implicated in corrosion processes.

BIOSYNTHESIS OF MINERAL SULFIDES

Large deposits of sulfides of iron, copper, silver, zinc, lead, cobalt, nickel, molybdenum, arsenic, cadmium, and mercury are mined. These deposits are either of magmatic or hydrothermal origin or are associated with sedimentary rock. In syngenetic theories, it is postulated that metal sulfides were deposited along with pyrite in sedimentary rock (Temple, 1964). Since hydrothermal ore-bearing fluids are common, many metal sulfide deposits are formed epigenetically by introduction of ore metals and replacement reactions with pyrite.

The scope and degree of biogenic H_2S production in nature has already been stressed. The surface muds of the littoral zones of oceans and lakes are the natural habitats for the sulfate-reducing microbe. A black mud is visible evidence of sulfide production and iron sulfide deposition.

The biosynthesis of metal sulfides requires two conditions and takes place in two steps.

$$SO_4^{2-} + 8\ H \leftrightharpoons 4\ H_2O + S_2^- + Energy$$
$$R + S^{2-} \leftrightharpoons RS$$

where R = metal.

The sulfate-reducing microbes produce sulfide which reacts with specific metals to form the corresponding metal sulfide. Metal sulfide production, the second reaction, is purely inorganic. For example, iron compounds, either in solution or colloidal form in mud, react with biogenic H_2S to form amorphous hydrotroilite ($FeS \cdot n\ H_2O$). The latter compound is unstable and is converted to the iron disulfides pyrite and marcasite.

Maddox and Perry (1964) have used these reactions to plug or reduce the porosity and permeability of underground formations. Controversy still exists as to how many other metal sulfides are formed in this manner.

Metals are toxic to many sulfate-reducing microbes, but this should not be considered a serious problem to the biogenesis of sulfide because microbes mutate and adapt easily. Sulfate-reducing microbes do not normally select and precipitate specified metals. They act as hydrogen sulfide generators, and if metals are carried into an anaerobic basin, certain metallic sulfides will precipitate quantitatively. Such metals react with H_2S strictly on their specific chemical affinity for S^{2-} ions. Baas-Becking and Moore (1961) and Miller (1950a,b,) have prepared mineral sulfides using H_2S generated by microbes. Ferrous sulfides were obtained from steel wool, covellite (CuS) from malachite [$CuCO_3 \cdot Cu(OH)_2$] and chrysocolla ($CuSiO_3 \cdot H_2O$), digenite ($Cu_2 \cdot n$ S) from cuprous oxide (Cu_2O), argentite (Ag_2S) from AgCl and $AgCO_3$, galena (PbS) from $PbCO_3$ and $2 PbO_3 \cdot Pb(OH)_2$, and sphalerite (ZnS) from zinc wire and smithsonite ($ZnSO_3$). Artificial seawater was used as the aqueous medium. Addition of Ni and Co yielded the corresponding sulfides but no identifiable minerals. Manganese and mercury carbonates did not yield sulfides. Addition of iron and copper oxide always gave iron sulfide and covellite. Low-temperature studies on the bioformation of mixed sulfides are needed.

Addition of Bi, Co, Cr, Mn, Pb, and Sb to culture media results in an increase in sulfide accumulation. Removal of free H_2S by the formation of metal sulfides has a favorable effect upon microbial growth. Addition of Ni, Zn, Cd, and Cu partly inhibits growth (Hata, 1960).

Microbial copper toxicity is not a serious problem in the syngenesis of copper sulfides. Copper sulfate must be added to concentrations of 0.25–0.29% to inhibit sulfate reduction by bacteria (Temple and LeRoux, 1964).

The relation of microbes to the migration of molybdenum in underground waters near rare sulfide deposits can be clarifide if is it postulated that desulfating bacteria precipitate molybdenum salts. Sulfur-oxidizing microbes increased the molybdenum content of underground water near the Kazakhstan molybdenum deposits (Kramarenko, 1962).

ELEMENTAL SULFUR

Datta (1946) used a microbe to produce elemental sulfur from the $MgSO_4$ present in van Delden's medium. The reaction was carried out in four weeks at 30°C.

Kaplan and Rittenberg (1963) using isotopic fractionation during sulfate reduction found only a small enrichment of ^{32}S as organic sulfur. Oxidation of H_2S using chemosynthetic and photosynthetic bacteria gave as products elemental sulfur, sulfate, and an unidentified polythionate. The amount of elemental sulfur formed in nature through microbial sulfate reduction is self-limiting (Fig. 5.1).

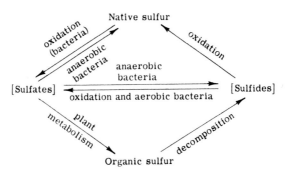

FIG. 5.1. The sulfur cycle.

DESULFURIZING PETROLEUM

There are conflicting reports concerning the anaerobic oxidation of hydrocarbons by sulfate-reducing bacteria. Rosenfeld (1947) obtained rapid decomposition of long-chain aliphatic hydrocarbons with *Desulfovibrio* sp. Fatty acids were formed and were in turn degraded. Hexadecane acted as a hydrogen donor for methylene blue reduction. A dehydrogenase is involved in these reactions. Other anaerobes were incapable of using hydrocarbons. Dutova (1961) was unable to obtain sulfate reduction with *Desulfovibrio desulfuricans* upon addition of hydrocarbons with a molecular weight less than that of decane. (M. W. 142). However, with mixed cultures sulfate reduction was observed in media containing heptane, nonane, and octane.

Harrison (1963) used a mixed culture of *Desulfovibrio aestuarii* and a mutant of *Pseudomonas fluorescens* to reduce the viscosity of crude oil to increase recovery of oil. Similar techniques have been used by Kolesnik and Shmonova (1957).

All tests made with the sulfate reducers do not show release of oil from petroleum-bearing sands. For example Updegraff and Wren (1954) did not observe oxidation of hexadecane or crude oil by 38 cultures of *Desulfovibrio*.

Sulfate-containing water in the Eastern Caucasian oil fields produces large amounts of H_2S. The primary microbial genera represented should be determined (Lindtrop, 1947).

Many of the sulfate reducers evidently can oxidize aliphatic hydrocarbons containing ten or more carbon atoms. Mixed cultures of *Desulfovibrio* and *Pseudomonas* are capable of oxidizing the lower molecular weight aliphatic hydrocarbons.

A strain of *Microsporum gypseum* reduced the sulfur content of crude petroleum from 3.0 to 2.2% (Brisou and Paquet, 1953). Tests with *Flavobacterium* and *Achromobacter* gave negative results. Direct splitting of the carbon-sulfur bonding is probably involved rather than sulfate reduction. The aerobic bacteria which desulfurize petroleum have been described by Kirshenbaum (1961).

CONTROL OF SULFATE REDUCERS

When practical the introduction of oxygen or air is used to control microbial sulfate reduction. Bactericidal action has been obtained with phenol, chlorophenols, chromates, zinc, copper, cadmium, and mercury ions (Gherardi and Bombara, 1960; Guynes and Bennett, 1958) and with monofluorophosphate (Bawn *et al.*, 1958).

The most active inhibitors of sulfate reduction are reported in Table 5.I. *Desulfovibrio desulfuricans* was used as the test organism. Normally it is less difficult to control than *Clostridium nigrificans*.

ECOLOGY OF SULFATE REDUCTION

Vast areas of the earth support an environment for the bioreduction of sulfate. Marine muds produce large volumes of hydrogen sulfide which reacts with any iron present to form iron sulfide ($2 FeS \cdot Fe_2S_3$). The remaining H_2S reacts with other metals or diffuses upward until it reaches an adequate oxygen supply where it is oxidized spontaneously or by sulfate-oxidizing microbes, e.g., *Chromatium*, and the photosynthetic bacteria (Kanwisher, 1962).

The sulfate-reducing environment should have an E_h of 600 and -400 mV. Sulfate-reducing microbes grow over a wide range of acidity and

TABLE 5.I

ANTIBACTERIAL AGENTS FOR SULFATE REDUCERS

Inhibitor	Concentration (ppm)
Group I[a]	
Phenol, 4-chloro-2-nitro-	50
Phenol, O-sec-amyl-	50
Phenol, 2-bromo-4-phenyl-	25
Phenol, 4-chloro-2-cyclohexyl-	25
Phenol, 2-chloro-4-nitro-	25
Group II[b]	
Methane, bromonitro-	50
Propanediol, 2-nitro-2-methyl-1,3-	50
Butanol, 3-chloro-3-nitro-2-	25
Butanol stearate, 2-chloro-2-nitro-1-	25
Butyl acetate, 2-chloro-2-nitro-	25
Hexanol, 4-chloro-4-nitro-3-	25
Pentanol, 1-chloro-1-nitro-2-	25
Pentanol, 3-chloro-3-nitro-2-	25
Propanol, 2-chloro-2-nitro-1-	25
Group III[c]	
Bis-p-chlorophenyl diguanidohexane	1
Quaternary ammonium compounds	10–50
Cetytrimethyl ammonium bromide	10
Proflavine	10
2, 2'-Dihydroxy-5, 5'-dichlorodiphenyl methane	15
Pentachlorophenol	50
Chloramphenicol	50
Potassium chromate	50

[a] From Bennett et al. (1958).
[b] From Bennett et al. (1960).
[c] From Bawn et al. (1958).

alkalinity although neutral or slightly alkaline conditions are preferred (Bass-Becking et al., 1960). Seawater and certain brine levels encourage growth. Growth inhibition has been observed in brines from Devonian strata (Kuznetsova, 1960). These brines contain 8.4 gm-equiv. of salt and a high bivalent : univalent cation ratio. The bivalent : univalent cation ratio becomes critical between 0.36–0.4 inless the gram-equivalence of salt decreases below 2.6. The growth of sulfate reducers in subsurface waters is favored by low mineralization (Dostalek and Kvet, 1964), and their concentrations may be used to determine the origin of subterranean waters.

Sulfate reduction in waterlogged soils increases the titrable alkalinity and decrease the available Cu^{2+} and Mg^{2+}. Hydrated Na_2CO_3 deposits in certain lakes of the Libyan Desert have resulted from bacterial sulfate reduction (Abd-El-Malek and Rizk, 1963).

The biogeochemical interest in sulfate reduction is widespread (Table 5.II). Laboratory yields of sulfide have been obtained as high as 3000 mg/liter (Butlin and Postgate, 1955). The amount of sulfate reduced was 24–31%. The addition of H_2 (Temple, 1964) and metals to precipitate sulfide will increase reduction.

The Black Sea is the classic example of a sulfate-reducing basin but a Libyan Lake, the Ain-ez-Zania, is far more active. Black Sea reactions are as follows:

$$2\,CaSO_4{}^+ + 2\,Cu^{2+} \leftrightarrow 2\,CuS + 2\,CO_2$$
$$CaS + 2\,CO_2 + 2\,H_2O \leftrightarrow Ca(HCO_3)_2 + H_2S$$
$$Ca(HCO_3)_2 \leftrightarrow CaCO_3 + CO_2 + H_2O$$

The Ain-ez-Zania contains 20 mg of H_2S/liter at the surface and 108 mg of H_2S/liter at the bottom (Butlin and Postgate, 1954). Lake Faro, a stagnant lake in Sicily, has a sulfide content similar to that of Lake Ain-ez-Zania (Genovese, 1962).

The rate of sulfate reduction in Lake Ain-ez-Zania provides enough sulfide so that 100 tons of sulfur/500,000 gal water is synthesized annually. The oxidation of the mineral sulfide to sulfur is mediated by the photosynthetic sulfur bacteria.

MICROFOSSILS AND RELICT ORGANIC COMPOUNDS

The sulfide present in the Kupferschiefer of Germany is arranged in small spheres 20 μ in diameter (Love, 1962). These are delicate microfossil remains of pyrite associated with *Pyritosphaera barbaria*. The principal forms were regarded as the remains of microbes which initiated the authigenic precipitation of the primary sulfides in the sediments of the foul-bottomed early Zechstein Sea of Europe. Similar formations are found in Marl Slate of northeast England which is a deposit from the same sea. Temple (1964) does not regard these small spheres as fossilized bacteria, although no question is raised as to their occurrence.

TABLE 5.II

Biochemistry of Sulfide Formation

Location	H$_2$S concentration (mg/liter)	Rate of synthesis (mg H$_2$S/liter/day)	Products	References
Copahue, Neuquen Territory, Argentina	28			Dalo (1955)
McMurdo Sound, Antarctica			Pyritic sediments	Barghoorn (1961)
Subterranean waters (Upper Tortonian Deposits Rabinsh limestone) (Russia)		0.01–2.0		Ivanov (1959b)
Volga Reservoir		0.09–13.1		Sorokin (1961)
Caspian Sea (oil-water flood)	629			Gasanov and Akhundov, (1959)
Water of Yazov S-Ved		3		Ivanov and Kostruba (1961)
Solenoe Lake (water-mud, water)		1.4 0.05		Ivanov (1959a)
Black Sea	100		Pyrite, marcasite	Volkov (1961) Sorokin (1962)

Russia
Northern
Near Baltic
Central
Urals Volga
Western Ukraine
Azov-Black Sea
Kerch
Northern Caucasus
Azerboidzhan
Sochi-Abkhaz
Western Turkmenia
Kopet-Dag
Surkhan-Dar'ya Fergana
Angara-Lema
Caucasian Mineral Springs

Yarotskii (1960)

Lacustrine milieu

Calcite ($CaCO_3$)
Aragonite($MgCO_3$)
Siderite ($FeCO_3$)

20

Volkova and Toshinskaya (1961)
Riviere and Vernhet (1962)

REFERENCES

Abd-El-Malek, Y., and Rizk, S. G. (1963). Bacterial sulfate reduction and the development of alkalinity. I. II. III. *J. Appl. Bacteriol* **26**. 7–26.

Agnihotri, V. P. (1963). Aspergilli. VIII. Sulphur requirements of some ascosporic members of the *Aspergillus nidulans* group. *Pathol. Microbiol.* **26**, 810–816.

Baas-Becking, L. G. M., and Moore, D. (1961). Biogenic sulfides. *Econ. Geol.* **56**, 259–272.

Baas-Becking, L. G. M., Kaplan, I. R., and Moore, D. (1960). Limits of the natural environment in terms of pH and oxidation-reduction potentials. *J. Geol.* **68**, 243–284.

Barghoorn, E. S. (1961). Sulfate-reducing bacteria and pyritic sediments in Antarctica. *Science* **134**, 190.

Barta, J. (1962). Decontamination of industrial effluents by means of anaerobic continuous action of sulfur-reducing bacteria. *Continuous Cultivation Microorganisms* **2**, 325–327.

Baumann, A., and Denk, V. (1952). The physiology of sulfate reduction. *Arch. Mikrobiol.* **15**, 283–307.

Bawn, C. E. H., Roffey, F., Topley, B., Melville, H., and Pratt, D. D. (1958). Natural Chemical Laboratory Steering Committee. Dept. Sci. Ind. Res., 97pp. Dawson and Goodall, London.

Bennett, E. O., Guynes, G. J., and Isenberg, D. L. (1958). The sensitivity of sulfate-reducing bacteria to antibacterial agents (phenolic compounds). *Producers Monthly* **23**, 18–19.

Bennett, E. O., Guynes, G. J., and Isenberg, D. S. (1960). The sensitivity of sulfate-reducing bacteria to antibacterial agents. III. The nitroparaffin derivatives. *Producers Monthly* **25**, 26–27.

Brisou, J., and Paquet, R. (1953). Attempt at desulfurizing petroleum by bacteria. *Compt. Rend.* **236**, 862–863.

Bromfield, S. M. (1953). Sulfate reduction in partially sterilized soil exposed to air. *J. Gen. Microbiol.* **8**, 378–390.

Butlin, K. R., and Postgate, J. R. (1954). The microbiological formation of sulfur in Cyrenaican lakes. *In* "Biology of Deserts" (J. L. Cloudsley-Thompson, ed.), pp. 113–122. Inst. Biol., London.

Butlin, K. R., and Postgate. J. R. (1955). Microbiological formation of sulfide and sulfur. *Proc. 6th Intern. Congr. Microbiol., Rome, 1953* pp. 126–143. Staderini, Rome.

Dalo, H. R. (1955). Sulfate-reducing bacteria. *Rev. Asoc. Bioquim. Arg.* **20**, 42–47.

Datta, S. C. (1946). Sulphate reduction and production of elemental sulfur by bacteria. *J. Sci. Ind. Res. (India)* **1B**, 28–30.

Dostalek, M., and Kvet, R. (1964). Utilization of osmotolerance of sulfate-reducing bacteria in a study of the origin of subterranean waters. *Folia Microbiol. (Prague)* **9**, 103–114.

Dutova, E. N. (1961). Significance of the presence of sulfate-reducing bacteria in oil exploration as it was observed during study of subsurface water in Central Asia. *Tr. Inst. Mikrobiol., Akad. Nauk SSSR* **9**, 101–104.

Gasanov, M. V., and Akhundov, A. R. (1959). The appearance of H_2S in water floods using Caspian Sea water. *Azerb. Neft. Khoz.* **38** 29–32.

Genovese, S. (1962). Sulle condizione fisico-chimiche dello stagno di Faro in seguito cell operatura di un nuovo canale. *Atti. Soc. Peloritana Sci. Fis. Mat. Nat.* **8**, 67–72.

Gherardi, D., and Bombara, G. (1960). Control of the sulfate-reducing bacteria. *Ann. Univ. Ferrara, Sez.* **2**, 653–669.

Ghose, T. K., and Basu, S. K. (1961), Bacterial sulfide production from sulfate-enriched spent distillery liquor. *Folio Microbiol.* (*Prague*) **6**, 335–341.

Ghose, T. K., Mukherjee, S. K., and Basu, S. K. (1964). Bacterial sulfide production from sulfate-enriched spent distillery liquor. III. *Biotechnol. Bioeng.* **6**, 285–298.

Goldner, B. H., Otto, L. A., and Carfield, J. H. (1963). Studies on sulfate-reducing bacteria. *1st Bien. Symp. Microbiol., Long Beach, Calif.*, 1962, pp. 16–26. Am. Petrol. Inst., Dallas, Texas.

Guynes, G. J., and Bennett, E. O. (1958). The sensitivity of sulfate-reducing bacteria to antibacterial agents (the mercurials). *Producers Monthly* **23**, 15–17.

Harrison, W. M. (1963), Bacterial treatment of media containing hydrocarbons and sulfides. U.S. Patent 3,105,014.

Hata, Y. (1960). Influence of heavy metals upon the growth and the activity of marine sulfate-reducing bacteria. *Norinsho Suisan Koshusho Kenkyu Hokoku* **9**, 363–375.

Hvid-Hansen, N. (1951). Sulfate-reducing hydrocarbon producing bacteria in ground water. *Acta. Pathol. Scand. Microbiol.* **29**, 314–334.

Ivanov, M. V. (1959a). The sulfur cycle in lakes, with use of radioactive sulfur (S^{35}). *Tr. 6-90* (Shestogo) *Soveshch. op Probl. Biol. Vnutren. Vod.* pp. 152–157.

Ivanov, M. V. (1959b). The role of sulfate-reducing bacteria in the formation of hydrogen. sulfide in subterranean waters of the Upper Tortonian deposits. *Mineralog. Sb., L'vovs. Geol. Obshchestvo pri L'vovsk. Gos. Univ.* **13**, 178–189.

Ivanov. M. V., and Kostruba, M. F. (1961). Microbiological investigation of the Carpathian sulfur deposits. III. Formation of hydrogen sulfide. *Mikrobiologiya* **30**, 130–134.

Kanwisher, J. (1962). Sulfur chemistry in marine muds. *Biochem. Sulfur Isotopes, Proc. Symp., Yale Univ., 1962*, pp. 94–103. Yale Univ. Press, New Haven, Connecticut.

Kaplan, I. R., and Rittenberg, S. C. (1963). Microbiological fractionation of sulfur isotopes. *Biochem. Sulfur Isotopes, Proc. Symp., Yale Univ., 1963.* pp. 80–93. Yale Univ. Press, New Haven, Connecticut.

Katagiri, H., Kanie, M., and Ishida, T. (1947). Sulfur bacteria. *Rept. Chem. Res. Inst., Kyoto Univ.* **16**, 25–26.

Kirshenbaum, I. (1961). Bacteriological desulfurization of petroleum. U.S. Patent 2,975,103.

Kolesnik, Z. A., and Shmonova, N. I. (1957). Change in petroleum composition by bacteria of the *Pseudomonas* species under anaerobic concentrations. *Dokl. Akad. Nauk SSSR* **115**, 1197–1199.

Kramarenko, L. E. (1962). Bacterial biocenoses in underground waters of beds of some useful fossils and their geological significance. *Mikrobiologiya* **31**, 694–701.

Kuznetsova, V. A. (1960). Occurrence of sulfate reducers in oil-bearing strata of the Kuibyshev field as related to salt composition of stratum brines. *Mikrobiologiya* **29**, 408–414.

Lindtrop. N. T. (1947). Reduction of sulfates in Grozny petroleum deposits. *Dokl. Akad. Nauk SSSR* **57**, 923–926.

Love, L. G. (1962). Biogenic primary sulfide of the Permian Kupferschufer and Marl Slate. *Econ. Geol.* **57**, 350–366.

Maddox, J., Jr., and Perry, R. B. (1964). Treatment of underground formations to render them less permeable. U.S. Patent 3,118,500.

Miller, L. P. (1950a). Tolerance of sulfate-reducing bacteria to hydrogen sulfide. *Contrib. Boyce Thompson Inst.* **16**, 78–83.

Miller, L. P. (1950b). Formation of metal sulfides through the activities of sulfate-reducing bacteria. *Contrib. Boyce Thompson Inst.* **16**, 85–89.

Mongait, I. L., Kulakov, E. A., Iskhanova, E. B., and Vandyuk, N. V. (1962). Biological purification of waste waters from synthetic fiber manufacture. I. purification of general industrial waste. *Ochistka Prom. Stochnykh. Vod Vses. Nauchn.-Issled. Inst. Voldosnabzh. Kanz. Gidrotekn. Soorenii i Inzh. Gidrogeol. Sb. No.* **3**, 154–166.

Murakami, E. (1952). Sulfate-reducing bacteria. I. *Symp. Enzyme Chem. (Tokyo)* **7**, 53–54.

Peck, H. D., Jr. (1962). Role of adenosine 5'-phosphosulfate in the reduction of sulfate to sulfite by *Desulfovibrio desulfuricans*. *J. Biol. Chem.* **237**, 198–203.

Pipes, W. O. (1960). Sludge digestion by sulfate-reducing bacteria. *Purdue Univ., Eng. Bull., Ext. Ser.* **106**, 308–319.

Postgate, J. R. (1959). Sulfate reduction by bacteria. *Ann. Rev. Microbiol.* **13**, 505–520.

Postgate, J. R. (1960). On autotrophy of *Desulfovibrio desulfuricans*. *Z. Allgem. Mikrobiol.* **1**, 53–56.

Riviere, A., and Vernhet, S. (1962). Euxinic regime and carbonate sedimentation in lacustrine milieu. *Compt. Rend.* **255**, 3013–3015.

Rosenfeld, W. D. (1947). Anaerobic oxidation of hydrocarbons by sulfate-reducing bacteria. *J. Bacteriol.* **54**, 664–668.

Rossmoore, H. W., Shearer, M. E., and Shearer, C. (1964). Growth studies on *Desulfovibrio desulfuricans*. *Develop. Ind. Microbiol.* **5**, 334–342.

Schiff, J. A. (1964). Sulfate utilization by Algae. II. Identification of reduced compounds formed from sulfate by *Chlorella*. *Plant Physiol.* **39**, 176–179.

Sealey, J. Q. (1951). The biology of *Clostridium nigrificans*. Ph.D. Thesis, University of Texas, Austin, Texas.

Segel, J. H., and Johnson, M. M. (1963). Intermediates in inorganic sulfate utilization by *Penicillium chrysogenum*. *Arch. Biochem. Biophys.* **103**, 216–226.

Senez, J. C. (1955). Consumption of molecular hydrogen by non-proliferating suspensions and by cell extracts of *Desulfovibrio desulfuricans*. *Bull. Soc. Chem. Biol.* **37**, 1135–1146.

Sigal, N., Senez, J. C., Gall, J. L., and Sebald, M. (1963). Base composition of the deoxyribonucleic acid of sulfate-reducing bacteria. *J. Bacteriol.* **85**, 1315–1318.

Sorokin, Y. I. (1954). Chemistry of the process of hydrogen reduction of sulfates. *Tr. Inst. Mikrobiol., Akad. Nauk SSSR* **3**, 21–34.

Sorokin, Y. I. (1961). The content and the rate of sulfides forming in the loams of Volga reservoirs in 1959. *Byul. Inst. Biol. Vodokhranilishch, Akad. Nauk SSSR* **4**, 44–48.

Sorokin, Y. I. (1962). Determinations of bacterial sulfate reduction in the Black Sea with S^{35}. *Mikrobiologiya* **31**, 402–410.

Temple, K. L. (1964), Syngenesis of sulfide ores: An evaluation of biochemical aspects. *Econ. Biol.* **59**, 1473–1491.

Temple, K. L., and LeRoux, N. W. (1964). Syngenesis of sulfide ores: Sulfate-reducing bacteria and copper toxicity. *Econ. Geol.* **59**, 271–278.

Updegraff, D. M., and Wren, G. B. (1954). Release of oil from petroleum-bearing materials by sulfate-reducing bacteria. *Appl. Microbiol.* **2**, 309–322.

Volkov, I. I. (1961). Free hydrogen sulfide and products of its transformation in sediments of the Black Sea. *Tr. Inst., Okeanol., Akad. Nauk. SSSR* **50**, 29–67.

Volkova, Y., and Toshinskaya, A. D. (1961). Biogenic formation of H_2S in deep mineral waters. *Mikrobiologiya* **30**, 693–698.

Yanagita, T., and Kogane, F. (1963). Site of biosyntheses of sulfur compounds in growing colonies of *Aspergillus niger. J. Gen. Appl. Microbiol.* (*Tokyo*) **9**, 331–336.

Yarotskii, L. A. (1960). Main patterns of the formation of hydrogen sulfide waters. *Vopr. Formirovan. i. Rasprostr. Mineral'n. Vod SSSR, Tsentr. Nauchn.: Issled. Inst. Kurortol. i Fiziotetapii, Tr. Soveshch., Moscow, 1958* pp. 141–168.

ZoBell, C. E. (1958). Ecology of sulfate-reducing bacteria. *Producers Monthly* **22**, 12–29.

6

BIOTRANSFORMATION
OF IRON

GEOCHEMISTRY

Metallic iron is isometric with distinct cubic cleavage. It is malleable and fuses at 1535°C. Its abundance has helped develop the space age. It is steel gray to iron black in color. It forms a mixed crystal with up to about 30% nickel.

The iron content of the lithosphere approximates 5.0%. This increases to 40% in meteorites. Cobalt and nickel are common constituents of iron deposits. The atomic radii of these three metals are quite close: iron—1.27Å; cobalt—1.26 Å; nickel—1.24 Å. All enter into the metallic lattice of platinum where they make up to 30% by weight of platinum ores found in a native state. Magmatic rocks contain the divalent metals Mg^{2+}, Fe^{2+}, and Mn^{2+} (8 : 1 : 0.002).

Iron has been produced from the Lake Superior area whose ultimate exploitation has depended upon Pennsylvania coal. Brazilian iron ore is richer than Lake Superior ore but the use of it has lagged because of the lack of coking coal. Many sedimentary strata supply metallic ores high in iron. Ores of this type are found in Alabama, Newfoundland, north-eastern France, and England. The ore beds in Birmingham, Alabama, date back to the Silurian Age. Deposits of the same age in the Appalachians extend from New York southward. Near Birmingham, the beds are

20 feet thick. They are separated from adjoining beds by sandstone and shale. The sandstone contains some hematite which cements the rock and which also coats and replaces many fossils. The hematite granules are often flattened indicating that formerly they were jellylike masses of iron rust on an ocean floor. Such ores are of commercial value only because (a) they are calcareous and calcite acts as a flux in a blast furnace, and (b) deposits are close to sources of coking coal.

Our present oceans are very low in iron, a condition differing quite drastically from that when the Silurian iron deposits were formed. The low amounts of ferric iron in the oceans are associated with the low solubility of ferric iron at a pH of 8.0. Ocean water has about 1 ppm iron. Iron concentration increases with depth. Oolitic oxidative iron ores are often deposited in concentric zones. This may be caused by seasonal variations in areas where sinking waters high in oxygen cause circular zones high in oxidizing and low in reducing conditions. Ferrous iron carried into these areas as well as that supplied by inflow from adjacent lagoons of stagnant bottom water supply the iron to these formations. Goethite and hematite are altered by ferrous carbonate and ferrous silicate, while secondary

TABLE 6.I

MINERALS OF IRON

Mineral	Formula
Pyrite	FeS_2
Lawrencite	$FeCl_2$
Hercynite	$FeAl_2O_4$
Brownmillerite	$Ca_4Al_2 \cdot Fe_2O_{10}$
Ilmennite	$FeTiO_3$
Pseudobrookite	Fe_2TiO_5
Siderite	$FeCO_3$
Rhomboclase	$FeH(SO_4)_2 \cdot 4\,H_2O$
Ferrinatrite	$Na_3Fe(SO_4)_3 \cdot 3\,H_2O$
Voltaite	$KFe''Fe'''(SO_4)_3 \cdot 4\,H_2O$
Szomolnokite	$FeSO_4 \cdot H_2O$
Melanterite	$FeSO_4 \cdot 7\,H_2O$
Siderotil	$FeSO_4 \cdot 5\,H_2O$
Kornelite	$Fe_2(SO_4)_3 \cdot 7\,H_2O$
Carposiderite	$(H_2O)Fe_3(SO_4)_2OH_5 \cdot H_2O$
Butlerite	$Fe(SO_4)(OH) \cdot 2\,H_2O$
Amarantite	$Fe(SO_4)(OH) \cdot 3\,H_2O$
Liminite (hematite)	$Fe_2O_3 \cdot nH_2O$

replacement of $CaCO_3$ occurs when solutions of ferrous bicarbonate permeate sedimentary rocks containing calcium carbonate. Siderite ($FeCO_3$) is deposited. An anaerobic environment is required.

Chemical oxidation of ferrous to ferric iron may occur under aerobic conditions or in brackish marine waters low in oxygen where microbes are involved in iron oxidation under anaerobic conditions. Nodules consisting of ferric oxide and manganese dioxide are found off the California and many other coastlines. The nodules contain 21.5% Fe_2O_3 and 29.0% MnO_2 (Sverdrup et al., 1942). Ferrugenous muds are high in aluminum, iron, and manganese (100 : 55 : 1). Of the many iron minerals the more important ones are listed in Table 6.I.

IRON AND LIFE PROCESSES

Before discussing the microbes involved in iron transformation, the importance of iron to microbial life can be best emphasized in the types of organic compounds and reactions with which it is associated. Since iron can be both rapidly oxidized and rapidly reduced, by donating or accepting electrons, it functions in the oxidation and reduction of many compounds. A discussion of this will be presented later.

Several parasitic protists and members of the bacterial genus *Hemophilus* require hemin for growth. The parasitic protozoan flagellates known as the hemoflagellates, including the trypanisome causing African sleeping sickness, also require heme compounds. Lwoff (1934) has calculated that one type of hemoflagellate cell requires 746,000 molecules of hemin/cell division. Each molecule of hemin contains 1 atom of iron. Iron bound in heme does not exchange with the surrounding medium.

Herpetomonas eulcidarum commonly isolated from the lumen of the gut of mosquitoes may be used as a test microbe for quantitatively determining heme and iron, biochemically.

Iron porphyrins are synthesized by *Propionibacterium shermanii* and numerous other microbes. The biosynthesis of porphyrins is stimulated by the presence of iron and α-aminolevulinic acid. Coproporphyrin III accumulates in cultures of *Candida guilliermondii* when grown under either aerobic or anaerobic culture conditions (Shavlovskii and Bogatchuk, 1960).

Organisms not requiring molecular oxygen for respiration do not require the heme compounds associated with electron transport mechanisms. Some clostridia which apparently possess some heminlike compounds also require iron for growth. This is unusual since the clostridia are obligate anaerobes. Waring and Werkman (1942) grew the facultative anaerobe *Aerobacter indologenes* in media very low in iron. No heme enzymes were found in these cells, and the formic dehydrogenase and hydrogenlyase enzymes were reduced to 3 % of their normal cellular level. A heme enzyme is believed to link dehydrogenation of formic acid and the action of hydrogenase, i.e.,

$$H_2 \leftrightarrow 2\,H \leftrightarrow 2\,H + 2\,e^-$$

One strain of *Hemophilus influenzae* needs hemin when grown anaerobically (Granick and Gilder, 1947) which indicates that hemin has a function other than electron transport as coupled with oxygen utilization.

Cytochromes *b*, *c*, and *a* have an iron molecule which helps tie together a peptide and/or tetrapyrrole structure. Apparently the iron is oxidized and reduced, thus assisting in electron transport. Cytochromes are active in thiosulfate reduction in the sulfate-reducing bacteria (Ishimoto and Koyama, 1955). A similar involvement may be found in sulfate reduction.

Toxin production by *Corynebacterium diphtheriae* is controlled by the amount of iron present. A fungus *Ustilago sphaerogena* produces ferrichrome a when adequate iron is added (Garibaldi and Neilands, 1956), and *Bacillus subtilis*, *Aspergillus niger*, and *Bacillus megaterium* can be induced to form similar compounds. The iron-binding component of *B. subtilis* has the properties of a polyhydric alcohol. A yellow ferric chelate of citric acid is formed by *A. niger*. *Bacillus subtilis* and *A. niger* both excrete large amounts of coproporphyrin III in addition to the other iron chelates mentioned. Coproporphyrin III is also synthesized by *Candida guilliermondii* (Shavlovskii and Bogatchuk, 1960).

Iron is required for nitrogen fixation by *Azotobacter vinelandii* (Bruemmer *et al.*, 1957), and both *Clostridium pasteurianum* and the blue-green algae *Nostoc muscorum* require iron salts and biotin for nitrogen fixation (Carnahan and Castle, 1958). The requirement for iron is greater when growth is on elemental nitrogen rather than ammonia. A mechanism is proposed in which a chelate of a variable-valent iron ion serves as a key reactant in nitride formation.

Many biological compounds act as chelating agents: α-amino acids, *o*-hydroxycarboxylic acids, polyalcohols (glycerol, mannitol), and organic acids (citric acid, maleic acid). Chelation probably is the basic mechanism in the binding and mobilization of metals in nature, and iron is important because it forms complexes with many compounds. The complexing of iron in humus for example, needs additional investigation. Iron is bound with lignin and proteins and forms ferrilignoproteins.

Iron increases the cobalt uptake in *A. niger* (Adiga *et al.*, 1962) but not zinc and nickel. It is also involved in pigment production in *Torulopsis pulcherrima* (Roberts, 1946). Iron is an absolute mineral requirement in the metabolism of higher plants and animals.

In examining the abundant literature on iron bacteria, the microbes involved in iron transformations present the same complexity as the minerals with which they are involved. Iron is metabolized two ways. It may be assimilated within the cell or attacked by dissimilating processes in which actual cellular needs are in a sense secondary to the reactions involved. A natural taxonomic division between the nonfilamentous and filamentous forms of bacteria should be made. No other taxonomic division of this ecological group will be made here.

For a review of the older literature on iron, see Pringsheim (1948). A comprehensive index of names of iron microorganisms and synonyms as well as an index to the historical development of this subject is given. Classic works of Lieske (1919) and Winogradsky (1888, 1922) should also be consulted.

NONFILAMENTOUS IRON BACTERIA

Thiobacillus ferrooxidans

Temple and Colmer (1951) isolated *T. ferrooxidans* from acid mine waters and showed it to be responsible for oxidizing ferrous iron.

$$6 FeSO_4 + 2 HNO_3 + 3 H_2SO_4 \rightarrow 3 Fe_2(SO_4)_3 + 2 NO + 4 H_2O$$

They carried out iron oxidation experiments under conditions where direct atmospheric oxidation or iron could be excluded. Except that it oxidizes iron, this microbe cannot be distinguished from *T. thiooxidans*. Nitrate has been coupled to iron oxidation in other biosystems (de Barjac, (1954).

Thiobacillus ferrooxidans oxidizes elemental sulfur (pH 5.0), thiosulfate (pH 4.0), and tri- and tetrathionate (pH 6.0). Polythionates accumulate during thiosulfate oxidation, but not during the oxidation of elemental sulfur. The optimal pH for oxidation of chalcopyrite, bornite, and pyrite oxidation were, respectively, 2.0, 3.0, and 2.0 (Landesman *et al.*, 1966). Both pyrite and marcasite are oxidized (Temple and Delchamps, 1953).

$$FeS_2 + 3\tfrac{1}{2} O_2 + H_2O \xrightarrow{\textit{T. ferrooxidans}} FeSO_4 + H_2SO_4$$

(Pyrite)   (Ferrous sulfate)

$$2 FeSO_4 + \tfrac{1}{2} O_2 + H_2SO_4 \xrightarrow{\textit{T. ferrooxidans}} Fe_2(SO_4)_3 + H_2O$$

(Ferric sulfate)

The rate of biooxidation in an acid medium of ferrous iron is thirty or more times greater than that of pure chemical oxidation. Once ferric sulfate is formed, one has an agent which itself is a good chemical oxidant. It will in turn oxidize pyrite.

$$FeS_2 + Fe_2(SO_4)_3 \rightarrow 3 FeSO_4 + 2 S$$

(Pyrite)

The sulfur moiety is subject to further microbial oxidation to sulfuric acid. Some ferric sulfate is hydrolyzed to basic ferric sulfate and sulfuric acid.

$$Fe_2(SO_4)_3 + 2 H_2O \rightarrow 2 Fe(OH)SO_4 + H_2SO_4$$

The degree of hydrolysis depends on the ratio of Fe^{2+} to Fe^{3+} ions, the presence on O_2, the presence of sulfate, and the degree of acidity.

Cultures of *T. ferrooxidans* are initially inhibited in media containing ferric chloride in place of ferric sulfate. The inhibition is associated with the requirement of this microbe for a high ratio of sulfate to chloride ions for iron oxidation. Adaptation to cloride is possible and spontaneous mutants capable of oxidizing $FeCl_2$ in low sulfate environments develop rapidly (Lazaroff, 1963).

The rate of oxidation of sulfides or an excess of Fe^{3+} ions does not affect the oxidation of pyrite and chalcopyrite by *T. ferrooxidans* (Ivanov, 1962). An excess of ferrous ions can inhibit the oxidation of these and other mineral sulfides. Ferric ions themselves are active in oxidizing sulfides.

Thiobacillus ferrooxidans grows well in solutions containing up to 12 gm of Fe^{2+}, 2.5 gm Cu^{2+}, or 2.5 gm Zn^{2+} (Marchlewitz *et al.*, 1961). After adaptation to 5.0 gm Cu^{2+} or 25 gm Zn^{2+} per liter, the actual leaching of copper by the microbe is improved by adding sulfides such as pyrite and marcasite to copper schists.

Beck (1960) found that 120 gm of $FeSO_4$ was required for *T. ferrooxidan* to synthesize 16 gm of carbon. From this he calculated the microbe's efficiency in utilizing the energy available from ion oxidation at pH 2.0–3.5:

$$4 FeSO_4 + 2 H_2SO_4 + O_2 \rightarrow 2 Fe_2(SO_4)_3 + 2 H_2O$$
Fixation 1 g-atom C = 115 kcal
Oxidation of iron = 11.3 kcal/g-atom
Efficiency = 3.2%

It was estimated that at least 30% efficiency could be expected. If adequate water is present, the ferric sulfate is hydrolyzed.

$$4 \, Fe_2(SO_4)_3 + 12 \, H_2O \rightarrow 2 \, Fe_2(OH)_6 + 6 \, H_2SO_4$$

In the oxidation of siderite, R. L. Starkey (1946) has calculated the free energy.

$$4 \, FeCO_3 + O_2 + 6 \, H_2O \rightarrow 4 \, Fe(OH)_3 + 4 \, CO_2$$
$$\Delta F° = -40 \, kcal$$

This value is high, and at 25°C the free energy value should be -2.9 kcal. From this calculation it can be seen that 55.8 gm of iron must be oxidized to yield sufficient energy to synthesize 0.209 gm of cell material. In the process, 107 gm of ferric hydrate is produced.

Ferrobacillus ferrooxidans

Ferrobacillus ferrooxidans is distinguished from *T. ferrooxidans* by its inability to utilize thiosulfate. This differentiation is weak, and the two should be combined into one generic classification. Both can be regarded as *true* iron bacteria possessing the potential to obtain energy from the oxidation of ferrous iron under very acid conditions. *Ferrobacillus sp.* and *T. ferrooxidans* are autotrophs.

The Calvin scheme for CO_2 fixation has been found to operate in iron-oxidizing cells of *F. ferrooxidans* (Maciag and Lundgren, 1964).

Ferrobacillus ferrooxidans oxidizes ferrous to ferric iron. It is distinguished from *Thiobacillus* organisms not only by its inability to oxidize thiosulfate but by the inability of some cultures to oxidize sulfur (Leathen *et al.*, 1956).

This species can undergo definite morphological variation which can be controlled by the degree of aeration (Silverman and Rogoff, 1961). Cells produced under excessive aeration possessed a high Q_{O_2} for oxidizing ferrous iron. Lower Q_{O_2} values were obtained with cells grown under normal or less vigorous conditions of aeration.

Pyrite found in the Kosaka mine in Japan is oxidized by *F. ferrooxidans* (Ito *et al.*, 1960).

Bacillus Forms

Although they are not true iron bacteria, many heterotrophic bacilli form precipitates and remove iron from solution. Ferric hydroxide is apparently a waste product, and iron salts are a necessary metabolite for these forms. *Bacillus bruntzii* assimilates $Fe_3(PO_4)_2$ and synthesizes a pigment containing 6.0% iron (Marchal and Marchal, 1959). A study on the relative concentration of microbial types in oil field injection waters containing iron has shown them to be high in *Desulfovibrio, Pseudomonas, Bacillus,* and *Clostridium* species. These microbes are involved in corrosion, slime formation, hydrocarbon oxidation, and plugging (Carlson *et al.*, 1961). Some bacterial forms may precipitate $Fe(OH)_3$ and MnO_2 by synthesizing high molecular weight polysaccharides which cause them to coalesce.

A less well-known bacterium *Siderocapsa* produces coccoid or spherical cells and forms small colonies of 1–30 cells. Highly mucoid capsules are formed. Ferrous and ferric hydrate accumulates in the capsule giving it a rust brown color (Fig. 6.1).

Cells of *S. treubii* are $0.6–1.6\mu$ in diameter. *Siderocapsa major* is a small rod $1.4 \times 1.0 \times 2.0\mu$. Cells are easily stained by an ammonium-oxalate crystal-violet solution. Certain cultures utilize organic materials which means that these microbes are probably heterotrophic.

The taxonomic position of the genus *Siderocapsa* is difficult to assess. They are nonfilamentous and if morphology is the primary criterion of classification, they should be completely separated from the Caulobacteriales and placed in the Eubacteriales. *Siderocapsa* is found in marine habitats. Its mode of metabolism is heterotrophic. It forms lobate brown deposits on leaves and water weeds. It is also frequently isolated from exposed iron piping in underground reservoirs.

Sphaerotilus

Sphaerotilus natans is a member of the Chlamydobacteriaceae. It was first described by Kutzing in 1833. Earlier work has been summarized by Pringsheim (1948) who describes it as a thread-forming bacterium without spores. Gram-negative rods 1μ wide and $2–6 \mu$ long are formed, often with refractive granules (oil and volutin). Young filaments are naked or have a sheath which is either slimy or sticky and difficult to observe. Older filaments are rigid and brittle with a thickness of up to 1.0μ (Fig. 6.2). Such filaments are refractive and vary from colorless to yellow. Reproduction is by fragmentation of the naked filaments or by the production of motile swarm cells having subpolar flagellae. Swarm cells are formed when the sheath is soft and are liberated by its swelling and disintegration. *Sphaerotilus* is found in an environment with low concentrations of organic matter and oxygen, but with small quantities of oxidizable iron and/or manganese. Mono- and disaccharide sugars, organic acids, alcohols, and amino acids are utilized as carbon sources. Both nitrates and amino acids are utilized for nitrogen.

Sphaerotilus natans is a nuisance organism, occurring as slimy growths in rivers and streams where sewage industrial wastes are dumped. These slime masses are unsightly and become numerous enough to interfere with fishing where nets are used. In the lower Williamette and Columbia Rivers,

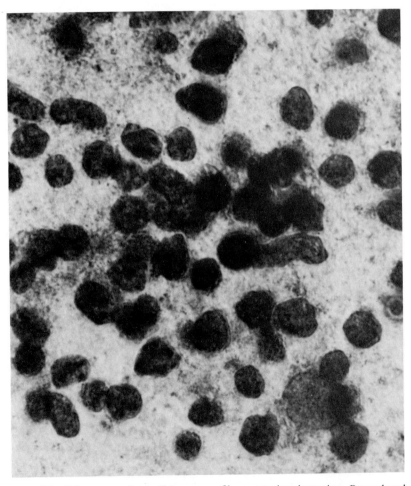

FIG. 6.1. *Siderocapsa*, a less well-known nonfilamentous iron bacterium. Reproduced with permission from Layne & Bowler, Inc., Memphis, Tennessee.

this characteristic pollution is caused by the dumping of sulfite waste liquors. In the production of 100 tons of pulp by the sulfite process Liebmann (1951) calculated that a yield of 70 tons of *Sphaerotilus* can be

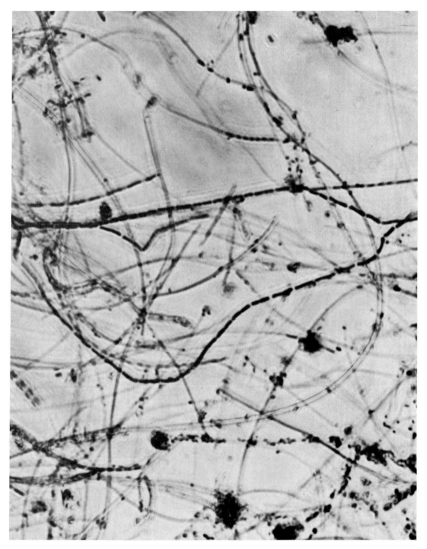

FIG. 6.2. *Sphaerotilus*, an aerobic microbe found in stagnant water with organic impurities. Reproduced with permission from Layne & Bowler, Inc., Memphis, Tennessee.

produced. Control is possible when the wastes are kept below 50 ppm. Growth varies with amount of nutrients and rate of stream flow. Profuse growth appears during the colder months, probably because competition for nutrients and oxygen by other microbes is less.

The predominant bacterium in Mill Creek near Walla Walla, Washington, which receives cannery waste, is *Sphaerotilus natans. Aerobacter aerogenes* is also present and accentuates slime growth. Also identified were species of *Beggiatoa, Leuconostoc,* and *Flavobacterium* (Nakamura and Dunstan, 1958).

Varying ratios of nutrients to ferrous salts result in different *Sphaerotilus* forms which cause definite confusion to those seeking uniformity in morphology. Pringsheim (1949) has shown organisms identified as *Leptothrix, Cladothrix,* and *Crenothrix* to be variant forms of *Sphaerotilus.* This raises a question as to whether the genus *Leptothrix* is purely autotrophic, since *Sphaerotilus* is known to be a heterotroph. Pringsheim (1948) states, however, that " observations are in favor of the utilization of energy from inorganic oxidation by members of the genus *Sphaerotilus* of the *Leptothrix* type." These observations were made by adding a small amount of organic matter to a culture, whereby a large increase in cellular material was obtained with an increased oxidation of magnanous and ferrous compounds (Pringsheim, 1949). Dondero (1961) suggests that because of the many chemical reactions which iron is known to undergo, the deposition of iron in the sheath of *Sphaerotillus* might well proceed without biochemical oxidation. For example, it is postulated that ammonia present in the cells and the polysaccharide sheath could cause precipitation of iron.

The sheath of *Sphaerotilus* is actually composed of a delicate inner sheath consisting of a hemicellulose membrane containing dextrose, galactose, arabinose, xylose, and ribose units; and an outer sheath contains ferric hydroxide and probably manganese oxide (Dondero, 1961). Winogradsky believed this outer sheath to be the result of autotrophic oxidation of ferrous carbonate, with ferric hydroxide precipitating out, according to the reaction:

$$4\ FeCO_3 + O_2 + 6\ H_2O \rightarrow 4\ Fe(OH)_3 + 4\ CO_2$$
$$\Delta F^\circ = -40\ cal$$

The status of this organism as an autotroph is doubtful (Halvorson and Starkey, 1927; R. L. Starkey and Halvorson, 1927; Halvorson, 1931; R. L. Starkey, 1945b). In microbial transformations, iron can be oxidized, reduced, precipitated, and dissolved by both biological and nonbiological reactions. These reactions take place in the presence of microoganisms without necessarily being a part of their metabolism. However, the sheathed bacteria are capable, through some unknown process, of assimilating iron and manganese from waters low in these elements and concentrating them in their sheaths. Whether or not this can be done by the organisms on a strictly autotrophic basis has yet to be proven.

Cladothrix is similar to *Sphaerotilus*; it will not be covered here.

Gallionella

Gallionella is present in natural waters and is hard to distinguish from the genus *Spirophyllum*. Many scientists regard these two genera as being identical for this reason. This is especially true of natural forms encrusted with iron. Spiral thread formation is characteristic of *Gallionella*. The thread is loosely or tightly twisted and often looks like a string of beads (Fig. 6.3).

Gallionella cells are asymmetrical, and ferric hydrate is secreted on one side to form long slender twisted ribbons. These ribbons or stalks support the bacterial cells which are kidney-shaped rods $0.5 \times 1.2 \mu$. As in *Sphaerotilus* the ferric hydrate is secreted from the concave side of the cells. Division is by transverse fission and is followed by the production of a secretion filament by each cell. At the point of division, secretion strand branches making it look as if the organism itself is branched. Filaments attain a length of 200 μ or more. *Gallionella* is believed to be a strict autotroph (R. L. Starkey, 1945a). This paper by Starkey is a classic on the biotransformation of iron.

Gallionella ferruginea is found in mine springs in Meltingen, Switzerland, where it removes siderite from water (Sartar and Meyer, 1948):

$$4 \, FeCO_3 + O_2 + 6 \, H_2O \rightarrow 4 \, Fe(OH)_3 + 4 \, CO_2$$

In only a few cultures can manganous carbonate or ferrous sulfate be substituted for ferrous carbonate (Lieske, 1919).

Gallionella cultures have been described which live and multiply freely in seawater. The filaments are much longer in seawater and there is less twisting (Sharpley, 1961). *Gallionella* induces corrosion in water-flood operations in the secondary recovery of oil by two mechanisms. First, it oxidizes ferrous iron to hydrated ferric hydroxide. Oxygen is consumed, anaerobic conditions are created by aerobic bacteria, and the environmental conditions change to favor growth of sulfate-reducers such as *Desulfo-vibrio*. Second, as ferrous deposits increase, anaerobic conditions prevail at the metal surface, and the microbial attack becomes more vigorous as a differential aeration cell is formed.

On an examination of 76 municipal wells in Wisconsin, 53 % contained species of *Gallionella* and *Leptothrix*. The concentration varied from 10–100 million/ml (Lueschow and Mackenthun, 1962).

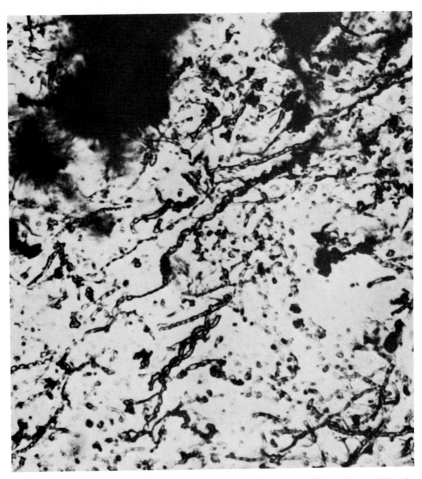

FIG. 6.3. *Gallionella*. Aerobic and filamentous forms resemble strings of beads. When actively growing, it forms masses one or more inches thick in two weeks time. Reproduced with permission from Layne & Bowler, Inc., Memphis, Tennessee.

Leptothrix

Leptothrix is probably more widespread than *Crenothrix*, a related genus; however, it is not a serious offender in water fouling. In the early stages of growth, *Leptothrix* forms transparent tubular threads which become coated with copious quantities of ferric hydroxide with increased growth (see Fig. 6.4). This coating has to be removed, e.g., with dilute

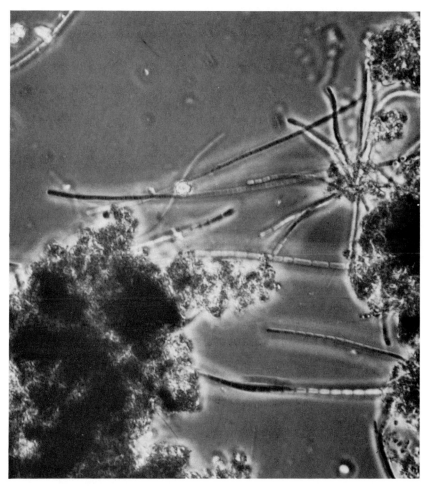

Fig. 6.4. *Leptothrix*. Cells in filaments are first cylindrical and colorless, becoming yellow or brown as they become encrusted with iron. Reproduced with permission from Layne & Bowler, Inc., Memphis, Tennessee.

hydrochloric acid, before the microbe can be identified. Cellular structure is not definite. The threads are normally straight, but curved forms do occur. Cells reproduce by regular cell division, by fragmentation of filaments, and by motile cylindrical swarm cells (0.7×2 to $14\,\mu$). True branching is seen in some species. The filaments are not commonly attached at the base with "hold fast cells." The inner diameter of $L.\ ochracea$ is about $1\ \mu$, and the sheath increases the thickness 0.5–$1\ \mu$, while filaments may attain a length of 1.0 cm. Actually, longer filaments are observed. The cells slip easily out of the ferric hydrate sheath leaving an empty cylindrical tube.

New cells immediately initiate sheath formation. $L.\ ochracea$ occurs in both pure water wells and springs which are low in organic matter. A related species, $L.\ crassa$, is more commonly found in slow flowing water containing higher quantities of organic matter. Some species of $Leptothrix$ are facultative autotrophs. $Leptothrix$ grows best in media containing ferrous carbonate, but ferric ions above a certain concentration inhibit growth and no growth occurs in the absence of ferrous ions. In cultures, added ferrous carbon is readily absorbed by the cell and oxidized and deposited in the sheath. The quantities of ferric hydrate deposited are several times the weight of the cell contained therein. Manganese carbonate improved growth in medium containing organic carbon, and is required when the microbe is cultivated as a strict autotroph.

$Leptothrix\ ochracea$ can utilize glucose and grow heterotrophically, but its morphology is atypical under these growth conditions. Iron oxidation occurs as follows:

$$2\ FeCO_3 + O_2 + 3\ H_2O \rightleftharpoons 2\ Fe(OH)_3 + 2\ CO_2$$
$$2\ C_3H_6O_3 + 2\ Fe(OH)_2 \rightleftharpoons 2\ CH_3COOH + 2\ Fe(OH)_3 + 2\ CO_2 + 3\ H_2$$
$$CH_3COOH + 2\ Fe(OH)_2 \rightleftharpoons 2\ CO_2 + 2\ Fe(OH)_3 + 3\ H_2$$

TABLE 6.II

CONDITIONS FOR GROWTH OF MAIN IRON BACTERIA[a]

| | Growth temperature (°C) | | | | |
	Min.	Opt.	Max.	pH	Metabolism
Gallionella	—	6	25	6.4–6.8	Autotroph
Leptothrix	5	25	40	5.8–7.8	Facultative autotroph
Crenothrix	—	18–24	—	5.5–6.2	Facultative autotroph

[a] From Duchow and Miller (1948).

The iron is both a source of energy and an acceptor of "oxygen" (Sartar and Meyer, 1948). A comparison between the types of metabolism and pH and temperature optima for *Gallionella*, *Leptothrix*, and *Crenothrix* are presented in Table 6.II.

Crenothrix

Crenothrix is the most important and the predominant filamentous form of iron bacteria occurring in dirty discolored water. Microscopically it is thin, colorless, and threadlike during early development, turning brown-black at maturity. The filaments normally contain a single row of cells with rounded ends (Fig. 6.5). Upon maturation the apex widens and

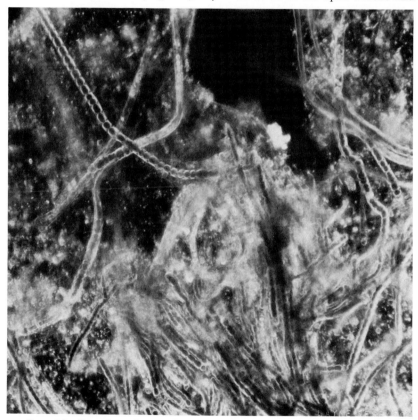

FIG. 6.5. *Crenothrix*. An aerobic iron bacterium with thin sheaths which are colorless at the tip and iron-encrusted at the base. Reproduced with permission from Layne & Bowler, Inc., Memphis, Tennessee.

spherical conidia or spores are expelled as seen in *C. polyspora*. Cells are 2 μ in diameter and are nonmotile. Chains of cells are formed similar to those of streptococci, however, the cells do become elongated upon division. Cells exude a muscilagenous tubular sheath (0.2 μ thick) which make it possible for them to cohere to solid objects. The sheath soon becomes impregnated with $Fe(OH)_3$. Cells dividing within the sheath provide the

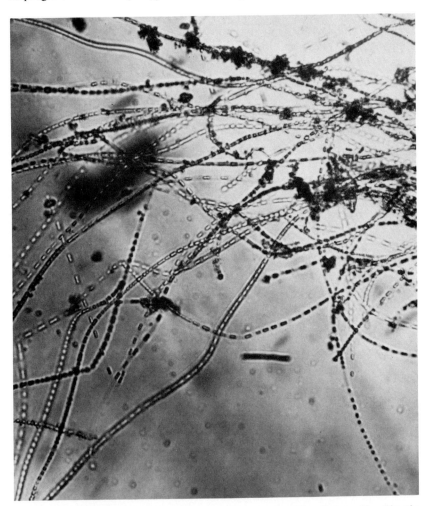

FIG. 6.6. *Clonothrix*. An aerobic iron bacterium resembling *Sphaerotilus*. Sheaths are iron- and manganese-encrusted. Reproduced with permission from Layne & Bowler Inc., Memphis, Tennessee.

force necessary for expelling the uppermost cells. They form a brownish hairy slime on rocks.

Crenothrix grows autotrophically and heterotrophically. The accumulation of fouling material tends to be far greater when organic constituents are present. If the water or stream is undergoing agitation, the cells or conidia are discharged and form masses and clumps rather than filaments. This is a type of zoogloea.

In classic studies hay infusion and ferrous carbonate or manganese carbonates were used to cultivate *Crenothrix*. Well water with ferric ammonium citrate and ferrous sulfate or ferrous carbonate gives the best growth.

Clonothrix

Clonothrix forms filaments and attaches firmly to particulate material (Fig. 6.6). False branching is common. The sheaths formed are probably polysaccharides and are encrusted with iron. The filaments are 5–7 μ wide at the base and taper to 2 μ at the tip. The cells are cylindrical, and reproduction is by spherical conidia which are formed in chains by transverse fission. *Clonothrix fusca* is the species found in most waterworks. Roze, who originally described this culture in 1896, stressed its greenish color and it was accordingly placed with the algae in the *Cyanophyceae*. Whether this genus should be separated from the other better-known iron bacteria is questionable.

SULFUR AND MANGANESE BACTERIA

The sulfur bacteria cannot be omitted from a discussion of iron metabolism because many microbes metabolize sulfur, sulfide, and related iron complexes. This is also true of the manganese bacteria. The sulfate reducers produce H_2S and precipitate FeS_2 and are always present in decaying organic matter. *Beggiatoa alba* is an important sulfur bacterium which can be confused with the iron bacteria. In natural waters it forms long tubular threads, or trichomes, of uniform thickness with rounded ends.

The diameter of the trichomes ranges from 1.4–1.6 μ and the length from 80–1500 μ. The individual cell length depends upon the strain and varies from 4.0–16 μ. Cell width is approximately that of the trichomes (Scotten and Stokes, 1962). Cells are motile possessing a slow, jerky movement or a gliding motion, depending upon the viscosity of the medium.

Cells contain both volutin and lipoidal inclusions, and sulfur granules are formed within the sheath. *Beggiatoa* is recognized easily in malodorous and discolored waters.

IRON REDUCTION

Iron reduction has been observed in a number of soils (Brydon and Heystek, 1958) where is it favored by a low pH. Iron reduction is more common in soils having sufficient organic matter. That is not always true since, in some cases, addition of organic acids and glucose has actually inhibited iron reduction (Kamura and Takai, 1961). Kalakutskii (1959) isolated a culture of *Pseudomonas* which produced 10–15 gm reduced Fe^{2+}/liter under anaerobic conditions. The culture was obtained from a gley-podzolic soil. Iron reduction was particularly associated with a butyric-acid-producing anaerobe. In some Russian soils, workers have observed that ferric hydroxide is reduced to ferrous oxide, under anaerobic conditions, and organic matter is consumed, (Kalakutskii and Duda, 1961).

Micrococcus lactilyticus uses molecular hydrogen to reduce a number of metals including ferric hydroxide and ferricyanide (Woolfolk and Whiteley, 1962):

$$Fe(OH)_3 + \tfrac{1}{2} H_2 \rightarrow Fe(OH)_2 + H_2O$$
$$Fe(CN)_6^{3-} + \tfrac{1}{2} H_2 \rightarrow Fe(CN)_6^{4-} + H^+$$

The microbial reduction of iron will receive much greater study in the years ahead.

BIOGEOCHEMISTRY

The ferrous-ferric system in lake and ground water is reversible. Under acid conditions the reversibility is independent of pH. At pH values near neutrality, which are normally encountered in the biosphere, ferric ions in equilibrium with (solid phase) ferric hydroxide are favored. Although trivalent iron in the ionic state is hard to detect with conventional techniques, it is present along with the ferrihydroxy ion ($FeOH^{2+}$) and possibly ferrite (H_2FeO_3) in lake and ground waters.

The oxidation of pyrite is catalyzed by both *Thiobacillus* and *Ferrobacillus* microbes.

$$FeS_2 + H_2O + 3\tfrac{1}{2} O_2 \rightarrow FeSO_4 + H_2SO_4$$
(Pyrite)   (Ferrous sulfate)

Ferrous sulfate is formed and is oxidized by these same microbes, as well as by the iron bacteria and many others.

$$FeSO_4 + \tfrac{1}{2} O_2 + H_2SO_4 \rightarrow Fe_2(SO_4)_3 + H_2O$$
$$\text{(Ferric sulfate)}$$

Ferric hydroxide, or ferric sulfate, is the end product. Ferric sulfate is one of the most important compounds formed in run-off or ground water. It not only serves as a chemical oxidant for pyrite, but it reacts with a host of mineral sulfides, many of which have been discussed.

$$FeS_2 + Fe_2(SO_4)_3 \rightarrow 3\ FeSO_4 + 2\ S$$

Depending upon the pH and other ground water conditions, hydrolysis of ferric sulfate proceeds as follows:

$$Fe_2(SO_4)_3 + 2\ H_2O \rightarrow 2\ Fe(OH)SO_4 + H_2SO_4$$
$$4\ Fe_2(SO_4)_3 + 12\ H_2O \rightarrow 2\ Fe_2(OH)_6 + 6\ H_2SO_4$$

Pyrite, chalcopyrite, marcasite, and bornite are oxidized by autotrophs (Table 6.III).

TABLE 6.III

OXIDATION OR SOLUBILIZATION OF IRON ORES

Microbe	Mineral	Reference
Thiobacillus thiooxidans	Pyrite (soil)	Quispel *et al.* (1952)
Thiobacillus ferrooxidans	Pyrite	Ivanov *et al.* (1961)
Thiobacillus ferrooxidans	Chalcopyrite (Utah)	Bryner *et al.* (1954)
Thiobacillus ferrooxidans	Chalcopyrite	Duncan and Trussell (1964)
Thiobacillus ferrooxidans	Bornite	Duncan and Trussell (1964)
Ferrobacillus ferrooxidans	Pyrite (coal)	Silverman *et al.* (1961)
Ferrobacillus ferrooxidans	Marcasite	Silverman *et al.* (1961)

Bornite is oxidized as follows:

$$2\ Cu_5FeS_4 + 18\ O_2 + 3\ H_2O \rightarrow 8\ CuSO_4 + Cu_2O + 2\ Fe(OH)_3$$

Chalcopyrite can be oxidized by microbes (*a*), or through the formation of ferric sulfate (*b*):

(*a*) $CuFeS_2 + 4\ O_2 \rightarrow CuSO_4 + FeSO_4$

(*b*) $CuFeS_2 + Fe_2(SO_4)_3 + 3\ H_2O + 3\ O_2 \rightarrow CuSO_4 + 5\ FeSO_4 + 2\ H_2SO_4$

Under acid conditions *T. ferrooxidans* maintains most of the iron in the ferric form. Molybdenum-pyrite ores from the Bingham Canyon in Utah are oxidized and both the molybdenum and iron are leached.

Ferric Sulfate as a Leaching Agent

The oxidation of ferrous solutions to ferric sulfate by microbes provides nature with a potent oxidant to oxidize many mineral sulfides. With "R" representing the metal, the following general reaction can be expected:

$$2 \, RS + Fe_2(SO_4)_3 + 2 \, H_2O + 3 \, O_2 \rightarrow 2 \, RSO_4 + 4 \, FeSO_4 + 2 \, H_2SO_4$$

Mineral sulfides which react with ferric sulfate are listed below in the order of decreasing activity (Ivanov et al., 1961):

Mineral	Formula
Pyrrhotite	$Fe_{1-x}S$
Tetrahedrite	$(Cu,Fe,Zn,Ag)_{12}Sb_4S_{13}$
Galena	PbS
Arsenopyrite	$FeAsS$
Sphalerite	ZnS
Pyrite	FeS_2
Enargite	Cu_3AsS_4
Marcasite	FeS_2
Chalcopyrite	$CuFeS_2$

Ferric sulfate can be carried to even the deepest ore deposits by ground water.

The basic concept which should be stressed, is that the geologist and engineer become alert to the fact that many ores which are not amenable to mining by conventional mining may be potentially leachable if they contain pyrite or some other mineral sulfide. *In situ* solution leaching will introduce new and needed technology in mining. Much of the chemistry will depend on controlling what already takes place biochemically in nature.

Deposition and Dissolution of Iron Minerals

Except in relation to corrosion or other minerals, specific studies on the deposition and dissolution of iron minerals are limited. It is quite obvious from the previous discussion that microbes catalyze the oxidation and reduction of iron through all its important valence states.

$$Fe \leftrightharpoons Fe^{2+} + 2 \, e^- \leftrightharpoons Fe^{3+} + e^-$$
$$(O_2) \qquad\qquad (O_2)$$

In microbial catalysis, the iron minerals which are deposited and solubilized will depend primarily on the pH of the ground minerals and the anions available for reaction. In natural water, iron is present as Fe^{3+}, Fe^{2+}, and $Fe(OH)^{2+}$. Ferrous iron remains in solution and precipitates upon oxidation to the ferric state. The transport of ferric hydroxide depends upon its solubility which is related to the pH. At pH 6.0, ferric hydroxide is 100 times more soluble than at pH 7.0 and is 100,000 times more soluble than at pH 8.5 (Cooper, 1937). The iron which is present in weakly acid streams flowing to oceans is precipitated in the alkaline sea. Fresh water contains about 100 times more iron than marine waters. Iron precipitated in the sea forms a positively charged colloid which chelates and entraps other ions from the sea as it passes to the bottom. Ions removed by this mechanism are arsenic, molybdenum, nickel, vanadium, phosphorous, antimony, copper, selenium, and lead. The ferric hydroxide gel also oxidizes manganese ion to form a precipitate of manganese and iron which then removes cobalt, zinc, thorium, tin, and silver from seawater (Arrhenius, 1959).

Phytoplankton and filamentous algae collected from the Columbia river have been found to concentrate iron-59 from river water 100,000–200,000 times (Lowman, 1963).

To supplement the reactions discussed in the text, a chart has been developed (Table 6.IV) in which the deposition and dissolution of iron minerals of carbonate, oxides, sulfides, sulfate, silicates, and miscellaneous groups are given. References for some of the reactions are cited; other reactions are based upon experience and need verification. However, there is little question about the utilization of siderite by the filamentous iron bacteria. Heterotrophs which produce copious quantities of carbon dioxide initiate carbonic acid formation which should lead to the deposition of carbonates. Goethite is also deposited by a number of microbes (Iwasa, 1963) as are the other oxides of iron lepidocroite, hemitite, and limonite. The chemistry of magnetite is a little more complex, but it also should be a good substrate for the iron oxidizers.

The iron sulfides have both cation and anion moieties which are readily utilized by the species *Thiobacillus* and *Ferrobacillus*. Ferrous iron is oxidized to ferric and sulfide is started on its way to sulfate or sulfuric acid. The sulfate reducers provide the sulfide for the deposition of pyrite, troilite, etc.

Thiobacillus cultures synthesize sulfuric acid to provide the sulfate for the deposition of melanterite and coquimbite. The iron oxidizers will supplement the chemical oxidation of iron to produce ferric ions in coquimbite deposits.

TABLE 6.IV

CHART OF PREDICTED DEPOSITION AND DISSOLUTION OF IRON MINERALS BY MICROBES[a]

Iron mineral and/or	Composition	Origin	Deposition	Dissolution	Reference
Carbonates					
Siderite	$FeCO_3$	S,H	B	B,F	—
Oxides					
Goethite	$Fe_2O_3 \cdot H_2O$	S	M	B	Iwasa (1963)
Lepidocrocite	$Fe_2O_3 \cdot H_2O$	S	M	B	—
Hematite	Fe_2O_3	S	M	B	—
Magnetite	Fe_3O_4	S	M	B	—
Limonite	$2\ Fe_2O_3 \cdot 3\ H_2O$	S	M	B	Berner (1962)
Sulfides					
Troilite	FeS	S	SO_4-Red	B(T)	Love and Zimmerman (1961)
Hydrotroilite	$FeS \cdot nH_2O$	I,H,S	SO_4-Red	B(F)	—
Pyrite, marcasite	FeS_2	I,H,S,	SO_4-Red	B(F)	Love and Zimmerman (1961)
Pyrrhotite	Fe_nS_{n+1}	I,H,S	SO_4-Red	B(F)	Quispel et al. (1952)
Sulfates					
Melanterite	$FeSO_4 \cdot 7\ H_2O$	S	—	S-Oxid	—
Coquimbite	$Fe_2(SO_4)_3 \cdot 9\ H_2O$	S	—	S-Oxid, Fe-Oxid	—

Silicates

	Formula / Type				Reference
Thuringite	$7\,FeO \cdot 3(Al,Fe)_2O_3 \cdot 5\,SiO_3 \cdot nH_2O$	S	M	M	—
Chamosite	$4\,FeO \cdot Al_2O_3 \cdot 3\,SiO_2 \cdot 3\,H_2O$	I,S	M	M	—
Greemalite (glauconite)	$Fe_9 \cdots Fe_2 \cdots (Si_4O_{11})(OH)_{12} \cdot 2\,H_2O$	I,S	M	M	—
Iron-siliceous incrustations		—	Lo	—	Pillai *et al.* (1947)
Miscellaneous					
Fe^{3+}, Mn^{3+} (oxides)	Alluvial and podzolized soil horizons	P$_P$	—		Aristovskaya (1963)
Pyritized sulfur oolite-pyrite	Koessen Marl	B$_P$	—		Fabricius (1962)
Fibrous iron precipitate	Crude tar	B	—		Garnett (1960)
Limestone	Incrustation: Calcite— 10–12% Fe	L	—		Paclt (1951)
Ferro-manganese concretions	Kars Sea	B$_P$	—		Kalinenko (1946)

[a] *Abbreviations:* B, Bacteria; B$_P$, *Bacterium precipitatum*; F, Fungi; M, Microbes; SO$_4$-Red, sulfate-reducing microbes; (T), *Thiobacillus*; (F), *Ferrobacillus*; S-Oxid, sulfur-oxidizing microbes; Fe-Oxid, iron-oxidizing microbes; Lo, *Leptothrix ochracea*; P$_P$, *Pedomicrobium podzolicum*; L, lichens; I, igneous; H, hydrothermal; S, sedimentary.

The iron silicates are complex and contain iron, aluminum silicate, and water. Microbes are thought to be active in both their deposition and dissolution. Pillai *et al.*, (1947) observed an iron-siliceous incrustation formed by *Leptothrix ochracea*. Insect larvae were present in the deposit.

Iron and manganese oxides are found in alluvial and podzolized soil horizons (Aristovskaya, 1963), and ferromanganese concretions are found in the seas (Kalinenko, 1946). Pyritized sulfur and oolite-pyrite have been found in Koessen Marl (Fabricius, 1962), and fibrous iron precipitates are formed in crude tars (Garnett, 1960). Lichens are active in depositing incrustations of calcite with an iron content as high as 12%.

REFERENCES

Adiga, P. R., Sastry, K. S., and Sarma, P. S. (1962). The influence of iron and magnesium on the uptake of heavy metals in metal toxicities in *Aspergillus niger*. *Biochim. Biophys. Acta* **64**, 546–548.

Aristovskaya, T. V. (1963). Decomposition of organomineral compounds in podzolic soils. *Pochvovedenie* pp. 30–43.

Arrhenius, G. (1959). Sedimentation on the ocean floor. *In* "Researches in Geochemistry" (P. H. Abelson, ed.), pp. 1–24. Wiley, New York.

Beck, J. V. (1960). A ferrous-iron oxidizing bacterium, I. Isolation and some general physiological characteristics. *J. Bacteriol.* **79**, 502–509.

Berner, R. A. (1962). Experimental studies of the formation of sedimentary iron sulfides. *Biochem. Sulfur Isotopes, Proc. Symp., Yale Univ.*, 1962. pp. 107–120. Yale Univ. Press, New Haven, Connecticut.

Bruemmer, J. H., Wilson, P. W., Glenn, I. L., and Crane, F. L. (1957). Electron-transporting particle from *Azotobacter vinelandii*. *J. Bacteriol.* **73**, 113–116.

Brydon, J. E., and Heystek, J. (1958). Mineral and chemical study of the dikeland soils of Nova Scotia. *Can. J. Soil Sci.* **38**, 171–186.

Bryner, L. C., Beck, J. V., Davis, D. B., and Wilson, D. G. (1954). Microorganisms in leaching sulfide minerals. *Ind. Eng. Chem.* **46**, 2578–2592.

Carlson, V., Bennett, E. O., and Rower, J. A., Jr., (1961). Microbial flora in a number of oilfield water-injection systems. *Soc. Petroleum Eng. J.* **1**, 71–80.

Carnahan, J. E., and Castle, J. E. (1958). Some requirements of biological nitrogen fixation. *J. Bacteriol.* **75**, 121–124.

Cooper, L. H. N. (1937). Some conditions governing the solubility of iron. *Proc. Roy. Soc.* **B124**, 229–307.

de Barjac, H. (1954). Reduction of sulfates and nitrates under anaerobic conditions, an occurrence in the iron cycle. *Congr. 8th Intern. Botan., Paris*, 1957, *Collog. Anal. Plantes Probl. Engrals. Mineraux* pp. 21–27. Inst. Rech. Huiles Oleagineux, Paris.

Dondero, N. C. (1961). *Sphaerotilus*, its nature and economic significance. *Advan. Appl. Microbiol.* **3**, 78–107.

Duchon, K., and Miller, L. B. (1948). The effect of chemical agents on iron bacteria. *Proc. Tech. Assoc. Pulp Paper Ind.* **126**, 47–58.

Duncan, D. W., and Trussell, P. C. (1964). Advances in the microbiological leaching of sulfide ores. *Can. Met. Quart.* **3**, 43–55.

Fabricius, F. (1962). The structure of " oolite-pyrite " of the Koessen strata (Rhaetion) as a contribution to the problem of " mineralized bacteria." *Geol. Rundschau* **51**, 647–657.

Garibaldi, J. A., and Neilands, J. B. (1956). Formation of ironbinding compounds by microorganisms *Nature* **177**, 526–527.

Garnett, P. H. (1960). Growth of iron bacteria in crude tar. *Nature.* **185**, 942–943.

Granick, S., and Gilder, H. (1947). Distribution, structure, and properties of the tetrapyrroles. *Advan. Enzymol.* **7**, 305–368.

Halvorson, H. (1931). Studies on the transformation of iron in nature. III. The effect of CO_2 on the equilibrium in iron solution. *Soil Sci.* **32**, 141–165.

Halvorson, H., and Starkey, R. (1927). Studies on the transformation of iron in nature. I. Theoretical considerations. *J. Physiol. Chem.* **31**, 626–631.

Ishimoto, M., and Koyama, J. (1955). Role of cytochrome in the thiosulfate reduction by sulfate-reducing bacteria. *Bull. Chem. Soc. Japan* **28**, 231–232.

Ito, J., Wakazono, Y., Konodo, M., Yagi, S., and Katsuki, H. (1960). Special bacteria in mine water and their utilization for mining operations. *Nippon Kogyo Kaishi* **76**, 524–529.

Ivanov, V. I. (1962). Iron oxidation by *Thiobacillus ferrooxidans*. *Mikrobiologiya* **31**, 795–799.

Ivanov, V. I., Nagirnyak, F. I., and Stepanov, B. A. (1961). Bacterial oxidation of sulfide ores. I. The role of *Thiobaccillus thiooxidans* in the oxidation of chalcopyrite and sphalerite. *Mikrobiologiya* **30**, 688–692.

Iwasa, Y. (1963). Iron hydroxides in soils. II. Effect of 5N·NaOH on iron hydroxides. *Nippon Dojo-Hiryogaku Zasshi* **34**, 335–338.

Kalinenko, V. O. (1946). Role of bacteria in formation of ferromanganese concretions. *Mikrobiologiya* **15**, 364–369.

Kalakutskii, L. V. (1959). The role of microorganisms in the processes of iron reduction in the soil. *Nauchn. Dokl. Vysskei Shkoly, Biol. Nauki* **1**, 225–229.

Kalakutskii, L. V. and Duda, V. I. (1961). Role of microorganisms in iron reduction in the soil. *Nauchn. Dokl. Vysshei Shkoly, Biol. Nauki* **1**, 172–176.

Kamura, T., and Takai, Y. (1961). The microbial reduction mechanism of ferric iron in paddy soils. *J. Sci. Soil Tokyo* **32**, 135–138.

Landesman, J., Duncan, D. W., and Walden, C. C. (1966) Oxidation of inorganic sulfur compounds by washed call suspensions of *Thiobacillus ferrooxidans*. *Can. J. Microbiol.* **12**, 957–964.

Lazaroff, N. (1963). Sulfate requirement for iron oxidation by *Thiobacillus ferrooxidans*. *J. Bacteriol.* **85**, 78–83.

Leathen, W. W., Kinsel, N. A. and Braley, S. A., Sr. (1956). *Ferrobacillus ferrooxidans*, a chemosynthetic autotrophic bacterium. *J. Bacteriol.* **72**, 700–706.

Liebmann, H. (1951). The action of sulfite cellulose waste liquors on smaller and larger streams. *Gesundh.-Ingr.* **72**, No. 17/18.

Lieske, T. (1919). On the nutritional physiology of the iron bacteria. *Zentr. Bakteriol. Parasitenk., Abt. II* **49**, 413.

Love, L. G., and Zimmerman, D. O. (1961). Bedded pyrite and micro-organisms from Mount Isa shale. *Econ. Geol.* **56**, 873–896.

Lowman, F. G. (1963). Iron and cobalt ecology. *Proc. 1st Natl. Symp. Radioecol., Fort Collins, Color., 1961* pp. 561–567. Reinhold, New York.

Lueschow, L. A., and Mackenthun, K. M. (1962). Detection and enumeration of iron bacteria in municipal water supplies. *J. Am. Water Works Assoc.* **54**, 751–756.

Lwoff, A. (1934). Die Bedeutung des Blatfarbstoffes fur die parasitischen Flagellaten. *Zentr. Bakteriol., Parasitenk., Abt. I. Orig.* **130**, 497–518.

Maciag, W. J., and Lundgren, D. G. (1964). Carbon dioxide fixation in the chemoautotroph *Ferrobacillus ferrooxidans. Biochem. Biophys. Res. Commun.* **17**, 603–607.

Marchal, J. G., and Marchal, A. (1959). Tentative explanation for the inability of *Bacillus bruntizii* and *Pseudomonas cattleyaecolor* cultured in rolling tubes to produce their characteristic pigments. *Trav. Lab. Microbiol. Fac. Pharm. Nancy* **19**, 37–44.

Marchlewitz, B., Hasche, D., and Schwartz, W. (1961). Investigation on the behavior of thiobacteria against heavy metals. *Z. Allgem. Mikrobiol.* **1**, 179–191.

Nakamura, M., and Dunstan, G. H. (1958). Studies on the control of slime in cannery waste water in the Walla Walla area. *D. W. Indust. Res. Inst. Technol. State Coll. Wash.* pp. 1–24.

Paclt, J. (1957). Origin of ferrous limestone coatings in the plain of Sharon in Isreal. *Israel Exploration J.* **1**, 141–145.

Pillai, S. C., Rajagopalan, R., and Subrakmanyan, V. (1947). Incrustations in water pipes as affected by filamentous iron bacteria. *Indian Med. Gaz.* **82**, 36–38.

Pringsheim, E. G. (1948). Iron bacteria. *Biol. Rev.* **24**, 201–245.

Pringsheim, E. G. (1949). The filamentous bacteria *Sphaerotilus, Leptothrix, Cladothrix* and their relation to iron and manganese. *Phil. Trans. Roy. Soc. (London)* **B233**, 453–482.

Quispel, A., Harmsen, G. W., and Otzen, D. (1952). The chemical and bacteriological oxidation of pyrite in soil. *Plant Soil* **4**, 43–55.

Roberts, C. (1946). The effect of iron and other factors on the production of pigment by the yeast *Torulopsis pulcherrima. Am. J. Botany* **33**, 237–244.

Sartar, A., and Meyer, J. (1948). A study of evolution of two iron-containing bacteria their factors of energy and synthesis. *Compt. Rend.* **226**, 443–445.

Scotten, H. L., and Stokes, J. L. (1962). Isolation and properties of *Beggiatoa. Arch. Mikrobiol.* **42**, 353–368.

Sharpley, J. M. (1961). The occurrence of Gallionella in salt water. *Appl. Microbiol.* **9**, 380–382.

Shavlovskii, G. M., and Bogatchuk, A. M. (1960). The cause of coproporphyrin accumulation in *Candida guilliermondii* cultures. *Biokhimiya* **25**, 1043–1048.

Silverman, M. P., and Rogoff, M. H. (1961). Morphological variation in *Ferrobacillus ferrooxidans related* to iron oxidation. *Nature* **191**, 1221–1222.

Silverman, M. P., Rogoff, M. H., and Wender, I. (1961). Bacterial oxidation of pyrites materials in coal. *Appl. Microbiol.* **9**, 491–496.

Starkey, R. L., and Halvorson, H. (1927). Studies on the transformation of iron in nature. II. Concerning the importance of microorganisms in the solution and precipitation of iron. *Soil Sci.* **24**, 381–402.

Starkey, R. L. (1945a) Precipitation of ferric hydrate by iron bacteria. *Science* **102**, 531.

Starkey, R. L. (1945b). Transformation of iron by bacteria in water. *J. Am. Water Works Assoc.* **37**, 963–985.

Sverdrup, H. U., Flemming, R. H., and Johnson, M. W. (1942). "The Oceans," 1087 pp. Prentice Hall, New York.

Temple, K. L., and Colmer, A. R. (1951). The autotrophic oxidation of iron by a new bacterium: *Thiobacillus ferrooxidans. J. Bacteriol.* **62**, 605–611.

Temple, K. L., and Delchamps, E. W. (1953). Autotrophic bacteria and the formation of acid in bituminous coal mines. *Appl. Microbiol.* **1**, 255–258.

Waring, W. S., and Werkman, C. H. (1942). Growth of bacteria in an iron-free medium. *Arch. Biochem.* **1**, 303–310.

Winogradsky, S. (1888). On iron bacteria. *Botan. Ztg.* **43**, 262.

Winogradsky, S. (1922). Iron bacteria as inorganic oxidants. *Zentr. Bakteriol., Parasitenk., Ast. II.* **57**, 1.

Woolfolk, C. A., and Whiteley, H. R. (1962). Reduction of inorganic compounds with molecular hydrogen by *Micrococcus lactilyticus. J. Bacteriol.* **82**, 647–658.

7

COPPER BIOGEOCHEMISTRY

COPPER MINERALS LEACHED BY MICROBES

Razzell (1962) has pointed out that the importance of bacterial action in leaching sulfide minerals is becoming widely recognized in the mining industry. Bacterial leaching has warranted thorough investigation because of the accumulation of immense tonnages of low-grade ores in the last 25 years and because the higher-grade ores are being depleted. Razzell advocates the bacterial leaching of covellite, chalcopyrite, chalcocite (all copper minerals), millerite (a nickel mineral), and molybdenite (a source of molybdenum). Bacterial activities associated with leaching are the formation of ferric from ferrous iron, sulfuric acid formation from sulfur, and soluble metal formation from insoluble sulfides. Some bacteria capable of oxidizing ferrous iron also grow well with other energy sources such as sulfur and metallic sulfides. Copper solutions containing 1.2 gm/liter were obtained by leaching chalcopyrite for 100 hours. Almost 100% dissolution of covellite was obtained in six months of leaching. Studies on 12 Canadian ores showed sulfuric acid requirements of 100–120 lb/ton of ore to give a pH of 2.5. By using bacteria to convert sulfur to sulfuric acid, a savings of $0.62–0.73/ton of ore treated was predicted.

The earliest studies on copper leaching, conducted by Kennecott, have shown the practicality of leaching waste dumps (Zimmerly et al., 1958). The rate of growth of bacteria on metallic sulfide is similar to the rate of growth on sulfur. Free sulfuric acid is synthesized under these conditions.

Microbial leaching of covellite (CuS) was quite high. Studies on chalcopyrite ($CuFeS_2$) showed that over 80% of the mineral was converted to acid-soluble copper in seven months. Leaching of chalcocite (Cu_2S) as well as millerite (NiS) and molybdenite (MoS_2) was observed. More recently Mariacher and Spence (1962) had some success in using bacteria to leach copper, molybdenum, manganese, and zinc from low-grade ore. The bacteria used in these tests were isolated from acid soils and acid mine waters.

The copper-silver ores of Mansfeld, Thuringia, in Germany have been known for more than 750 years (Stradlinger, 1950). Deposits of this type originated from marine sediments of bituminous marls and slates; they were once high in microbial activity. Beds of copper near Sangerhausen in Germany containing 2–5% copper may be used for copper production in the future. The genesis of these copper ores is one of the most fascinating problems of chemical geology, but the literature is too extensive and too complex for development here. There is a definite controversy over whether epigenetic or syngenetic formation occurred. The ore-bearing solutions were assumed to have originated from the Paleozoic Hartz Mountains, the peripheries of the Halle Region, or the basic eruptive rocks of the Vogtland and the Erzgebirge.

Evidence for the syngenetic origin of copper ores is seen in the fluviatile migration of the typical ore elements including copper, molybdenum, manganese, antimony, vanadium, gold, and selenium in a highly diluted solution into the marine sediments of the Zechstein Basin of Mansfeld. There may also have existed small amounts of copper compounds in the older sediments which were flooded by the Zechstein Sea; they are generally called "copper marsh ores." The role of bacteria in the deposition and especially in the local copper enrichments of bornite and chalcocite is of great chemical importance. Areas oxidized down to the typical red rock are very low in copper. The general chemical conditions which existed during the formation of the Mansfeld ores may have been similar to those actually observed in the black muds of the Black Sea. The red rock may indicate the formation of ferric sulfate solutions which redissolved copper. The ionized copper (Cu^{2+}) probably migrated to the marginal parts of the sediments and became enriched by precipitation. On the basis of this theory new rich copper deposits were discovered near Sangerhausen and in Greater Hessia.

The principal copper ore minerals are pyrite and chalcopyrite (Sirel, 1949). Accessory minerals are sphalerite, millerite, pyrrhotite, pentlandite, cobaltite, arsenopyrite, and magnetite. Cobaltite and chalcopyrite occur in the zone of oxidation. Microscopic studies support the view that copper is deposited by syngenetic sedimentary processes from the marine environment. This theory is based largely on the gellike characteristics of many pyrite deposits.

Chemistry of Biochemical Leaching of Copper

The chemistry involved in the microbial leaching of copper ores is reviewed by Sutton and Corrick (1963).

Ferrous iron is oxidized by many microbes. In the presence of sulfuric acid, ferric sulfate is normally formed (Zimmerly *et al.*, 1958).

$$2 \text{ FeSO}_4 + \text{H}_2\text{SO}_4 + \tfrac{1}{2} \text{O}_2 \rightarrow \text{Fe}_2(\text{SO}_4)_3 + \text{H}_2\text{O}$$

If sulfuric acid is limiting, basic iron sulfate is formed:

$$4 \text{ FeSO}_4 + 2 \text{ H}_2\text{O} + \text{O}_2 \rightarrow 4 \text{ Fe(OH)SO}_4$$

Pyrite (FeS_2) is oxidized to either ferrous or ferric sulfate and sulfuric acid.

$$2 \text{ FeS}_2 + 7 \text{ O}_2 + 2 \text{ H}_2\text{O} \rightarrow 2 \text{ FeSO}_4 + 2 \text{ H}_2\text{SO}_4$$
$$2 \text{ FeS}_2 + 7\tfrac{1}{2} \text{ O}_2 + \text{H}_2\text{O} \rightarrow \text{Fe}_2(\text{SO}_4)_3 + \text{H}_2\text{SO}_4$$

Oxidation of ferrous sulfate to ferric sulfate usually follows.

If the sulfuric acid is depleted, ferric sulfate hydrolyzes to sulfuric acid, and a brownish ferric hydroxide precipitates.

$$\text{Fe}_2(\text{SO}_4)_3 + 6 \text{ H}_2\text{O} \rightarrow 2 \text{ Fe(OH)}_3 + 3 \text{ H}_2\text{SO}_4$$

The degree of hydrolysis of ferric sulfate varies with the reaction temperature. Under certain conditions, which are not well defined, ferric sulfate hydrolyzes to form sulfuric acid and ferric oxide.

$$\text{Fe}_2(\text{SO}_4)_3 + 3 \text{ H}_2\text{O} \rightarrow 3 \text{ H}_2\text{SO}_4 + \text{Fe}_2\text{O}_3$$

This reaction may be reversed under atmospheric conditions in the presence of high acid concentrations. These reactions are not detrimental to microbial activity. Excess iron often causes precipitation of ferric salts, and it may become desirable to remove such salts from leaching solutions. Prevention of hydrolysis of ferric sulfate to ferric hydroxide is important because the former reacts with pyrite to give high yields of ferrous sulfate and sulfuric acid.

$$7 \text{ Fe}_2(\text{SO}_4)_3 + \text{FeS}_2 + 8 \text{ H}_2\text{O} \rightarrow 15 \text{ FeSO}_4 + 8 \text{ H}_2\text{SO}_4$$

In addition chalcocite reacts chemically with ferric sulfate to give copper sulfate, ferrous sulfate, and elemental sulfur.

$$Cu_2S + 2\ Fe_2(SO_4)_3 \rightarrow 2\ CuSO_4 + 4\ FeSO_4 + S$$

Elemental sulfur itself is oxidized by *Thiobacillus concretivorus* to sulfuric acid.

$$2\ S + 3\ O_2 + 2\ H_2O \rightarrow 2\ H_2SO_4$$

The oxide of copper reacts with ferric sulfate to form copper sulfate and iron oxide.

$$3\ CuO + Fe_2(SO_4)_3 \rightarrow 3\ CuSO_4 + Fe_2O_3$$

In the precipitation of copper from solution by cementation the chemical reaction is:

$$CuSO_4 + Fe \rightarrow FeSO_4 + Cu$$

The ferrous sulfate solution is recycled through the system. For a discussion of the problems and economics of the microbial leaching of copper, see Johnson (1965).

LEACHING TECHNIQUES

Both heap leaching and hillside leaching techniques are used in leaching copper. Once the copper is solubilized it is precipitated. In copper precipitation the copper in solution displaces the iron of the cans to form cement copper. The copper content of cement copper runs 16% or higher.

Wood-lined launders may be used in the precipitating plant. The plant-tails water is fed into the settling pond where a final filtration process removes the last of the copper. Copper precipitation requires 1.5 lb of iron/1 lb of copper.

To avoid hauling, low-grade ores may also be bulldozed over a hillside onto areas which contain impervious surface rocks to minimize solution losses. Sites should be selected where drainage can be controlled. The procedures for leaching such areas are similar to those used for heap leaching. Tremendous tonnages of low-grade copper ores can be effectively leached in this manner.

REVIEW OF MICROBES USED TO LEACH COPPER ORES

As high-grade copper ores are depleted, a need for more effective methods for recovering valuable metals has developed. A problem of major importance in the hydrometallurgical field is the improvement of metal extraction in the leaching of low-grade sulfide ores. Pioneer work in this subject was carried out by Rudolfs and Helbronner (1922) at the New Jersey Agriculture Experiment Station. In these studies, it became evident that certain unidentified sulfur-oxidizing microbes were able to change iron and zinc sulfides into the more soluble sulfate forms. Bryner and Anderson (1957), Bryner et al. (1954), and Bryner and Jameson (1958) proposed the use of the biochemical reactions of microorganisms to extract copper ores. Until recently little attention had been focused on these bacteria in mining and metallurgical processes.

DeCuyper (1964) suggested the use of bacteria belonging to the genus *Thiobacillus* as potential aids in leaching copper. The best known species, *Thiobacillus thiooxidans*, was first isolated from soil by Waksman and Joffe (1922). It is a strict autotroph which utilizes elemental sulfur, dioxide, and thiosulfate as sources of energy and derives all its carbon from carbon dioxide. Much work on this bacterium has been done by Starkey (1925), Vogler and Umbreit (1942), Volger et al. (1942), and others. This microbe possesses some outstanding physiological characteristics, such as resistance to high concentrations of acid and rapid growth at pH 2 to 3. It facilitates leaching by producing sulfuric acid from sulfur.

In West Virginia and Pennsylvania the effluent of bituminous coal mines was reported to be polluted by high levels of acid and high concentrations of iron. Investigations of this problem by Colmer et al. (1950) led to the isolation of a new autotrophic iron-oxidizing bacterium of the genus *Thiobacillus*. The species name *Thiobacillus ferrooxidans* was therefore suggested by Temple and Colmer (1951). Temple and Delchamps (1953) characterized this microbe by its ability to oxidize ferrous iron very rapidly in acid solution and its failure to grow on elemental sulfur. It is, however, able to grow by oxidizing thiosulfate as the sole energy source and it does synthesize acid from museum grade pyrite. Microbes similar to *Thiobacillus ferrooxidans* were isolated from acid waters in Scotland by Ashmead (1955).

In 1954 Leathen and Braley isolated *Ferrobacillus ferrooxidans*. This microbe (Leathen et al., 1956) oxidizes ferrous iron to the ferric state within the pH range of 2.0–4.5, but does not oxidize acid thiosulfate and elemental sulfur. Sulfur oxidation may be a variable characteristic, since Silverman and Lundgren (1959) observed a slow but significant oxidation of sulfur by *Ferrobacillus ferrooxidans*.

Ferrobacillus sulfooxidans was isolated in 1960 by Kinsel. This bacterium was characterized by its ability to utilize both ferrous iron and elemental sulfur as energy sources.

Chemosynthetic iron- and sulfur-oxidizing autotrophic bacteria have been isolated and identified in effluent water of several copper mines (Sutton and Corrick, 1961a,b) where leaching has occurred in the south-western United States. The natural presence of iron and sulfur bacteria in copper ore deposits and their survival in acid environments support the bioenergetic and microbiological data collected for these microbes. The most important microorganisms were those capable of producing solutions of ferric sulfate and sulfuric acid since solutions of this type are excellent solvents for many minerals. The acid ferric sulfate formed in leaching areas was first believed to be formed by atmospheric oxidation of sulfur-containing materials associated with the copper deposits. The later theory that microbial activity was primarily responsible for this oxidation was strengthened by the establishment of the presence of these bacteria at copper leaching areas. The indigenous presence of these microbes also advanced possibilities for the controlled use of microbial action in metal extraction processes (Corrick and Sutton, 1961). Objectives were: (*a*) to determine if pure strains of the bacteria *Ferrobacillus ferrooxidans*, *Thiobacillus concretivorus*, and *T. ferrooxidans* could utilize the iron and sulfur occurring in sulfide minerals to produce appreciable quantities of ferric sulfate and sulfuric acid for dissolving copper and (*b*) to develop the chemistry involved in the microbial oxidation of sulfide minerals. These objectives were of particular interest as they related directly to the feasibility of employing microorganisms in leaching operations.

A detailed study of the morphologically similar bacteria *Thiobacillus thiooxidans*, *T. concretivorus*, *T. neopolitanus*, *Ferrobacillus ferrooxidans*, and *T. ferrooxidans* was completed. The principal distinguishing character-istics were: (*a*) *Thiobacillus thiooxidans* oxidized elemental sulfur and thiosulfate to sulfuric acid using ammonium ions as a source of nitrogen, but could not oxidize ferrous iron or utilize nitrate; (b) *Thiobacillus concretivorus* was similar to *Thiobacillus thiooxidans* except that it used nitrate and ammonium ions, but not nitrites, as a nitrogen source; (c) *Thiobacillus neopolitanus* was similar to *Thiobacillus concretivorus* except that it used nitrites as well as nitrates and ammonium ions as a nitrogen source; (*d*) *Ferrobacillus ferrooxidans* oxidized ferrous iron in an acid environment to ferric iron, but did not oxidize elemental sulfur or thiosulfate; and (e) *Thiobacillus ferrooxidans* oxidized ferrous iron to ferric iron and thiosulfate to sulfuric acid, but did not oxidize elemental sulfur appreciably. These data agree with other studies (Bryner *et al.*, 1954).

An important step in the possible applications of *Thiobacillus* and *Ferrobacillus* species in leaching low-grade ores was made with the discovery of the presence of bacteria in mine waters in Bingham Canyon, Utah. Davis (1953) found bacteria to be at least partially responsible for the copper found in solution. Bryner *et al.* (1954), Bryner and Anderson (1957), and Bryner and Jameson (1958) identified two strains of autotrophic bacteria, *Thiobacillus thiooxidans* and *T. ferrooxidans*, in the leach streams issuing from the waste rock dumps. Investigations showed that the iron-oxidizing bacterium found in these acidic leach waters was physiologically different from reported cultures of *Thiobacillus ferrooxidans* or *Ferrobacillus ferrooxidans*. Beck (1960) demonstrated that this microbe obtained its energy by oxidizing either ferrous iron to ferric iron or sulfur to sulfate. It grew slowly on an acid thiosulfate medium.

Other studies have been reported on the existence of microorganisms in copper mine waters. Audsley *et al.* (1961), working in the Department of Scientific and Industrial Research in England, firmly established the presence of both *Ferrobacillus ferrooxidans* and *Thiobacillus thiooxidans* in such waters.

The autotrophic bacteria oxidize ferrous to ferric iron more rapidly than does atmospheric oxygen, especially under acidic conditions. Strains of iron autotrophic bacteria were found in the acid mine waters from Kennecott Copper Corporation mines at Bingham Canyon, Utah (Zimmerly *et al.*, 1958) and Chino, New Mexico. The ability of these bacteria to oxidize many sulfide minerals was noted. Among the minerals oxidized were chalcopyrite, covellite, chalcocite, bornite, tetrahedrite, molybdenite, and sphalerite. Bacteria were active in the dissolution of these minerals. Bacteria which occur naturally in acid mine waters and tolerate high concentrations of copper have a low tolerance for other metals. For example, bacteria taken directly from Bingham Canyon will not grow in solutions containing more than 150 ppm of zinc, even though they thrive in solutions containing appreciable higher concentrations of copper. Adapting successive generations of these bacteria to media containing increasingly higher concentrations of other metals has produced strains which will tolerate higher concentrations of these metals. By gradual adaptation, bacteria were obtained which tolerated up to 17,000 ppm of zinc as compared to an original tolerance of 150 ppm. Also, the tolerance of these bacteria for copper has been increased to approximately 12,000 ppm. Other metallic ion concentrations to which tolerance has been developed in particular strains are as follows:

Metal	Maximum tolerance (ppm)
Aluminum	6,290
Calcium	4,975
Magnesium	2,400
Manganese	3,280
Molybdenum	160

The bacteria involved in this process are motile, nonsporeforming rods approximately $0.5-1.0\mu$ wide and $1.0-2.0\mu$ long. They are autotrophic in character, deriving their energy from the oxidation of ferrous iron and using carbon dioxide as a carbon source. Organic materials are not required for growth, but acid conditions are. They appear to be substantially identical with the *Thiobacillus ferrooxidans* except for a natural tolerance for copper in concentrations of 1–1.5 mg/liter.

Additional research indicates the possibility of enhancing bacterial leaching activity by the addition of nutrients to the leach solutions. Working with mine waters taken from the Bingham Canyon operation of Kennecott Copper Corporation, it was found that bacterial activity was increased significantly in the initial phase of the leaching operation by the addition of up to 200 ppm of nitrogen in the form of nitrate or ammonium ion. Although bacterial growth increased, no benefit to leaching was observed beyond 100 ppm of nitrogen. In long-term experiments the latter stages of leaching are not benefited by nitrogen addition.

Lazaroff (1963) found that the growth of *Thiobacillus ferrooxidans* was initially inhibited in media containing ferrous chloride in place of ferrous sulfate. This inhibition of growth was found associated with a requirement of a high proportion of sulfate to chloride ions or certain other anions for iron oxidation. Adaptation was attributed to the selection of spontaneous mutants capable of oxidizing iron in media high in chloride and low in sulfate.

Temple and Delchamps in the United States (1953), Audsley *et al.* in England (1961), and Ivanov *et al.* in the U.S.S.R. (1961) have indicated that the role of these bacteria consists chiefly of the oxidation of ferrous sulfate to ferric sulfate after the ferrous sulfate had been formed, e.g., by a nonbiological chemical oxidation of moist iron sulfides. Razzell in Canada (1962) indicated that bacterial leaching of metallic sulfides such as chalcopyrite, covellite, and nickel sulfides could operate directly on the mineral, although ferrous ions improved the rate of leachings.

Silverman *et al.* (1961) used manometric techniques for studying the oxidation of pyritic material in the presence of bacteria. Resting cells of *Ferrobacillus ferroxidans* accelerated the oxidation of coal, pyrite, and coarsely crystalline marcasite, but not of coarsely crystalline pyrite. Oxidation rates in the presence of *F. ferrooxidans* were increased by reducing the particle size of the pyrite sample and, in one instance, by removing the calcium carbonate from a calcite-containing sample.

Rogoff *et al.* (1960) studied the role of microorganisms in the oxidation of sulfur-containing materials. Mine waters containing iron-sulfur and thiosulfate-oxidizing, and sulfate-reducing bacteria were examined. A manometric study of the biological oxidation of pyrites showed that pyrite oxidation was increased 8–13 times over the controls in the presence of *Ferrobacillus ferrooxidans. Thiobacillus thiooxidans* apparently played no role in pyrite oxidation. The rate of pyrite oxidation in the presence of the bacteria was affected by particle size of the crushed material, pyrite content of the material, crystalline form of the pyrite, pH, etc. Other experiments were concerned with microbial oxidation of organically combined sulfur.

Shouten (1946) also studied the role of sulfur bacteria in the formation of so-called sedimentary copper ores and pyritic ore bodies. Small sulfide spheres in the ore were believed to have originated from sulfur bacteria. This is now thought to be physiologically improbable.

Factors affecting the rate and extent of biological leaching are particle size, temperature, pH, aeration, size of bacterial inoculum, and exposure to ultraviolet light (Bryner *et al.*, 1954; Bryner and Anderson, 1957; Silverman and Lundgren, 1959; Malouf and Prater, 1961; Silverman *et al.*, 1961; Razzell and Trussell, 1963a,b; Sutton and Corrick, 1963). Optimal leaching conditions include particle size of -325 mesh, a temperature near 35°C, a pH between 2.0 and 3.0, rapid aeration, a large inoculum, and protection from sunlight (ultraviolet light). Microbial leaching of sulfide ores is most commonly studied with airlift percolators (Bryner *et al.*, 1954), Warburg respirometers (Beck, 1960), and stationary leach bottles (Razzell and Trussell, 1963b). Percolators and stationary leach bottles normally meet all the requirements for optimal leaching with the possible exception of aeration. A practical method which produces rapid aeration and permits evaluation of many experimental variables is the use of the Erlenmeyer flask on a standard shaking apparatus. Almost all conditions for optimal leaching can be evaluated by this technique.

Duncan *et al.* (1964) found that the rate of leaching of chalcopyrite by *Thiobacillus ferrooxidans* was greatly accelerated by using shaker flasks in place of stationary bottles or percolators. Under conditions where *T. ferrooxidans* attacks the crystal lattice mineral sulfides directly, the rate of leaching may be increased by a mechanical system which provides a more intimate contact between the microbe and the surface of the mineral.

A further increase in rate and extent of leaching was obtained by the use of surface active agents such as Tween 20, 40, 60, and 80, Triton X-100, Quaker TT5386, and Hyamine 2389 (Audsley *et al.*, 1961). Tween 20 was the most effective. No individual component of the Tween molecule was responsible for the improved leaching. The Tween-to-chalcopyrite ratio was more important than the Tween-to-medium ratio. The increased rate of biochemical leaching of copper achieved with surfactants and mechanical agitation was comparable to that obtained with acidified ferric sulfate leaching.

Starkey *et al.* (1956) found the rate of oxidation of sulfur by *Thiobacillus thiooxidans* in a liquid medium was increased by shaking with surface-active agents. More recently, Schaeffer and Umbreit (1963) showed that *T. thiooxidans* produces phosphatidylinositol, a surface-active agent, that wets the surface of the mineral particle to be oxidized.

The increased rate and extent of biological leaching of chalcopyrite accomplished by the use of surfactants and shaking has improved the commercial attractiveness of this process. Utilization of these techniques with *Thiobacillus ferrooxidans* has produced 85% copper recovery from chalcopyrite in 24 days. Initial rates of leaching were essentially higher. Release of copper over the first 5 to 7 days was 10% per day. These rates and yields are much greater than the previously reported maximums of 60% copper in 470 days by Malouf and Prater (1961) and 40% copper in 60 days by Razzell and Trussell (1963b); they are comparable to the results of Columbo and Frommer (1962) in chemical leaching of chalcocite.

Oxygen deficiencies occur in percolator systems even though solutions entering the top are saturated with oxygen. From the descriptions and illustrations of percolators given in the literature (Bryner *et al.*, 1954; Bryner and Anderson, 1957; Malouf and Prater, 1961; Sutton and Corrick, 1963), it should be noted that when columns are flooded with nutrient solutions to a height of 3–5 cm above the surface of the sand and ore, the oxygen becomes depleted as minerals are oxidized and anaerobic conditions develop in the lower regions of the column. These conclusions are in agreement with the results obtained by Audsley and Daborn (1962) during the leaching of uranium ore in percolators. In deep columns of uranium ore high in pyrite, over 83% of the uranium was leached from the upper two-fifths of the column and only 32% from the lower one-fifth. When the frequency of wetting the mineral surfaces in deep columns was decreased from daily to weekly intervals, an increased rate of leaching was obtained. This was attributed to a greater volume of air in the voids.

The rates of copper leaching often depend on the type of ore and the system used for leaching. Using acidified ferric sulfate in a batch system, Columbo and Frommer (1962) leached 50% of the copper from Michigan ores in 4 hours and over 90% in 2–3 weeks. Using percolator systems inoculated with microbes, Bryner et al. (1954) recovered only 2.8% of the copper from one chalcopyrite sample in 70 days and 6.6% from another sample in 56 days. Malouf and Prater (1961) leached 50% of the copper present in their samples of chalcopyrite in about 170 days using airlift percolators; a total of 60% was leached in 470 days. Razzell (1962) was able to recover 25% of the copper from chalcopyrite in 60 days with percolators and 35% in 100 days in stationary bottles. In the Razzell studies 80% of the copper in chalcopyrite was converted to acid-soluble copper, but only 45% of this was soluble at pH 2.5. Later Razzell and Trussell (1963a) improved this yield to 40% by leaching 55 days.

Ehrlich (1965) studied the capacity of *Ferrobacillus* to leach low-grade copper sulfide ore with ferrous or ferric solutions in sterile sets of glass columns containing ground ore. One set of columns inoculated with *Ferrobacillus* and an uninoculated set were leached daily with solutions at pH 2.5 containing 960 μg/ml of either ferrous or ferric iron. A third uninoculated set was leached daily with iron-free solutions adjusted to pH 2.5. Effluents from each column were examined at weekly intervals for bacterial content, pH, and copper and iron content. Leaching of acidic ore with or without bacteria was more rapid than leaching of basic ore, especially with ferrous iron leaching solutions. Ferric was more effective than ferrous iron in leaching copper.

Thiobacillus ferrooxidans increase the leaching rate of pyrite and covellite in the absence of iron salts, i.e., by direct oxidation (Ivanov and Nagirnyak, 1962). With the addition of iron salts the bacterium acted to regenerate ferric sulfate. A pure culture of *Thiobacillus thiooxidans* had no effect on the leaching of these minerals. Iron sulfate was required for maximum leaching of covellite and pyrite with both *Thiobacillus thiooxidans* and *Thiobacillus ferrooxidans* being required. The addition of iron salts markedly accelerated the leaching of chalcopyrite by *Thiobacillus ferrooxidans*. Ferrous sulfate and ferric sulfate had almost the same effect and the addition of pyrite had little effect. Bacterial action was maximum at pH 1.7–2.0.

The bacteria were not destroyed in 24 hours by adding either 18.3 gm of sulfuric acid/liter simultaneously to the same culture or by adding 10–20 gm of iron, copper, and zinc salts/liter. Mild aeration aided bacterial regeneration. Nutrients other than those present in natural waters were not necessary. Periodic percolation leaching of copper sulfides was twice as effective as continuous leaching.

Duncan and Trussell (1964) found that various sulfide minerals and ores could be leached rapidly, and with a good yield, by *Thiobacillus ferrooxidans*. The addition of ferrous sulfate and ferric sulfate was not necessary for biological leaching, but its addition may have been beneficial for the maintenance of the pH in the case of minerals or ores low in iron. *Thiobacillus ferrooxidans* leached 72% of the copper from museum-grade chalcopyrite in 12 days, 100% in 26 days. With covellite, about 50% of the copper was removed in 76 days. In tests with chalcocite more than 90% was leached in 30 days, and with bornite, 100% was leached in 20 days. In leaching chalcopyrite ore, 45–90% of the copper was dissolved from different samples in 10 days.

Sutton and Corrick (1961a,b) leached the copper minerals azurite, chalcopyrite, chrysocolla, covellite, cuprite, and digenite individually, using the indigenous microbial population present and some *Thiobacillus* cultures. The leaching was accomplished with 50 ml of nutrient solution which was continuously circulated by compressed air through percolators charged with 10 gm of ore and 150 gm of Ottawa sand. The liquid medium was extracted and analyzed, fresh solution was added, and the cycle repeated for ten 7-day periods. Appreciable amounts of copper and cobalt were extracted from these minerals when elemental sulfur was present. Sulfides were not oxidized to acid in these ores. Ores containing copper minerals in the silicate, carbonate, and oxide forms were more susceptible to bacterial action than ores containing sulfide copper minerals. *Thiobacillus thiooxidans* would not tolerate nor would it react with certain ores until strains of the bacterium had been progressively adapted to increase concentrations of the individual ores. The acid produced by the bacterium was effective in extracting copper from the ores and, on analysis, proved to be sulfuric acid.

Zinc and other minerals may be recovered from low-grade ores. Leaching of copper in Bingham Canyon has been increased one thousand times by the presence of certain iron-oxidizing becteria, such as *Thiobacillus thiooxidans* and *T. ferrooxidans*, which are common in various mine waters. These bacteria make waste dumps more pervious to leach water.

The bacteria active in this leaching require oxygen to live and become inactive at depths where oxygen is limiting. The temperature of optimum activity is 35°C. The leaching efficiency of the microbe decreases as the temperature rises and inactivation occurs at temperatures above 50°C. The optimum acidity for most leach solutions is at pH 2.0–3.5. This range of acidity is most favorable for the growth of the sulfur-oxidizing bacteria. Low acidity also keeps slime deposits from forming in waste dumps and pipelines, while the waste dump becomes more porous and pervious to water and air. Bacterial action is inhibited at pH of 6.9; at a pH of 9.0 the desirable autotrophic bacteria are destroyed.

Zimmerly *et al.* (1958) found that the autotrophic bacteria *Thiobacillus ferrooxidans*, which resists relatively high concentrations of metal ions, can be used to oxidize ferrous sulfate leach solutions to the ferric state. By exposing the bacterial strains to successively higher concentrations, it was possible to increase their tolerance up to 17,000 ppm of zinc, 12,000 ppm of copper, and 6290 ppm of aluminum. Certain nutrients in the form of nitrates or ammonium ions in amounts up to 200 ppm increased the bacterial activity significantly. Mine waters from copper ore tailings which contained 0.26 gm of copper/liter increased its copper content to 11.74 gm/liter in 119 days of using bacterial leaching solutions. The process is particularly applicable to leaching of sulfide minerals. A feature of the process is that the iron pyrite in the ore continuously and automatically supplements the original iron and acid which is utilized.

For heaps low in sulfide, a " breeder " solution of the bacteria in sulfuric acid and ferric sulfate must be prepared before it is pumped to the top of the heaps. Where an appreciable amount of sulfide ores is present, it is necessary only to control the acidity of the solution. As the solution flows down through the dumps, *Thiobacillus thiooxidans*, which oxidizes sulfur, and *T. ferrooxidans*, which oxidizes iron, initiate complementary chemical reactions. *Thiobacillus ferrooxidans* produces ferric sulfate which, in acid solution, leaches copper, chalcocite, covellite, and chalcopyrite. At the same time, the *T. thiooxidans* strains produce more sulfuric acid to maintain an acid pH. These chemical reactions are exothermic and at leaching dump depths of 250 ft the temperature of the leaching solution increases 14°C over the initial temperature. Although the increased temperature may kill some of the active bacteria it improves the leaching capacity and offsets any detrimental effect.

The studies reported herein show that copper sulfides of both primary and secondary origin are easily leached by microbial processes. The copper sulfides of primary origin formed at higher temperatures are probably a little more difficult to leach.

Ores which contain 0.8 % copper in the form of chalcopyrite, chalcocite, or covellite will normally yield, on complete oxidation of the sulfide, 49.0, 12.3, or 24.5 lb of sulfuric acid/ton of ore, respectively. The presence of other sulfide minerals, such as pyrite, will increase the yields of sulfuric acid. Such ores are easily leached of copper. Certain oxygen-containing minerals such as copper carbonates (azurite, malachite) and chrysocolla (copper silicate) and some cuprites (copper oxide) are more difficult to leach biochemically. If sulfur is added, or if other mineral sulfides are present, the sulfur-oxidizing autotrophs will synthesize sufficient sulfuric acid to solubilize the copper.

MICROBIAL LEACHABILITY OF COPPER ORES

The copper present in copper ores never appears to be directly utilized by even the autotrophic microbes. The explanation may lie in that (*a*) copper has normal valence states of Cu^+ and Cu^{2+}, and oxidation of copper does not provide a high energy value for autotrophic microbes, (*b*) the metal is quite toxic to most biological systems, and (*c*) it is not normally an essential element of life processes.

A leachability chart is shown in Table 7.I. Copper-containing minerals

TABLE 7.I

PREDICTED MICROBIAL LEACHABILITY OF COPPER ORES

Chemical composition	Mineral	Moiety oxidized biochemically	Oxidation by Fe^{3+} [a]
Cu_2S	Chalcocite	S^{2-}	+
$CuFeS_2$	Chalcopyrite	S^{2-}, Fe^{2+}	—
CuS	Covellite	S^{2-}	+
$Cu_3(CO_3)_2(OH)_2$	Azurite	—	+
Cu_2O	Cuprite	—	+
Cu_3AsS_4	Enargite	As^{2+}, S^{2-}	—
$Cu_4(OH)_6SO_4$	Brochantite	—	—
$Cu_3(OH)_4SO_4$	Antbrite	—	—
$Cu_2Cl(OH)_3$	Atacamite	—	—
$Cu_2CO_3(OH)_2$	Malachite	—	+
Cu_5FeS_4	Bornite	Fe^{3+}, S^{2-}	+
Cu_9S_5	Digenite	S^{2-}	+
$CuSiO_3 \cdot 2\,H_2O$	Chrysocolla	—	—

[a] $Fe_2(SO_4)_3$.

such as chalcocite, chalcopyrite, covellite, enargite, bornite, and digenite all contain a sulfide moiety and are leached as the sulfide is oxidized microbially to sulfuric acid. Chalcopyrite and bornite also contain reduced ron which autotrophs rapidly oxidize to the ferric state. In these latter minerals the microbes attack two moieties both of which are required in cellular metabolism.

The arsenic present in enargite is also attacked by specific autotrophic bacteria, thus offers a second line of attack for biochemical solubilization of this mineral.

The other copper minerals listed are more difficult to leach, probably because they are more toxic to biolgoical systems. Minerals in this category are: brochantite, antbrite, and atacamite. Chrysocolla is difficult to leach primarily because it is a silicate of copper.

BOTANICAL PROSPECTING

Microbial prospecting technology for copper deposits has not been developed. Where the copper content in surface ores is sufficiently high, it will definitely influence the microbial population. These habitats have not been studied.

Higher plants have been evaluated in relation to botanical prospecting. Since this area is indirectly related to microbial prospecting it is only briefly reviewed. Since botanical prospecting has definite merit, from a comparative biochemical viewpoint, analysis of the surface ores should yield specific microbial populations and cultures.

It has been possible to detect abnormal copper concentration in soil since the uptake of trace elements by plants is influenced by the pH of the soil. Warren *et al.* (1949) found that twigs representing a full season's growth are preferred in the sampling of deciduous trees for copper. Replicate analyses are needed because deviations of up to 50% occur (Millman, 1957).

Douglas fir is used as a reliable index for copper evaluation except when it is found growing in limestone or calcite regions. The copper prospects are good for the area if the copper-zinc ratio in Douglas fir is greater than 0.25 and/or there are more than 10 ppm copper in plant materials or 350 ppm copper in their ash. The same relation holds for Jack pine (lodgepole pine), and Englemann spruce; in the latter the Cu/Zn ratio decreases to 0.20 where mineralization of copper occurs. Data obtained from dwarf and rock mountain junipers should be used cautiously and only if other species are unavailable. More than 8 ppm copper in plant material and copper-zinc ratio greater than 0.40 are indicative of copper mineralization.

Analysis of willows should be avoided. There are too many species and a large variation in analyses of copper and zinc uptake by suckers and young twigs is encountered. Poplars appear to be particularly sensitive to copper and zinc, so that twigs representing growth over a period of 2 to 3 years can be used for analysis. Red alder and green and mountain alder are useful for copper searches. These trees contain lower quantities of zinc. A copper-zinc ratio greater than 0.40 suggests an underlying copper deposit. Sagebrush twigs concentrate copper, when it is available, with 25 ppm copper and copper : zinc ratios of 1.0 being found in mining areas.

Three plant groups—the pink family, the mint family, and the mosses— have been used as copper indicators, since they grow only in copper-bearing soils. Basil, a member of the mint family, grows only in soil containing more than 100 ppm of copper.

Geobotanical prospecting also involves finding ore by the lack of vegetation in soils which are toxic to a particular plant. Soils containing 8–14% copper in some generally forested areas in the Congo have no tree growth, but are covered by grass and shrubs. Bare areas have also been used in finding copper in Armenia and Rhodesia.

Botanical prospecting has several disadvantages, namely: (a) absorption by plants depends on many factors including soil pH and age, (b) the depth of effectiveness is determined by root penetration, and (c) the sampling and analysis itself can become complicated.

Advantages, however, are many: (a) roots often penetrate into inaccessible areas, not reached by surface soil sampling, (b) errors created by halos are eliminated, (c) larger areas can be covered by one sample, (d) interferences from transported surface soil are eliminated, and (e) use of indicator plants is simpler than other methods (Cannon, 1960).

REFERENCES

Ashmead, D. (1955). The influence of bacteria in the formation of acid mine waters. Colliery Guardian **190**, 694–698.

Audsley, A., and Daborn, G. R. (1962). Natural leaching of uranium ores. II. A study of the experimental variables. *Trans. Inst. Mining Met.* **72**, 235–246.

Audsley, A., Daborn, G. R., and Pearson, D. (1961). Sci. Rept. N.C.L./CON. 2 Dept. Sci. Ind. Res., London.

Beck, J. V. (1960). A ferrous-iron-oxidizing bacterium. I. Isolation and some general physiological characteristics. *J. Bacteriol.* **79**, 502–509.

Bryner, L. C., and Anderson, R. (1957). Micro-organisms in leaching sulfide minerals. *Ind. Eng. Chem.* **49**, 1721–1724.

Bryner, L. C., and Jameson, A. K. (1958). Microorganisms in leaching sulfide minerals. *Appl. Microbiol.* **6**, 281–287.

Bryner, L. C., Beck, J. V., Davis, D. B., and Wilson, D. G. (1954). Micro-organisms in leaching sulfide minerals. *Ind. Eng. Chem.* **46**, 2587–2592.

Cannon, H. L. (1960). Botanical prospecting for ore deposits. *Science* **132**, 591–597.

Colmer, A. R., Temple, K. L., and Hinkle, M. E. (1950). An iron-oxidizing bacterium from the acid drainage of some bituminous coal mines. *J. Bacteriol.* **59**, 317–328.

Columbo, A. F., and Frommer, D. W. (1962). Leaching Michigan copper ore and mill tailings with acidified ferric sulfate. *U.S., Bur. Mines, Rept. Invest.* 5924.

Corrick, J. D., and Sutton, J. A. (1961). Three chemosynthetic autotrophic bacteria important to leaching operations at Arizona copper mines. *U.S. Bur. Mines, Rept. Invest.* 5718.

Davis, D. B. (1953). Biological oxidation of copper sulfide minerals. Master's Thesis, Brigham Young University, Provo, Utah.

DeCuyper, J. A. (1964). Bacterial leaching of low grade copper and cobalt ores. *Metallurgical Soc. Conf., Dallas, Texas,* 1963. Vol. 24.

Duncan, D. W., and Trussell, P. C. (1964). Advances in the microbiological leaching of sulfide ores. *Can. Met. Quart.* **3**, 43–55.

Duncan, D. W., Trussell, P. C., and Walden, C. C. (1964). Leaching of Chalcopyrite with *Thiobacillus ferrooxidans:* Effect of surfactants and shaking. *Appl. Microbiol.* **12**, 122–126.

Ehrlich, H. L. (1965). Bacterial leaching of low-grade copper sulfide ore with and without ferrous or ferric iron. *Ann. Meeting. Am. Soci. Microbiol., Atlantic City, New Jersey,* 1965 (private communication).

Ivanov, V. I., and Nagirnyak, F. I. (1962). Intensification of leaching of copper sulfide minerals with thionic bacteria. *Tsvetn. Metal.* **35**, No. 8, 30–36.

Ivanov, V. I., Nagirnyak, F. I., and Stepanov, B. A. (1961). Bacterial oxidation of sulfide ores. I. The role of *Thiobacillus ferrooxidans* in the oxidation of chalcopyrite and spahlerite. *Mikrobiologiya* **30**, 688–692.

Johnson, P. H. (1965). Acid-ferric sulfate solutions for chemical mining. *Soc. Mining Eng. Eng.* **2**, 401–410.

Kinsel, N. A. (1960). New sulfur oxidizing iron bacterium: *Ferrobacillus sulfooxidans.* *J. Bacteriol.* **80**, 628–632.

Lazaroff, N. (1963). Sulfate requirements for iron oxidation by *Thiobacillus ferrooxidans* *J. Bacteriol.* **85**, 78–83.

Leathen, W. W., and Braley, S. A. (1954). A new iron oxidizing bacterium. *Bacteriol. Proc.* **44**, 37–40.

Leathen, W. W., Kinsel, N. A., and Braley, S. A. (1956). *Ferrobacillus ferrooxidans:* A chemosynthetic autotrophic bacterium. *J. Bacteriol.* **72**, 700–704.

Malouf, E. E., and Prater, J. D. (1961). Role of bacteria in the alteration of sulfide minerals. *J. Metals* **13**, 353–356.

Mariacher, B. C., and Spence, W. C. (1962). A look into the future of mineral benefici-ation. *Mining Eng.* **14**, 57–60.

Millman, A. P. (1957). Biogeochemical investigations in areas of copper-tin mineraliz-ation in southwest England. *Geochim. Cosmochim. Acta* **12**, 85–93.

Razzell, W. E. (1962). Bacterial leaching of metallic sulfides. *Can. Mining Met. Bull.* **55**, 190–191.

Razzell, W. E. and Trussell, P. C. (1963a). Microbiological leaching of metallic sulfides. *Appl. Microbiol.* **11**, 105–110.

Razzell, W. E., and Trussell, P. C. (1963b). Isolation and properties of an iron-oxidizing *Thiobacillus. J. Bacteriol.* **85**, 595–603.

Rogoff, M. H., Silverman, M. P. and Wender, I. (1960). Elimination of sulfur from coal by microbial action. *Am. Chem. Soc., Div. Gas Fuel Chem.* **2**, 25–36.

Rudolfs, W., and Helbronner, A. (1922). Oxidation of zinc sulfide by microorganisms. *Soil Sci.* **14**, 459.

Schaeffer, W. I., and Umbreit, W. W. (1963). Phosphatidylinositol as a wetting agent in sulfur oxidation by *Thiobacillus thiooxidans. J. Bacteriol.* **85**, 492–493.

Shouten, C. (1946). The role of sulfur bacteria in the formation of the so-called sedimentary copper ores and pyritic ore bodies. *Econ. Geol.* **41**, 517–538.

Silverman, M. P., and Lundgren, D. G. (1959). Studies on the chemoautotrophic iron bacterium *Ferrobacillus ferrooxidans.* II. Manometric studies. *J. Bacteriol.* **78**, 362–331.

Silverman, M. P., Rogoff, M. H., and Wender, I. (1961). Bacterial oxidation of pyritic materials in coal. *Appl. Microbiol.* **9**, 491–496.

Sirel, M. A. (1949). The copper deposit of Ergani-Maden, Turkey, *Neues Jahrb. Mineral, Abhandl.* **80A**, 36–100.

Starkey, R. L. (1925). Concerning the carbon nitrogen nutrition of *Thiobacillus thiooxidans,* and autotrophic bacterium oxidizing sulfur under acid conditions. *J. Bacteriol.* **10**, 165–175.

Starkey, R. L., Jones, G. E. and Frederick, L. R. (1956). Effects of medium agitation and wetting agents on oxidation of sulfur by *Thiobacillus thiooxidans. J. Gen. Microbiol.* **15**, 329–334.

Stradlinger, H. (1950). Mansfeld copper ores. *Chemiker Ztg.* **74**, 722–725.

Sutton, J. A., and Corrick, J. D. (1961a). Possible uses for bacteria in metallurgical operations. *U.S., Bur. Mines, Inform. Circ.* **8003**, 8 pp.

Sutton, J. A., and Corrick, J. D. (1961b). Bacteria in mining and metallurgy. Leaching selected ores and minerals. Experiments with *T. thiooxidans. U.S., Bur. Mines., Rept. Invest.* **5839** 16 pp.

Sutton, J. A., and Corrick, J. D. (1963). Leaching copper minerals by means of bacteria. *Mining Eng.* **15**, 37–40.

Temple, K. L., and Colmer, A. R. (1951). The autotrophic oxidation of iron by a new bacterium, *Thiobacillus ferrooxidans. J. Bacteriol.* **62**, 605–617.

Temple, K. L., and Delchamps, E. W. (1953). Autotrophic bacteria and the formation of acid in bituminous coal mines. *Appl. Microbiol.* **1**, 255–258.

Vogler, K. G., and Umbreit, W. W. (1942). Studies on the metabolism of autotrophic bacteria. *J. Gen. Physiol.* **26**, 157–167.

Vogler, K. G., LePage, G. A., and Umbreit, W. W. (1942). Studies on the metabolism of autotrophic bacteria. I. The respiration of *Thiobacillus thiooxidans* on sulfur. *J. Gen. Physiol.* **26**, 89–102.

Waksman, S. A., and Joffe, J. S. (1922). Microorganisms concerned in the oxidation of sulfur in the soil. II. *Thiobacillus thiooxidans,* a new sulfur-oxidizing organism isolated from the soil. *J. Bacteriol.* **7**, 239–256.

Warren, H. V., Delavault, R. E., and Irish, R. I. (1949). Biogeochemical researches on Copper in British Columbia. *Trans. Roy. Soc. Can., Sect. IV* [3] **43**, 119–137.

Zimmerly, S. R., Wilson, D. G., and Prater, J. D. (1958). Cyclic leaching process employing iron-oxidizing bacteria. Kennecott Copper Corporation. U.S. Patent 2,893,964.

8

MICROBES AND ZINC

Zinc was not distinctly recognized as an element by the ancients, but it was in use long before it was identified as a constituent of copper alloys and brass. The term "cadmia" was applied to ores of zinc and to the oxides of zinc obtained from the furnace, and cadmium itself is present in many ores. The Chinese were probably the first to extract zinc metal. It was originally sold as "spialter" or spelter which refers to a metallic product containing zinc with cadmium, iron, arsenic, and lead as impurities. The earth's crust contains an estimated 0.004% which is less than the values estimated for zirconium, vanadium, titanium, and strontium.

Zinc is bipositive in its most active state in nature. It complexes easily with sulfide to form the mineral sphalerite which contains 67% zinc. Sphalerite ores are also an important source of cadmium and, to a lesser degree, of indium, gallium, and germanium.

Igneous rocks contain up to 200 gm/ton of zinc. The affinity of both zinc and cadmium for sulfur is less than that of iron, nickel, cobalt, and copper. Sphalerite is a common mineral and is found associated with galena, pyrite, marcasite, chalcopyrite, smithsonite, calcite, and dolomite. The amount of iron present in sphalerite gives an indication of the temperature of formation, and is used in this context as a geological thermometer. Low iron and manganese contents are proof of a low temperature of formation, whereas, high levels of iron, manganese, and cobalt indicate hydrothermal formation at high temperatures.

Zinc as well as cadmium became highly enriched in hydrothermal minerals formed at relatively low temperatures.

Mineral	Formula
Sphalerite and wurtzite	ZnS
Zincite	ZnO
Smithsonite	$ZnCO_3$
Willemite	$Zn_2(SiO_4)$
Hemimorphite	$Zn_4(OH)_2(Si_2O_7) \cdot H_2O$

The similarity in ionic radii of zinc and iron explains the presence of zinc in magnetite and ilmenite. Zinc also is present in the ion-dependent minerals gahnite ($ZnAl_2O_4$) and franklinite ($Zn, Mn)Fe_2O_4$ which occur in metamorphic rocks. Zinc is not normally present in feldspar.

With weathering, divalent zinc goes readily into solutions as a chloride or sulfate. In solution it is easily transported in surface or ground waters. In enrichment zones it is redeposited as a sulfide, oxide, carbonate, or silicate.

Zinc has an atomic number of 30, an atomic weight of 65.38, and several isotopes. ^{65}Zn (half-life, 250 days) and ^{63}Zn (half-life, 38 minutes) are used as biological tracers (Schneider and Disselkamp, 1958).

BIOGEOCHEMISTRY OF ZINC

Zinc in low concentrations is known to be a growth stimulant for plants; and some plants even concentrate the element. Ashed material from the plant *Thlaspi calaminare* contained 16% Zn. Large quantities of zinc are normally quite toxic to plants and animals.

Zinc is an essential nutrient for both plants and animals; the growth requirement for zinc by *A. niger* was demonstrated by Raulin (1869). Zinc is an active component of the enzyme carbonic anhydrase and makes up 0.33% of the molecule (Keilin and Mann, 1944). Zinc is present in several dehydrogenases (Vallee, 1959), certain peptidases, transphorylase, and it forms a stable complex with antibiotics such as bacitracin. Zinc is bound to the imidazole groups encountered in proteins and enzymes, e.g., albumin, α-lipoprotein, α-glycoprotein, alkaline phosphatase, and serum esterase. Since up to 66% of the total zinc complexed with proteins is loosely bound, it may be leached, transported, and become available for mineral deposition.

The crystallization of insulin requires the presence of Zn^{2+}, Ca^{2+}, Co^{2+}, or Ni^{2+}. One zinc atom probably chelates two imidazole residues of amorphous insulin (Hallas-Moller *et al.*, 1952; Scott, 1934).

Rudolfs and Helbronner (1922) using zinc-blend ores high in ZnS leached up to 0.37 gm Zn/gm of ore using sulfur-oxidizing bacteria.

With sphalerite and wurtzite ores, under limiting aeration, microbes will oxidize the sulfide moiety to sulfate with the formation of zinc sulfate.

$$ZnS + 2\,O_2 \rightarrow ZnSO_4$$
$$\Delta F_{25°C} = -167 \text{ kcal}$$

If adequate oxygen is present, the zinc sulfide will be oxidized to zinc sulfate and sulfuric acid.

$$2\,ZnS + 4\,O_2 + H_2O \rightarrow ZnSO_4 + ZnO + H_2SO_4$$
$$\Delta F_{25°C} = -319 \text{ kcal}$$

The greatest solubilization was observed where the most sulfuric acid was formed.

Smithsonite ($ZnCO_3$) is most easily solubilized oxidatively by microbes synthesizing sulfuric or other organic acids.

$$ZnCO_3 + 1\tfrac{1}{2}\,O_2 + H_2O + S \rightarrow ZnSO_4 + H_2CO_3$$
$$\Delta F_{25°C} = -130 \text{ kcal}$$

The pH in these solutions decreased to 2.9. For good zinc solubilization, a pH of 2.0 or less is required. The addition of elemental sulfur increases the recovery of zinc in the above studies since it is oxidized by microbes to H_2SO_4.

Ono *et al.* (1959) conducted a spectrographic analysis of the ash of Filicales, Musci, Hepticae, certain lichens, and fungi. The alkaline earths and silicon were detected in microscopic quantities. *Parmelia tinctorum* contained 1.46% Zn^{2+}; *Cyclophorus lingua*, 1.34% Mn^+; and *Vittaria frexuosa*, 1.86% Mn^+. Aluminum, iron, manganese, zinc, copper, lead, phosphorus, and boron were found in most samples with silver, chromium, titanium, molybdenum, vanadium, and tin being detected in a few samples.

MICROBIAL METABOLISM IN ZINC

Aspergillus niger requires zinc for the synthesis of phenylalanine, tryptophan, and tryosine (Bertrand and de Wolf, 1957a,b, 1958, 1959, 1960, 1961a,b). Zinc is also required for the synthesis of the enzymes phosphofructokinase, glyceraldehydephosphate dehydrogenase,

6-phosphogluconic acid dehydrogenase, and glucose-6-phosphate dehydrogenase and is involved in nucleic acid synthesis in *A. niger*.

Zinc affects the organic acid and mycelium synthesis in such fungi as *Aspergillus*, *Rhizopus*, *Penicillium*, and *Citromyces* (J. W. Foster, 1939, 1949). The presence of zinc decreases organic acid synthesis and increases the amount of mycelium formed in what is known as the "zinc shunt" (Table 8.I). A zinc concentration of a few parts per million will initiate this reaction. With increased mycelial synthesis, CO_2 production increases. Zinc probably functions as a coenzyme in the oxidation of glucose.

TABLE 8.I

ZINC METABOLIC SHUNT

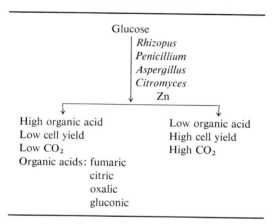

The addition of zinc to soil (Sulochana, 1952) did not produce an increase in the number of fungi, although in order of decreasing importance Li^{2+}, Mn^{2+}, Mo^{2+}, Co^{2+}, and Al^{3+} did. Li^{2+} caused a fivefold increase in the number of fungi, over controls whereas, Mn^{2+}, Mo^{2+}, Co^{2+}, B^{3+}, and Al^{3+} increased the fungal population approximately threefold. However, zinc is still regarded as essential for fungal growth.

Growth of penicillin-producing fungi is greatly retarded on zinc-deficient media; penicillin production is slowed by a low pH which occurs in the absence of zinc, possibly due to the buildup of gluconic acid. Zinc probably acts as a catalyst in glucose oxidation preventing the buildup of the gluconic acid as a by-product (J. W. Foster, 1949). A low pH value is favorable for the formation of the antibiotic notatin and unfavorable for penicillin production. Optimal levels of zinc enhance the production of characteristic yellow pigments present in "virtually all penicillin-producing fungi." Early workers on penicillin noted that a rapid decrease in the pH of the broth in the first 5 days of production, followed by an equally rapid increase in pH, was associated with good production of penicillin. When the pH failed to rise, notatin was formed at the expense of penicillin. These events are controlled to a large degree by the zinc content of the medium (J. W. Foster, 1949).

At relatively high concentrations, zinc is known to increase cytochrome synthesis in *Ustilago sphaerogens* with a simultaneous decrease in ferrichrome level. The concentration of zinc required for fungal growth varies from 0.001 to 0.5 ppm. Concentrations higher than 1.0 ppm are often toxic (Cochrane, 1958). In certain instances, toxic substances such as oxine are produced in fungi. Horsfall (1959) has been able to decrease the toxicity of such substances adding zinc.

Some of the more important physiological and metabolic effects of zinc on fungi are (Cochrane, 1958):

1. Accumulation of organic acids at zinc levels too low for normal growth. At optimal levels zinc decreases the percentage conversion of carbohydrate to citrate. Associated with this is a positive effect of the metal on carbon dioxide evolution and its action upon carbonic anhydrase activity.
2. Lowering of specific enzyme concentrations in cells grown in low zinc media.
3. Decrease in the accumulation of certain antibiotics.
4. Increase in cytochrome synthesis in *Ustilago sphaerogena* at high zinc concentrations with an accompanying decrease in ferrichrome accumulation.
5. Changes in cell composition.
6. Stimulation of the *in vitro* formation of vesicles by *Puccinia coronata*.
7. A shift in the proportions of related ring compounds produced by *Penicillium urticae*. In a low-zinc medium, 6-methylsalicylic acid is formed, whereas higher concentrations of zinc induces the production of gentisyl alcohol, patulin, and toluquinone.
8. Inhibition of sporulation.
9. Induction of mutants or cultural variants.
10. Acceleration of the 6-β-hydroxylation of progesterone by *Aspergillus ochraceus*.
11. Activation of an aminopeptidase in *Aspergillus parasiticus*.
12. Increase in diphosphopyridine nucleotidase formation by *Neurospora crassa*.

Zinc is required in several enzymatic reactions. Zinc, manganese, and cobalt are required for oxalacetate decarboxylase activity in *Azotobacter vinelandii* (Plaut and Lardy, 1949). Zinc is also required for phosphatase and polypetidase activity in certain species of *Clostridium, Propionibacterium*, and fungi. The mineral requirement for the formation of subtilin, a polypeptide antibiotic synthezised by *Bacillus subtilis*, is greater than that required for cell growth. Zinc is a primary mineral requirement along with K^+, Mg^{2+}, Fe^{2+}, and Mn^{2+}. Zinc also increases the yield of bacitracin synthesis in *Bacillus licheniformis*.

It has been shown in an experiment using *Rhizopus MX* that two to three times as much glucose was required to produce 1 gm of cell substance in the absence of zinc than in its presence (J. W. Foster, 1949), and also that two to three times as much lactic acid was synthesized. Thus, the amount of lactic acid obtained was dependent on the amount of glucose consumed, irrespective of the total amount of cell substance synthesized. Obviously, zinc catalyzed the oxidation of that portion not giving rise to lactic acid.

The profound effect of zinc on the metabolism of *R. nigricans* is shown in Tables 8.II and 8.III (J. W. Foster, 1949) when zinc is added to the medium. The zinc-deficient cultures attained growth only through the utilization of zinc present as a contaminant in the medium.

A similar study of Sulochana (1952) showed a small increase in number of bacteria by the addition of zinc, although there was a decrease in the actinomycete population. Enormous increases in bacterial populations were noted when manganese and boron were added to soil, while slight increases actinomycetes were shown with the addition of Li^{2+}, Co^{2+}, Al^{3+}, Mn^{2+}, and B^{3+}.

TABLE 8.II

Effect of Zinc on Metabolism of *R. nigricans*[a]

Zinc added (ppm)	Glucose con- sumed (mg)	Fumaric acid formed (mg)	Percent con- version	NH$_3$-N con- sumed (mg)	Total calcium in solution (mg)	Calcium due to fumaric acid (mg)	(%)
0.0	4702	2349	50.0	31.0	760	716	113
0.6	5088	452	8.9	82.1	338	220	71

[a] From J. W. Foster (1949).

TABLE 8.III

EFFECT OF ZINC ON CARBON METABOLISM
OF *R. nigricans*[a]

Zinc added (ppm)	Glucose carbon consumed (mg)	Carbon synthesized into cell substance (mg)
0.0	1881	279
0.6	2035	739

[a] From J. W. Foster (1949).

Besides *A. niger* and *R. nigricans*, zinc affects the metabolism of *Penicillium glaucum, P. notatum, Trichothecium roseum*, and others in a like manner (J. W. Foster, 1949).

As shown in Table 8.IV *Azotobacter chroococcum* requires zinc for nitrogen fixation (Banfi, 1963). This metal is also complexed as zinc coproporphyrin in *Corynebacterium diphtheriae* (Coulter and Stone, 1938). The enzymes tryptophan desmolase and pyruvic carboxylase of *Neurospora crassa* and *Rhizopus nigricans* require zinc in the cell surface (Bachmann and Odum, 1960). The lichen *Stereocaulon nanodes* contains up to 3300 ppm of this metal (Maquinay *et al.*, 1961).

TABLE 8.IV

ZINC REQUIREMENTS OF VARIOUS MICROBES

Microbe	Zn^{2+} Required for	Reference
Azotobacter chroococcum	Nitrogen fixation	Banfi (1963)
Corynebacterium diphtheriae	Zinc coproporphrin	Coulter and Stone (1938)
Aspergillus niger	Aldolase	Robinson (1951)
Neurospora crassa	Tryptophan desmolase	Nason (1950)
Rhizopus nigricans	Pyruvic carboxylase	J. W. Foster and Denison (1950)
Algae	Algal surface	Bachmann and Odum (1960)
Stereocaulon nanodes (lichen)	Zinc concentration (3300 ppm)	Maquinay *et al.* (1961)

MICROBIAL LEACHING OF ZINC

Bipositive zinc found in sphalerite, zincite, and smithsonite is probably not attacked directly by microbes. Not having several common valence states zinc per se does not offer the oxidative microbe an attractive bundle of energy. Zinc sulfide, however, has a sulfide moiety which could easily be oxidized by microbes to sulfate, and zinc sulfate is quite mobile in ground water. An experiment by the Russians Ivanov *et al.* (1962) demonstrated that the leaching of sulfide ores could be accomplished microbiologically through the use of the organism *Thiobacillus ferrooxidans*. Up to twelve times more zinc was leached from sphalerite (ZnS) using *T. ferrooxidans* than was leached in sterile controls. The oxidation of the sulfides takes place slowly. The addition of soluble iron increased the leaching of zinc dramatically; both ferrous and ferric iron gave good results. These results were attributed to the ability of *T. ferrooxidans* to regenerate $Fe_2(SO_4)_3$ faster than it was utilized in oxidizing ZnS.

Thus current evidence does not support a direct bacterial oxidation of the zinc moiety of ZnS. The primary bacterial oxidation is probably through generation of $Fe_2(SO_4)_3$. The ferric sulfate formed actively promotes the dissolution of sulfides including the sulfide of zinc, e.g., sphalerite (ZnS).

$$ZnS + 2\,Fe_2(SO_4)_3 + 2\,H_2O + O_2 \rightarrow ZnSO_4 + 4\,FeSO_4 + 2\,H_2SO_4$$
$$\Delta F_{25°C} = -122 \text{ kcal}$$

As the ferric ion is reduced during oxidation of zinc sulfides it is reoxidized by *T. ferrooxidans*. The production of sulfuric acid also aids in leaching and provides an acidic system with the low pH necessary for the growth of *Thiobacillus ferrooxidans*.

Basically, then, the process is as follows (see Fig. 8.1):

1. Bacterial and chemical oxidation of pyrite to yield ferrous sulfate.
2. Bacterial oxidation of ferrous sulfate to yield ferric sulfate.
3. Oxidation by ferric sulfate, because it dissolves sulfides, of ZnS to $ZnSO_4$ and ferrous sulfate. The zinc sulfate can then be recovered and the ferrous sulfate recycled to be again oxidized to ferric sulfate. To increase leaching, ferrous or ferric iron can be added to the liquor to greatly enhance dissolution of ZnS. Another reaction which is theoretically possible, is the direct microbial oxidation of ZnS. This is accomplished in metallurgical procedures by roasting ZnS:

$$ZnS + 2\,O_2 \xrightarrow{\Delta} ZnSO_4$$

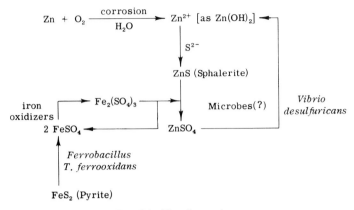

FIG. 8.1. The zinc cycle.

Zinc oxide and zinc carbonate can provide very little energy to the autotrophic microbe. Thus, bioenergetically, zinc per se appears to have no ostensible function. Its main value appears to be in its organizational and structural aspects as a metabolite. Its chelation by proteins and its function in enzymatic systems are the means by which it may be bound and carried through subterranean waters.

The possibility exists that sphalerite could be produced in nature by microorganisms, namely, the sulfate reducers. This was accomplished in the laboratory by Baas-Becking and Moore (1961). Sulfate reducers were shown to produce sphalerite from zinc metal and zinc carbonate. Zinc metal does not occur naturally, but zinc carbonate, known as smithsonite ($ZnCO_3$), does.

ZINC IN ALGAE AND IN MARINE ENVIRONMENTS

Zinc is found in ionic and sedimentary forms in lakes. Ionic zinc in the liquid phase is continually removed by microbes and ends up in solid materials such as seston and bottom sediments. In the aqueous phase it is present as a hydrated ion or complexed with organic materials. In the solid phase zinc is precipitated as zinc sulfide or zinc hydroxide.

$$Zn^{2+} + H_2O \rightleftharpoons ZnOH^+ + H^+ \underset{(H_2O)}{\longleftrightarrow} Zn(OH)_2 + H^+$$

The zinc cation enters into two different reactions with clay. It may occupy a position within the crystal lattice, or it may attach to a negatively charged site at the surface.

The amount of zinc-65 absorbed by algae is related to the zinc concentration in solution as measured by the Freundlich (1926) adsorption isotherm:

$$X = mkc^n$$

where X = amount removed; c = concentration in solution; m = mass of cells or adsorbent; and k and n = constants depending upon the chemical composition of the water and the type of cell material used. The addition of other cations to the solution decreases zinc uptake. The order of effectiveness of ions which reduce zinc uptake in *Golenkinia* cells was $H^+ > Ca^{2+} > Mg^{2+} > Na^+ > K^+$. Oxidized sediments remove as much as 96% of the zinc added to water, whereas, reduced sediments take up about 20% as much (Bachman 1963).

Many species of flagellates require zinc for growth (Ondratschedk, 1941), and zinc must be added as a micronutrient to *Chlorella pyrenoidosa* cultures (Myers, 1951). Zinc is also required by higher plants and the benefits of its use have been observed in corn, maze, deciduous fruits, pecans, tung, tree, etc.

Zinc is essential to the growth of freshwater phytoplankton, and it accumulates in high levels in silt (R. F. Foster, 1959; Nielsen and Perkins, 1957). In the studies by Davis *et al.* (1958) on the concentration of radionuclides by river organisms in the Columbia River below the Hanford reactor, zinc-65 was found in algae, sponges, insect larvae, snails, crayfish, and minnows (Chipman *et al.*, 1958; Ketchum and Bowen, 1958; Parker, 1961). Algal cultures of *Pelvetia canaliculata*, *Ascophylum nodosum*, and *Fucus vesiculosus* all concentrate zinc one thousand times over the surrounding seawater (Black and Mitchell, 1952).

In general microbes do not distinguish between stable and radioactive nuclides of zinc (Bennett, 1963). Marine microbes concentrate Zn^{65} from 23 to 67 thousand times over the external medium (Table 8.V). Cesium-137 is not concentrated appreciably by microbes. The highest concentration factor was 980 for an *Achromonas* sp. *Sphaerotilus natans* involved in iron corrosion concentrated copper-64 3900 times over the environment. No changes in the morphological or physiological behavior of these microbes were observed.

ZINC CORROSION

The corrosion of zinc in water proceeds in the following manner:

$$Zn + 2 H_2O \rightarrow Zn(OH)_2 + 2 H^+ + 2 e^-$$

TABLE 8.V

CONCENTRATION OF RADIONUCLIDES OF ZINC, CESIUM,
AND COPPER BY MARINE MICROBES[a]

Radionuclide uptake	Concentration factor over medium
Zinc-65	
Rhodomonas	0
Navicula confervacea	23,000
Nitschia spp.	42,000
Platymonas elliptica	67,800
Cesium-137	
Bacteria (several species)	15–26
Chlamydonomas spp.	28
Platymonas elliptica	50
Nitzschia sp.	97
Ochromonas sp.	980
Copper-64	
Sphaerotilus natans	3,900

[a] From Bennett (1963).

Zinc is often added to iron because it forms a good protective coating against further corrosion when oxidized. Atmospheric oxidation of the hydrogen produced by corrosion takes place, and it also furnishes energy for growth for the hydrogen-oxidizing bacteria.

$$2 H + \tfrac{1}{2} O_2 \rightarrow H_2O + 6.9 \text{ kcal}$$

In a paper by von Wolzogen Kuhr and van der Vlugt (1953) it was shown that the hydrogen oxidizers increased corrosion of zinc by removing hydrogen which acts to depolarize the system.

ZINC CYCLE

A zinc cycle involving microbes is not too complex. Zinc metal undergoes corrosion by oxidative action. Whether microbes obtain energy from this oxidation is not known. Bivalent zinc complexes easily with sulfide, oxide, and carbonate. As zinc sulfide it would be oxidized directly by microbes to zinc sulfate. The only definite method of oxidation of ZnS by microbes is by those bacteria which generate ferric sulfate which is a good oxidant. Zinc sulfate has been used as a substrate for the sulfate-reducing bacteria in which sphalerite is an end product. The sulfide moiety of ZnS is also amenable to microbial oxidation.

REFERENCES

Baas-Becking, L. G. M., and Moore, D. (1961). Biogenic sulfides. *Econ. Geol.* **56**, 259-276.

Bachmann, R. W. (1963). Zinc-65 in studies of the fresh water zinc cycle. *Proc. 1st Natl. Symp. Radioecol., Fort Collins, Colo.*, 1961, pp. 485-495. Reinhold, New York.

Bachmann, R. W., and Odum, E. P. (1960). Uptake of Zn^{65} and primary productivity in marine benthic algae. *Limnol. Oceanog.* **5**, 349-355.

Banfi, G. (1963). Fixing activity of a strain of *Azotobacter chroococcum* in the presence of zinc. *Ann. Microbiol.* **13**, 19-25.

Bennett, C. F. (1963). Microorganisms in environments contaminated with radioactivity. *Proc. 1st Natl. Symp. Radioecol., Fort Collins, Colo.*, 1961, pp. 175-177. Reinhold, New York.

Bertrand, D., and de Wolf, A. (1957a). Necessity of zinc as oligoelement, for glucose-6-phosphate dehydrogenase and 6-phosphogluconic acid dehydrogenase of *Asperigillus niger. Compt. Rend.* **245**, 1179-1184.

Bertrand, D., and de Wolf, A. (1957b). Zinc, a specific coenzyme of the aldolase of *Aspergillus niger. Compt. Rend.* **247**, 888-890.

Bertrand, D., and de Wolf, A. (1958). Zinc, dynamic trace element indispensable for the synthesis of phosphofructokinase and glyceraldehydephosphate dehydrogenase in *Aspergillus niger. Compt. Rend.* **246**, 2537-2539.

Bertrand, D., and de Wolf, A. (1959). Role of zinc in the metabolism of *Aspergillus niger. Bull. Soc. Chim. Biol.* **41**, 545-554.

Bertrand, D., and de Wolf, A. (1960). Zinc as a necessary trace element for the synthesis of tyrosine by *Aspergillus niger. Compt. Rend.* **250**, 2951-2952.

Bertrand, D., and de Wolf, A. (1961a). The necessity of zinc as a vital element in the synthesis of phenylalanine in *Aspergillus niger*, and its possible partial replacement by cadmium. *Compt. Rend.* **252**, 799-804.

Bertrand, D., and de Wolf, A. (1961b). Zinc requirement for nucleic acid synthesis in *Aspergillus niger. Compt. Rend.* **252**, 2613-2615.

Black, W. A. P., and Mitchell, R. L. (1952). Trace elements in the common brown algae and in sea water. *J. Marine Biol. Assoc. U.K.* **30**, 575-584.

Chipman, W. A., Rice, T. R., and Price, T. J. (1958). Uptake and accumulation of radioactive zinc by marine plankton, fish and shell fish. *U.S. Fish Wildlife Serv., Fishery Bull.* **50**, 279-292.

Cochrane, V. W. (1958). "Physiology of Fungi." Wiley, New York.

Coulter, C. B., and Stone, F. M. (1938). Identification of the porphyrin compound found in cultures of C. diphtheriae and Mycobacteria. Proc. Soc. Exptl. Biol. Med. 38, 423.

Davis, J. J., Perkins, R. W., Palmer, R. F., Hanson, W. C., and Cline, J. F. (1958). Radioactive materials in aquatic and terrestrial organisms exposed to reactor effluent water. Proc. U.N. 2nd Intern. Conf. Peaceful Uses Atom. Energy, Geneva, 1958. 18, 423–433. United Nations, New York.

Foster, J. W. (1939). Heavy metal nutrition of fungi. Botan. Rev. 5, 207–239.

Foster, J. W. (1949). "Chemical Activities of Fungi." Academic Press, New York.

Foster, J. W., and Denison, F. W. (1950). Role of zinc in metabolism. Nature 166, 833.

Foster, R. F. (1959). The need for biological monitoring of radioactive waste streams. 5th Nucl. Eng. Sci. Conf., Cleveland, 1959. Preprint V-6, 1–20. Engrs. Joint Council, New York.

Freundlich, H. (1926). "Colloid and Capillary Chemistry." Dutton, New York.

Hallas-Moller, K., Peterson, K., and Schlichtkrull, T. (1952). Crystalline and amorphous insulin-zinc compounds with prolonged action. Science 116, 394.

Horsfall, J. G. (1959). "Principles of Fungicidal Action." Chronica Botanica, Waltham, Massachusetts.

Ivanov, V. I., Nagirnyak, F. F., and Stepanov, B. A. (1962). Bacterial oxidation of sulfide ores. Microbiology (USSR) (English Transl.) 30, 575–578.

Keilin, D., and Mann, T. (1944). Activity of purified carbonic anhydrase. Nature 153, 107.

Ketchum, B. H., and Bowen, V. T. (1958). Biological factors determining the distribution of radioisotopes of the seal. Proc. U.N. 2nd Intern. Conf. Peaceful Uses At. Energy, Geneva, 1958. 18, 429–433.

Maquinay, A., Lamb, I. M., and Ramaut, J. L. (1961). Amount of zinc in a Belgian Calamine lichen, Stereocaulon nonades. Physiol. Plantarum 14, 284–289.

Myers, J. (1951). Physiology of the algae. Ann. Rev. Microbiol. 5, 157–180.

Nason, A. (1950). Effect of zinc deficiency on synthesis of tryptophan by Neurospora extracts. Science 112, 117.

Nielsen, J. M., and Perkins, R. W. (1957). The depletion of radioisotopes from the Columbia River by natural processes. U.S. At. Energy Comm. Rept., Hanford At. Prod. Operation HW–52908.

Ondratschedk, K. (1941). Uber den Mineralsalzbedarf heterotropher Flagellaten. Arch. Mikrobiol. 12, 241–253.

Ono, T., Kawai, H., and Kawai, M. (1959). Inorganic constituents in plants. I. Qualitative determination by spectrographic analysis and quantitative determination by polarography. Gifu Yakka Daigaku Kiyo 9, 41–54.

Parker, P. L. (1961). Zinc in a Texas bay. Texas, Univ. Inst. Marine Sci. Progr. Rept. Appendix II.

Plaut, G. W. E., and Lardy, H. A. (1949). The oxalacetate decarboxylase of Azotobacter vinelandii. J. Biol. Chem. 180, 13.

Raulin, J. (1869). Etudes clinques sur la vegetation. Ann. Sci. Nat. Botan. Biol. Vegetale 11, 93.

Robinson, G. N. (1951). Zinc requirements of Aspergillus niger. Nature 168, 210.

Rudolfs, W., and Helbronner, A. (1922). Oxidation of zinc sulfide by microorganisms. New Jersey Agr. Expt. Sta. Bull. 95, 459–465.

Schneider, K. H., and Disselkamp, C. (1958). Investigations on the zinc uptake of *Aspergillus niger* by means of Zn^{65}. *Landwirtsch. Forsch.* **11**, 133–137.

Scott, D. A. (1934). Crystalline insulin. *Biochem. J.* **28**, 1592.

Sulochana, C. B. (1952). Soils conditions and root diseases. IV. The effect of micro-elements on the occurrence of bacteria, actinomycetes and fungi in soils. *Proc. Indian Acad. Sci.* **B33**, 19–33.

Vallee, B. L. (1959). Biochemistry, physiology and pathology of zinc. *Physiol. Rev.* **36**, 443–490.

von Wolzogen Kühr, C. A. H., and van der Vlugt, L. S. (1953). Aerobic and anaerobic corrosion in water mains. *J. Am. Water Works Assoc.* **42**, 33–46.

9

MICROBES AND MANGANESE

The earth's crust contains approximately 0.1 % manganese as $MnAl_2O_4$. Plants and animals utilize this element in many biochemical processes. One of the largest single sources of manganese is the Nsutu mine on the Gold Coast of Africa. Since its opening in 1914, it has yielded 7.5 million tons of manganese. Mining is by opencast methods. Iron- and manganese-depositing bacteria are responsible for the deposition of the oxides of these two elements.

Obut (1962) hypothesizes the "manganese-oxidizing bacteria partici-pated in the formation of the Karce-Alma (Fergana) deposit which is the largest manganese deposit in Russia. Favorable conditions for an oxidizing-type flora existed in the Tethys basin in the late Cretacean period resulting in the formation of MnO_2. Similar paleobiological rationale may be used to explain the formation of manganese deposits in the Ural, Chiatura, Nikopol, and the Varna Region (Bulgaria) and in Asia Minor.

The most important manganese deposits in the world are those in the Chiatry district of Caucasia in Russia. They consist essentially of Pyrolusite (MnO_2). In deep-sea sediments rhodochrosite ($MnCO_3$) is common. Not all deep-sea manganese deposits form from direct precipitation from seawater. Submarine volcanic eruptions also contribute manganese to the marine environment. Other common deposits of manganese are shown in Table 9.I.

The important minerals of manganese are given in Table 9.II. Although manganese possesses many properties similar to iron and is found closely

TABLE 9.I

Manganese Deposits

Location	Deposit	Reference
Dzhetym-Too Range (Ordovician red silica)	Hematite, braunite, pyrolusite	Medvedev (1960)
	0.5–1-cm iron-manganese concretions Fe_2O_3 4–6% MnO_2 28–62% Jasper deposits per se Fe_2O_3 15–20% MnO_2 10–12%	
Atlantic and Pacific Oceans	Mn-nodules Fe 15%	Mero (1959)
Atlantic Ocean	Mn-nodules Mn 20%	Ehrlich (1963)
Indian Ocean (Calcareous formaniniferal muds)	Microconcretions of colloidal Fe and Mn	Kalinenko et al. (1962)
Africa Nsutu Manganese Mine	Fe, Mn	Service (1944)

associated with this element in nature, manganese sulfides such as albandite are not as common as pyrite. The oxides, hydroxides, silicates, and carbonate manganese complexes are most common.

Bivalent forms of manganese have the greatest mobility in nature. Where environmental conditions act to favor reduction, with weak acids being formed, manganese is reduced and leached from soil and ore deposits. Deposits high in organic matter, if saturated with water, induce microbial decomposition of the organic residium and initiate conditions favorable for leaching manganese. Proteolytic conditions yield basic solutions high in amines and ammonia which cause leaching of manganese. As carbon dioxide and oxygen are depleted from ground waters containing bicarbonate, precipitation of manganese carbonate occurs. Carbon dioxide fixation and oxygen depletion are initiated by the autotrophic population. If the manganese is in an organic colloidal state, the protective organic substance associated with it may be metabolized, and the manganese precipitated as a carbonate or hydroxide. Manganese deposited in nature by this process is found as slabs, nodules, dendritic infiltrations, or a black coating in mud. It is also found in manganiferous springs.

TABLE 9.II

IMPORTANT MANGANESE MINERALS AND MICROBIAL LEACHABILITY

Mineral	Formula	Microbial leachability
Manganosite	MnO	$+$
Pyrochroite	$Mn(OH)_2$	$+$
Manganite	$Mn_2O_3 \cdot H_2O$	$+$
Hausmannite	Mn_3O_4	$-$
Polianite, pyrolusite	MnO_2	$+$
Braunite	$3\ Mn_2O_3MnSiO_2$	$-$
Tephroite	$Mn_2(SiO_4)$	$-$
Manganfayalite, knebelite	$(Mn,Fe)_2(SiO_4)$	$-$
Rhodonite	$(Mn,Fe,Ca)(SiO_3)$	$-$
Pyromangite	$(Mn,Fe)(SiO_3)$	$-$
Spessartite	$Mn_3Al_2(SiO_4)_3$	$-$
Psilomelane	$BaMn\ Mn_8O_{16}(OH)_4$	$-$
Franklinite	$(Fe,Zn,Mn)(Fe,Mn)_2O_4$	$-$
Albandite	MnS	$+$
Rhodochrosite	$MnCO_3$	$+$
Galaxite	$MnAl_2O_4$	$-$
Jacobsite	$MnFeO_4$	$+$
Lithiophilite	$LiMnPO_4$	$+$

The iron and manganese present in such minerals as pyromangite, rhodonite, and jacobsite can be separated by natural weathering action. Solutions containing ferrous sulfate and manganese dioxide will yield ferric sulfate, ferric hydroxide, and manganous sulfate:

$$6\ FeSO_4 + 3\ MnO_2 + 6\ H_2O \rightarrow Fe_2(SO_4)_3 + 3\ MnSO_4 + 4\ Fe(OH)_3$$

The ferric sulfate precipitates except under highly acid conditions while the manganous sulfate follows the moving aqueous phase. Manganese dioxide is soluble under alkaline but not acidic conditions.

Manganese is related to iron in its chemical properties. In the periodic table it occurs next to the triad of Fe-Co-Ni. Unlike iron it forms a number of independent minerals in igneous rocks. Since sulfides of manganese are not common it differs from iron in this respect.

In silicate melts, $3\ MnO \cdot 2\ SiO_3$ is formed. Tephroite (Mn_2SiO_4), a manganese silicate, is found entrapped in certain feldspars. It is a rare mineral.

The sedimentary minerals of manganese are oxides, hydroxides, and carbonates. The most common are manganite, pyrolusite, manganomelane, and rhodochrosite. These deposits usually contain tetravalent forms of manganese. However, trivalent manganese is also found in manganite, braunite, and hausmannite. Divalent forms are more common in the silicate minerals.

MANGANESE NODULES

Nodular deposits of pyrolusite, the most important manganese ore, are found on the sea bottom. Nests and beds of manganese ores are also enclosed in residual clays derived from the decay of manganiferous limestones. Ehrlich (1963) found 20% manganese in nodules obtained from the Atlantic.

The radium content of manganese nodules is $2-330 \times 10^{-12}$ gm/gm, whereas radium in seawater averages 0.078×10^{-12} gm/liter. The rate of radial growth of manganese nodules calculated from the radium content is less than 1.0 mm/1000 years on the lower side of the nodule; it is 50% higher on the upper or convex side (Pettersson, 1955).

The genesis of the manganese rock nodules on the sea floor is still a moot question, although most data support involvement of a biochemical mechanism. The presence of organic materials in nodules lends additional supports to a biochemical origin. Graham (1959) has postulated that heavy metals are protected from hydrolysis by organic complexes in seawater. This organic complex may be on the cell surface of the polysaccharide slime surrounding a microbe. Part of the complex is metabolized and the remaining ions are precipitated.

Certain hairlike, sheath-forming bacteria remove iron and manganese from inland and ocean waters. Dried deposits of these bacteria contain 28% iron and manganese with a Fe/Mn ratio of 0.9 (Wolfe, 1960). High concentrations of these two metals are not required by bacteria in the aqueous environment since they can be extracted and concentrated from large volumes of water. Some forms of filamentous bacteria are highly specific and retain either iron or manganese in mucilaginous sheaths. The manganese in water gradually replaces the iron present in these sheaths. Biochemical procedures to concentrate either iron or manganese show a definite potential for use in the mining industry.

Manganese deposition causes fouling in certain domestic water lines. It is removed from water by the manganese zeolite process or by oxidation and precipitation with air, chlorine, or permanganate or by microbial catalysis.

ACTION OF MICROBES ON MANGANESE

Bacteria

Thiobacillus cultures are used to recover manganese from low-grade cores (Imai *et al.*, 1962). The sulfur oxidizers produce enough acid sulfate to reduce MnO_2 ores to soluble $MnSO_4$. Sixty-seven percent of the MnO_2 was leached from these ores in 9 days at 30°C. The ore was supplemented with sulfur, ammonium sulfate, and potassium phosphate prior to leaching.

Biochemical leaching of manganese dioxide is based upon the following chemical reaction:

$$MnO_2 + 2 S + 2\tfrac{1}{2} O_2 + H_2O \rightarrow MnSO_4 + H_2SO_4$$
$$\Delta F_{25°C} = -237 \text{ kcal}$$

Free hydrogen or some organic hydrogen donor will supplement the free energy available for driving this reaction.

$$2 MnO_2 + 2 H_2 + 2 S + 3 O_2 \rightarrow 2 MnSO_4 + 2 H_2O$$
$$\Delta F_{25°C} = -347 \text{ kcal}$$

When the supply of sulfuritic material and oxygen is plentiful, sulfuric acid is also formed in this latter reaction.

The oxidation of MnO per se is not a high energy yielding reaction.

$$2 MnO + O_2 + H_2O \rightarrow 2 MnO_2 \cdot H_2O$$
$$\Delta F^{\circ}_{25°C} = -40 \text{ kcal}$$

Since there is less potential energy physiologically available from this oxidation than from iron and sulfur oxidation, the manganese-oxidizing microbes are probably more dependent upon organic matter as a source of energy. This may also be the reason why the filamentous iron microbes concentrate both iron and manganese.

Manganous chloride has been reported to be a good mutagen (Demerec and Hanson, 1951). The mutagenic action of $MnCl_2$ can be correlated with how much is absorbed by the cell.

Manganese is involved in pigment and catalase synthesis in *Streptococcus faecalis*. Iron and zinc are also required for optimal pigment synthesis (Jones *et al.*, 1963).

Manganese is required for sporulation in *Bacillus megaterium* (Weinberg, 1964). It is also involved in enzymatic decarboxylations, polyglutamic synthetase, kinase action, and biosynthesis of various metabolites in other microbes. Manganese is usually required for nitrate reduction, though it may be replaced by ammonia.

The bacterial oxidation of manganese was demonstrated by Mann and Quastel (1946) in experiments with a Liss percolator. Beijerinck (1913) identified *Bacillus manganicus* as capable of oxidizing manganese. Repetition of this research, using Beijerinckj medium, led to the discovery by Bromfield and Skerman (1950) of two other bacterial synergists oxidizing manganese. These were species of *Corynebacterium* and *Chromobacterium*. In this study, Petri dishes filled with leached agar were covered with an "enamel" of manganese carbonate. The enamel was prepared by mixing solutions of equal weights of manganese sulfate and sodium bicarbonate. As cells grew they became intimately mixed with insoluble manganese oxides.

A culture of *B. manganicus* was isolated by Zavarzin (1963) using these methods; this culture would not grow on silica gel. Manganese oxidation here was stimulated with maleic acid. Zavarzin's study raises the question as to the purity of previous cultures of *B. manganicus*. Zavarzin isolated two species of *Pseudomonas* from a fixed culture of *B. manganicus*. These were identified as *P. eisenbergii* and *P. rathonis*, both of which required for manganese oxidation. Thus a symbiotic process is involved in at least one mechanism of manganese oxidation. Another culture identified as *Metallogenium symbioticum* also oxidizes manganese.

In soil or complex geological deposits of manganese, sulfur oxidation affects the degree of leaching of manganese. Using soil perfusion techniques, Vavra and Frederick (1952) added thiosulfate or elemental sulfur to Brookston silt loam or Maumee soils in amounts equivalent to 5 tons/acre and observed 10 100 times more solubilization in treated soil than in untreated soil. Treatment with calcium carbonate decreased the amount of soluble manganese released. In pure culture studies *T. thiooxidans* catalyzed the release of manganese. Manganese solubilization was attributed to acid production and a decrease in pH. Evidently MnO_2 was not formed under these conditions. Added H_2SO_4 was one-tenth as effective in Mn^{2+} release as comparable amounts of the acid produced biochemically from sulfur by *Thiobacillus* species. This demonstrates the catalytic effects of the microbe. Separation of the MnO_2 from the sulfur-oxidizing bacteria by a collodion membrane did not prevent the reduction of manganese.

Neither the microbe nor the sulfur have to be in contact with MnO_2 to release the soluble manganese ion. Manganese dioxide may be acting as an electron acceptor for the oxidation of either sulfur or organic compounds.

$$RH_2 + MnO_2 \rightarrow Mn(OH)_2 + R^{2+}$$
R = organic compound

Biochemical reduction of MnO_2 may actually be catalyzed by *T. thiooxidans*. Sulfur oxidation and manganese dioxide reduction do not occur normally but are possibly coupled in certain biological systems:

$$3\ MnO_2 + 4\ H^+ + S \rightarrow 3\ Mn^{2+} + SO_4^{2-} + 2\ H_2O$$

The leaching of manganese ores has been studied by Perkins and Novielli (1962). Manganiferous ores containing 4% MnO_2 were used in soil enrichment studies. Four mixed cultures of bacteria were obtained which leached manganese. Upon additional purification, four *Bacillus* species were isolated. Using airlift percolators 77.7–99.9% of the manganese was extracted in 60 days from ores obtained from Three Kids, Nevada, Charleston Hill, Nevada, and Aroostook, Maine. The mechanism of leaching manganese from the ores by microbes is probably different from the leaching of copper in the presence of sulfide minerals.

Some organic acid-producing bacteria, for example, those synthesizing 2-ketogluconic acid, will solubilize mineral salts such as $MnSiO_3$, manganese phosphate, and manganese-iron hard pans (Duff *et al.*, 1962). The organic acid acts as a chelating agent. Seventeen percent of the phosphate released was in the form of organic phosphate. The silicate released was found associated with three fractions: (1) ammonium molybdate reactive, (2) colloidal polymerized, and (3) amorphous forms. The ketogenic bacteria utilized were *Pseudomonas fluorescens*, *Bacterium herbicola*, and *Erwinia* spp. Some of the bacteria synthesizing hydroquinone type compounds may also chelate manganese. Hydroquinone is used to extract manganese sulfate from soil (Reid and Miller, 1962).

Species of *Corynebacterium*, *Chromobacterium*, and *Flavobacterium* oxidized manganese in a citrate medium. Fungi of the genera *Cladosporium*, *Trirgschemia*, and *Pleospora* also oxidize Mn^{2+} (Bromfield and Skerman, 1950).

Accretions of oxides of manganese are common in mollusc shells (*Macoma baltica*). Microscopic examinations of the surface show the presence of filamentous bacteria (Zavarzin, 1964). Bacteria catalyze the precipitation of manganese in seawater to yield rhodochrosite.

$$MnCl_2 + CaCO_3 \rightarrow MnCO_3 + CaCl_2$$

Accumulation of $MnCO_3$ is favored. Only *Mycoplana dimorpha* was successfully isolated and implicated in this reaction. This microbe is known to oxidize bivalent manganese.

Sphaerotilus discophora accumulates large quantities of divalent manganese (Johnson and Stokes, 1966). Cell suspensions readily oxidize manganese sulfate buffered at pH 7.2 to manganic oxide which is dark brown. The enzyme(s) catalyzing this reaction is inducible since it was not present except when cells were grown on media containing manganese. Autooxidation of Mn^{2+} takes place only at pH 10 or above, although it can be catalyzed at pH 8.0 by the addition of citric and butyric acids (Mulder, 1964).

Fungi

Some fungi are capable of oxidizing bivalent manganese in a calcium citrate-manganous sulfate medium. Those reported are *Helminthosporium victoriae*, *Culvularia lunata*, *C. brachyspora*, *C. inaequalis*, *Periconia circinata*, and *Cephalosporium* sp. (Timonin, 1950). Citrate was not required for the oxidation of the manganous ion. Salts of manganous sulfate and manganous chloride were easier to oxidize than was manganous carbonate.

Fungi such as *Hypoxylon punctulatum* and species of *Chaetomium* appear to have rather high manganese requirements (Barnett, 1964) and show only meager growth unless the medium contains 0.1 mg of manganese per liter. *Chaetomium globosum* branches excessively in a manganese-deficient medium and forms short thick cells. *Neocosmospora vasinfecta* grows well in a manganese-deficient medium, but will not form perithecia unless this cation is added.

Bertrand and Silberstein (1956) determined the manganese content of 60 Basidiomycetes and Ascomycetes species. The manganese varied from 3–89 mg of dried tissue/kg.

BIOGEOCHEMICAL PROSPECTING

Data are available on biogeochemical prospecting for manganese. The Mn/Fe ratio in leaves is a sensitive indicator of the manganese content of soil. The Mn/Fe ratio of leaves of the plant *Chamaecyporis obtusa* was 18.4 in mineralized and 3.5 in nonmineralized districts of Japan (Yamagata and Yamagata, 1957). Excess manganese in soil or surface minerals initiates changes in plant physiology. Iron deficiency follows which results in chlorosis of leaves. Gamelin (1937) noticed that the iron and manganese contents of wild trees vary universally. Plants found in hot dry areas contain practically no iron and manganese, whereas plants obtained from swamps or areas of heavy rainfall are rich in manganese (Warren et al., 1952). The amount of iron and manganese within a given plant species varies more than the amounts of zinc and copper which are used in biogeochemical prospecting (Warren and Delavault, 1948). Species of silver birch (Betula papyrifera) and black spruce (Picea marina) may be considered manganese rich, but, under specific conditions, they also may be rich in iron. Definite rules or patterns for the use of manganese and iron in biogeochemical prospecting are just emerging. All water organisms absorb minerals from water. A bacterial symbiont identified as Bacterium precipitatum aids in the selective concentration of manganese from natural waters which support the growth of sago pondweed. This microbe causes the deposition of calcite on the surface of the plant. It is identical to the microbe precipitating iron and calcite in the Kara Sea (Kalinenko, 1949). The intercellular concentration of manganese in sago pondweed, pond scum, and stonewort, all from Tempie Big Spring (Osborn, 1964), is several times greater than that deposited by land plants. Foliose lichen and black moss contain several times more manganese than crustose lichens. Micrococci spp. were isolated by Osborn (1964) from both manganese and phosphorite nodules taken from the ocean floor.

In aqueous environments, diatoms have been examined critically for both manganese and zinc. Surface seawater contains 0.023–0.099×10^{-12} ^{54}Mn, whereas the concentration of ^{65}Zn is 10% of this value (Folsom et al., 1963). Some dinoflagellates concentrate ^{54}Mn over one thousand times. Zinc is concentrated in some dinoflagellates.

Both manganese and iron promote the growth of diatoms and are absorbed to their hard surface. Manganese is deposited in the calcarious shells of foraminiferous species feeding upon diatoms. The calcium is replaced diadochically in the calcite structure of the shells. Manganese (calculated as MnO) is often concentrated in the shells by a factor of about 100,000.

MANGANESE CYCLE

The general pattern of movement and the interrelationship of manganese with various life processes is shown schematically in Fig. 9.1. Manganese-enriched sediments are formed in oceans which, after upheaval, will be leached to yield manganese appearing in soils, streams, and

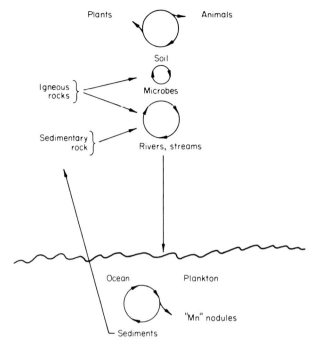

FIG. 9.1. Biogeochemical manganese cycle.

rivers, etc. The microbe is active in catalyzing both the oxidations and reductions in which manganese may be precipitated or solubilized. In soils, plants and animals require this metal for their life processes. Needles from fir trees are used in biogeochemical prospecting for manganese (Ljungren, 1951). They have higher concentration of manganese than broadleaf plants, and since 300 needles are used in each analysis a good average value can be obtained. Manganese varies from 1.2 to 7.6% of the ash. Spruce is also used in biogeochemical prospecting. The present data do not show that either of these plants are good indicator plants for manganese. The presence of magnesium and its similarity in biochemical function to manganese often marks or inhibits manganese assimilation.

Microbes precipitating manganese are universally spread throughout the biosphere (Thiel, 1925). A fungus *Oidium manganifera alpha* precipitates manganese from a solution of manganese tartrate. A heavy growth of sulfate-reducing bacteria develops in media containing the following manganese salts: $MnSO_4$, $MnCO_3$, $MnCl_2$, manganese peptone, manganese acetate, and manganese tartrate. On solid media, brown granules of manganese are observed surrounding colonies of bacteria. In liquid media, a dark gray flaky precipitate of manganese carbonate forms. Iron bacteria from the Cuyuna Range collected from spring deposits and mine waters also precipitate manganese. Iron bacteria precipitating manganese are: *Crenothrix polyspora, Leptothrix ochracea, Cladothrix dichtoma,* and *Clonothrix fusca.*

Since these organisms have universal distribution they are probably instrumental in precipitating the manganese found in many sediments. The sulfate-reducing bacteria precipitate manganese under anaerobic and reducing conditions and are abundantly present in marine muds. Iron bacteria remove manganese under aerobic and possibly micro-aerophilic conditions and are most active in iron springs, streams, rivers, etc.

Present data clearly indicate that a biochemical manganese cycle exists in nature which involves di-, tri-, and tetravalent manganese, and, potentially, other oxidative states of this element. The oxidative state encountered depends on several factors, e.g., the pH and the presence of sulfur, sulfate, O_2, and organic matter.

$$\text{Mn}^{2+} \xrightleftharpoons[\substack{\text{Biochemical or} \\ \text{chemical oxidation}}]{\substack{\text{Biochemical reduction} \\ \text{(pH5.5)}}} \text{Mn}^{4+}$$

The solubilization of manganese in soil is not just a chemical reaction. Reaction rates involving purely chemical reactions alone are too limiting to account for the amounts of manganese released.

Once manganese dioxide is formed, it is an adequate oxidant for oxidizing ferrous to ferric iron (Fig. 9.2). Thus the ferrous iron may be

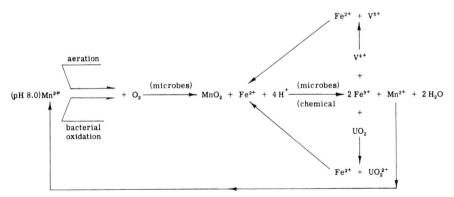

FIG. 9.2. The possible role of manganese in the oxidation of uranium and vanadium.

oxidized directly by the iron-oxidizing microbes or by oxidative products synthesized by microbes, e.g., Mn^{4+}, peroxides, etc. The natural role of ferric iron in mediating electron transfer in metal oxidation is beginning to be appreciated. Both quadrivalent uranium and vanadium are oxidized by ferric iron, respectively, to hexavalent or pentavalent states. The higher oxidized states of these metals are the soluble forms which are required in leaching or encountered in ground water.

REFERENCES

Barnett, H. L. (1964). High manganese requirements and deficiency symptoms in certain fungi. *Phytopathology* **52**, 746.

Beijerinck, M. W. (1913). Oxidation of manganous carbonate by microbes. *Koninkl. Ned. Akad. Wetenschapp., Verslag, Gewone Vergader. Afdel. Nat.* **22**, 415–420.

Bertrand, G., and Silberstein, L. (1956). The manganese content of fungi. *Compt. Rend.* **242**, 37–40.

Bromfield, S., and Skerman, V. (1950). Biological oxidation of manganese in soils. *Soil Sci.* **69**, 337–348.

Demerec, M., and Hanson, J. (1951). Mutagenic action of manganese chloride. *Cold Spring Harbor Symp. Quant. Biol.* **16**, 215–228.

Duff, R. B., Webley, D. M., and Scott, R. O. (1963). Solubilization of minerals and related materials by 2-ketogluconic acid-producing bacteria. *Soil Sci.* **95**, 105–114.

Ehrlich, H. L. (1963). Bacteriology of manganese nodules. I. Bacterial action of manganese nodule enrichments. *Appl. Microbiol.* **11**, 15–19.

Folsom, T. R., Young, D. R., Johnson, J. N., and Pellai, K. C. (1963). Manganese-54 and Zinc-65 in coastal organisms of California. *Nature* **200**, 4904.

Gamelin, H. (1937). Manganese and iron in the conifers of the province of Quebec. *Contrib. Lab. Ecole Hautes Etudes Com. Montreal* No. 8.

Graham, J. W. (1959). Metabolically induced precipitation of trace elements from sea water. *Science* **129**, 1428–1429.

Imai, K., Tano, T., and Noro, H. (1962). Recovery of manganese salt from its low-grade ore with microorganisms. Japanese Patent 15,208.

Johnson, A. H., and Stokes, J. L. (1966). Manganese oxidation by *Sphaerotilus discophorus. J. Bacteriol.* **91**, 1543–1547.

Jones, D., Deibel, R. H., and Niven, C. F., Jr. (1963). Apparent pigment production by *Streptococcus faecalis* in the presence of metal ions. *J. Bacteriol.* **86**, 171–172.

Kalinenko, V. O. (1949). The origin of ferro-manganese concretions. *Mikrobiologiya* **18**, 528–532.

Kalinenko, V. O., Belokopytova, O. V., and Nikolaeva, G. G. (1962). Bacteriogenic iron-manganese concretions in the Indian Ocean. *Okeanologiya* **2**, 1050–1059.

Ljungren, P. (1951). Some investigations of the bio-geochemistry of manganese. *Geol. Foren. Stockholm Forha.* **73**, 639–652.

Mann, P. J. G., and Quastel, J. H. (1946). Manganese metabolism in soils. *Nature* **158**, 154–156.

Medvedev, L. D. (1960). Iron-manganese concretions in the Ordovician red silica rocks of the Dzhetym-TOO Range. *Izv. Akad. Nauk Kirg. SSR., Ser. Estestv. i Tekhn. Nauk* **2**, 117–126.

Mero, J. L. (1959). Will deep sea bottom make tomorrow's manganese mines? *Mining World* **24**, 39.

Mulder, E. G. (1964). Iron bacteria, particularly those of the *Sphaerotilus*-Leptothrix group, and industrial problems. *J. Appl. Bacteriol.* **27**, 151–173.

Obut, A. M. (1962). Paleobiological bases of the formation of certain sedimentary manganese ore deposits. *Znachenie Biofpery Geol., Protsessakh* pp. 26–29.

Osborn, E. T. (1964). Intracellular and extracellular concentrations of manganese and other elements by aquatic organisms. *Geol. Sect. Water Supply, Paper* **1667-C**, 18 pp.

Perkins, E. C., and Novielli, F. (1962). Bacterial leaching of manganese ores. *U.S. Bur. Mines. Rept. Invest.* **6102**, 1–11.

Pettersson H. (1955). Manganese nodules and oceanic radium. *Deep-Sea Res.* **3**, 335–345.

Reid, A. S. J., and Miller, M. H. (1962). The manganese cycle in soil. II. Forms of soil manganese in equilibrium with solution manganese. *Can. J. Soil Sci.* **43**, 250–259.

Service, H. (1944). The Nsuta manganese mine. Largest single producer in the world. *S. African Mining Eng. J.* **55**, 177–183.

Thiel, G. A. (1925). Manganese precipitated by microoroganisms. *Econ. Geol.* **20**, 301–310.

Timonin, M. I. (1950). Soil microflora in relation to manganese deficiency. *Sci. Agr.* **30**, 324–325.

Vavra, J. P., and Frederick, L. (1952). The effect of sulfur oxidation on the availability of manganese. *Soil Sci. Soc. Am. Proc.* **16**, 141–144.

Warren, H. V., and Delavault, R. E. (1948). Biochemical investigations in British Columbia. *Geophysics* **13**, 609–624.

Warren, H. V., Delavault, R. E. and Irish, R. I. (1952). Preliminary studies on the biogeochemistry of iron and manganese. *Econ. Geol.* **47**, 131–145.

Weinberg, E. D. (1964). Manganese requirement for sporulation and other secondary biosynthetic processes of *Bacillus. Appl. Microbiol.* **12**, 436–441.

Wolfe, R. S. (1960). Microbial concentration of iron and manganese in water with low concentration of these elements. *J. Am. Water Works Assoc.* **52**, 1335.

Yamagata, N., and Yamagata, T. (1957). Fundamental studies on the biochemical prospecting for manganese. *Bull. Chem. Soc. Japan* **30**, 900–904.

Zavarzin, G. A. (1963). Symbiotic oxidation of manganese by two species of *Pseudomonas. Microbiology (USSR) (English Transl.).* **31**, 481–482.

Zavarzin, G. A. (1964). The mechanism of manganese deposition on mollusc shell. *Dokl. Akad. Nauk SSSR* **154**, 944–945.

10

MICROBES AND COBALT

COBALT BIOCHEMISTRY

Of the rarer heavy metals cobalt has an essential role in animal and plant nutrition and is an acid in food fermentation in the rumen of cattle and sheep. Cobalt deficiencies in livestock and ruminant animals still exist in many places throughout the world where soils and vegetation contain limiting quantities of this vital element.

A cobalt requirement in life processes is usually related to vitamin B_{12}. Anemias in higher animals have been greatly reduced by B_{12} therapy.

A simple cobalt cycle exists in nature (Scheme 10.I).

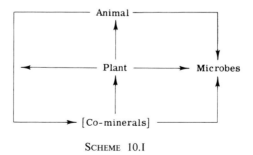

SCHEME 10.I

This cycle is similar to the carbon and nitrogen cycles except there is a notable absence of a gaseous phase. The comparative cobalt content of

169

the soils of the world has been thoroughly studied by Young (1956, 1960). The cobalt in most soils normally does not exceed a few parts per million.

GEOCHEMISTRY

Cobalt is not a common metal, but it is widely distributed in rocks. Geochemists estimate the concentration in the earth's crust at 0.001% which makes it the thirty-third most abundant element. The concentration is higher in basic formations than in acidic rocks such as granite. The simple salts of cobalt are bivalent, although trivalent and coordinate compounds are common. Valence is easily measured by magnetic susceptibility. With certain exceptions, cobaltous forms are paramagnetic while cobaltic forms are diamagnetic. The atomic radius of cobalt is similar to that of iron, magnesium, nickel, and zinc, and cobalt and nickel are regarded as members of the iron family. Cobalt is less siderophilic than nickel. The high affinity of cobalt for sulfur results in its frequent occurrence in sulfide bodies, which segregate from magmas, and in sulfide-rich muds formed biochemically through sulfate reduction. Magmatic rocks contain 4–200 ppm of cobalt. The cobalt in the pyrrhotite is concealed in small bodies which may be totally absent in some ores. Unlike nickel, cobalt does not form hydrosilicates during weathering; bicarbonates or colloidal hydroxides are formed instead. These are deposited along with manganese in the cobaltian wad (asbolan) which contains 4–34% cobalt.

The more common cobalt minerals are shown in Table 10.I. The chemical composition of these minerals is such that they are susceptible to microbial attack. The sulfide complexes of cobalt readily undergo microbial oxidation to yield bieberite $[Co(SO_4) \cdot 7 H_2O^{2-}]$ and sulfuric acid. These reactions have not been elucidated. Cobaltite (CoAsS) smaltite ($CaAs_{2-3}$), skutterdite ($CoAs_3$), and the arsenic-containing forms are probably oxidized by microbes to bieberite or erythrite $[CO_3(AsO_4)_2 \cdot 8 H_2O]$.

$$CoS_2 + 4 O_2 \longrightarrow Co(SO_4) \cdot 7 H_2O + H_2SO_4$$
$$(H_2O) \text{ (Bieberite)}$$

$$CoAsS + O_2 \longrightarrow Co(AsO_4) \cdot 8 H_2O + CO_3(AsO_4)_2 \cdot 8 H_2O$$
$$\text{(Cobaltite)} \quad (H_2O) \qquad\qquad \text{(Erythrite)}$$

The energy liberated from the oxidation of bivalent cobalt in the form of cobalt carbonate (spherocobalt) to trivalent cobalt hydroxide (heterogenite) is ample to support autotrophic growth.

$$2 \, CoCO_3 + 3 \, H_2O + 1\tfrac{1}{2} \, O_2 \rightarrow 2 \, Co(OH)_3 + 2 \, CO_3$$
$$\Delta F^{\circ}_{25^{\circ}C} = -54.7 \text{ kcal}$$

Autotrophs catalyzing this reaction have not been reported.

The iron complexed in glaucodot and in some forms of cobaltite may be oxidized by microbes thus increasing mineralization. Microbial oxidation and mineralization of cobalt minerals compete with the physico-chemical interrelationships of cobalt to other minerals. For example, the cobalt found in nonporous meteorites would be difficult to leach. The cobalt present in all minerals formed at high temperatures, e.g., pyrrhotites, pyrites, and sphalerites are more difficult to oxidize. These same minerals would be difficult to leach biochemically unless ground to a small

TABLE 10.I

COMPOSITION OF IMPORTANT COBALT MINERALS

Mineral	Formula
Cobalt minerals in magmatic deposits	
Cattierite	CoS_2
Linnaeite	Co_3S_4
Cobaltite	$CoAsS$
Safflorite	$CoAs_2$
Glaucodot (danaite)	$(Co,Fe)AsS$
Smaltite	$CoAs_{2-3}$
Skutterudite	$CoAs_3$
Weather and oxidation reaction products of cobalt	
Stainierite	$\alpha\text{-}CoOOH$
Heterogenite	$Co(OH)_2 + Co(OH)_3$
Spherocobaltite	$CoCO_3$
Bieberite	$Co(SO_4) \cdot 7 \, H_2O$
Erythrite	$Co_3(AsO_4)_2 \cdot 8 \, H_2O$
Igneous rocks high in cobalt	
Peridotite	
Gabbro	
Diorite	
Granite	
Gabbropierite	
Granophyre	

particle size. The bulk of cobalt found in igneous rocks is concealed in silicate minerals and these are more difficult to leach microbially.

Sutton and Corrick (1961) have leached cobalt from natural ores using airlift percolators. The rate of leaching was increased upon inoculation with *Thiobacillus thiooxidans*. Sulfur, but not mineral sulfides, could be added as substrates to increase leaching in these particular ores.

Cobalt and gersdorffite (NiAsS) form a complete solid-solution series. The crystal form of cobaltite resembles pyrite but may be distinguished from the latter by its silver color and perfect cleavage. It is found in high-temperature deposits as disseminations in metamorphosed rocks or as vein deposits with other cobalt and nickel minerals. Related cobalt minerals are skutterudite $(Co,Ni,Fe)As_3$, linnalite (Co_3S_4), and erythrite $Co_3(AsO_4)_2 \cdot 8 H_2O$. Large deposits of cobaltite occur at Tunaberg, in Sweden, and Cobalt, Ontario, in Canada. The largest production of cobalt today is in the Belgian Congo where cobalt is found associated with copper ores. The small percentages of oxidized cobalt present in the nickel-containing laterite deposits of Celebes, Cuba, and New Caledonia are largely untouched.

COBALT IN SOIL

Some soils are deficient in cobalt. Many pastures the world over are too low in cobalt to be good grazing land for cattle and sheep. Addition of cobalt salts from 0.5–2 lb/acre every three years usually is adequate to correct this condition. Cobalt chlorides or nitrates are soluble and are utilized rapidly by plants and microbes in the soil. Complexes of carbonate and phosphate become slowly available to plants (Young, 1956). Forage crops should contain 0.08 ppm for the maintenance of good health, while soil should contain 3–5 ppm of cobalt. In some instances this will not provide an adequate cobalt level for plant assimilation. Cobalt and cesium are not rapidly leached from soil since they are often bound in the bio-complex. Fungi such as *Pellicularia filamentosa*, *Rhizoctonia solani*, *Helminthosporum sativum*, *Phycomyces blakesleeanus*, *Phytophthora cactorum*, and *Rhizopus stolonifer* possess a translocation mechanism for transporting ^{60}Co and ^{137}Cs is agar media. Translocation of these elements in soil is not high. Cobalt and cesium in soil are retained close to the locus inoculated (Grossbard and Stranks, 1959).

Commercial fertilizers are usually deficient in cobalt. With the exception of basic slag, phosphorite meal, superphosphate or sewage sludge, the cobalt content of commercial fertilizers is less than 3 ppm. Basic slag may contain as high as 145 ppm cobalt whereas other fertilizers do not exceed 9 ppm.

COBALT METABOLISM BY MICROBES

Actinomycetes

The importance of cobalt in vitamin B_{12} synthesis in microorganisms is well known (Hendlin and Ruger, 1950). Increased synthesis of vitamin B_{12} by *Streptomyces griseus* is obtained by adding 2 ppm Co^{2+} to the culture medium, as has been shown by incorporation of radioactive cobalt into vitamin B_{12} (Rosenblum and Woodbury, 1951).

A dose of 2 μg of $CoO_2 \cdot 6 H_2O$/ml was sufficient to inhibit growth of certain cobalt-sensitive cultures of Actinomycetes (Kojima and Matsuki, 1956). The tolerance to cobalt was increased to 100 μg/ml by adaptation in a glucose-bouillon medium. Glucose oxidation itself was inhibited by addition of higher levels of cobalt. Purine and pyrimidine bases reversed cobalt inhibition only slightly, whereas casein hydrolysate is much more effective. Two amino acids, L-histidine and L-cysteine, could be substituted for casein hydrolyzate. The amino acids chelate the cobalt-forming metal-amino acid complex.

Fungi

Yeast cultures can be adapted to cobalt. Perlman and O'Brien (1954) adapted a cobalt-sensitive *Saccharomyces cerevisiae* culture to a medium containing 50 times the quantity of cobalt which inhibited the parent culture. Cells contained 9.9% cobalt when cultivated in a medium containing 750 ppm of cobalt. The cobalt complexed in these cells was utilized more efficiently for B_{12} synthesis in *Streptomyces griseus* than was cobalt nitrate. Of the cobalt assimilated by *Saccharomyces cerevisiae*, 10% is absorbed at the surface while the remainder is bound chemically. Between 40–63% is combined with proteins of which 20% can be liberated by autolysis (Bass and Zizuma, 1956). Addition of 5.6–560 μg% of $CoCl_2$ did not increase the rate of growth; however, the cobalt content of the yeast increased 11–24 mg% (dry weight). Over 90% of the cobalt combined with the cell biologically rather than by adsorption (Bass and Zizuma, 1955).

Torula utilis, growing on 2–17 μg cobalt/ml, accumulated 200–2000 μg cobalt/gm (dry weight) without any inhibition of growth (Drews *et al.*, 1952). Over 80% of the cobalt was taken up from the medium. At 19 μg cobalt per ml, or over, the amount of cobalt accumulating in cells decreased, and the cells were damaged.

At 0.002 M concentrations of Co^{2+}, growth and protein synthesis are inhibited in *Candida utilis*. At lower levels, the yeast contained more soluble nitrogen and nucleotide nitrogen (Erkama and Enari, 1956).

Bacteria

Nonsymbiotic Nitrogen Fixation

Molybdenum, calcium, and iron are involved in nonsymbiotic nitrogen fixation; other metals are involved to a lesser extent. Lead partially inhibits nitrogen fixation, whereas Cu^{2+} has no effect. Iron salts have been implicated in nitrogen fixation by *Azotobacter* and *Clostridium*, but their function is not clearly understood. Vanadium, uranium, and manganese stimulate nitrogen fixation; however, greater stimulation is obtained with Mo^{2+}, Co^{2+}, and Ni^{2+} (Iswaran and Sundara Rao, 1960). Calcium is required for atmospheric nitrogen assimilation in some species of *Azotobacter* and algae. Under some conditions strontium may be substituted for calcium.

Ionic cobalt in marine species or the cobalt present in vitamin B_{12} are not readily assimilated by most marine algae (Tsuchiya and Makuta, 1951). *Polysiphonia nigrescens* is an exception, where the concentration factor is approximately 10,000-fold in 7 days. *Polysiphonia fastigiata* and *Pelvetia canaliculata* concentrate lesser amounts of cobalt (Ericson, 1952).

METABOLIC REACTIONS INVOLVING COBALT

Pseudomonas cattleyaecolor synthesizes a violet pigment which contains cobalt (Marchal and Boussard, 1957). No metabolic function for this pigment is known.

Cobalt fixation by cells of *E. coli* K12 is suppressed by addition of Cu^{2+} or Ni^{2+}; either may share a common receptor with cobalt. Reserve amino acids may act as cobalt receptor sites (Katayama, 1961). The amount of cobalt fixed and the fixation rate increase between pH 5.6–7.7.

Trace amounts of cobalt (0.2 mg/ml) produces a marked inhibition of oxygen uptake in cells of *Salmonella pullorum* (Pital *et al.*, 1953). This selective inhibition of respiration enhances the therapeutic effect of penicillin against this organism. Inhibition of respiration by cobalt was observed in several other bacteria and yeast.

Trace amounts of Co^{2+}, Cu^{2+}, and Zn^{2+} added to Dorsetts' medium increased the growth of *Mycobacterium tuberculosis avium* threefold. A reddish-brown pigment, coproporphyrin III, increased sixty-two-fold. When tested separately, Cu^{2+} (1.07 ppm) had no effect, while Zn^{2+} (6.04 ppm) and Co^{2+} (0.27 ppm) gave half the yield of cells obtained by adding all three metals at the same time (Patterson, 1959).

Solutions of 1×10^{-5} M of Co^{2+}, Cr^{3+}, and Mn^{2+} do not affect growth of *Corynebacterium diphtheriae*, but they do reduce toxin production; this was not true, when Ag^{2+}, Zn^{2+}, Cu^{2+}, Cd^{2+}, Al^{3+}, Pb^{3+}, or Ni^{2+} (Clarke, 1958) was added.

A cobalt-activated acid pyrophosphatase is active in *Streptococcus faecalis* (Oginsky and Rumbaugh, 1955). The optimum pH was 5.3. A second enzyme hydrolyzing inorganic pyrophosphate was also present in *Streptococcus faecalis*; it had an optimum pH of 8.5 and was activated by Mg^{2+}. The acid pyrophosphatase was also activated by Mn^{2+} and Zn^{2+}, but the effect was less than that of Co^{2+}. Histidine, cysteine, lysine, and histamine also increased pyrophosphatase activity. Acid pyrophosphatase activated by Co^{2+} was found in *Escherichia coli, Aerobacter aerogenes, Proteus vulgaris, Pseudomonas aeruginosa, Lactobacillus lactis, Lactobacillus arabinosus, Streptococcus faecalis, Staphylococcus aureus, Leuconstoc mesenteroides*, and *Bacillus subtilis*.

Cobalt ions activate the enzymes dipeptidase (Smith, 1948) polypeptidase (Anderson, 1945), and arginase (Edlbacher and Baur, 1938). *Bacillus natto* and *Aspergillus oryzae* used in the natto and miso protein fermentations accumulate small amounts of cobalt. This cobalt was not complexed as vitamin B_{12} (Sawada *et al.*, 1955).

In some lactobacilli cultures, vitamin B_{12} is replaceable by one or more purine or pyrimidine nucleosides occurring as deoxyribose nucleic acid (Hoff-Jorgensen, 1949). These nucleosides must be supplied at one hundred or more times the amount of vitamin B_{12} concentration. It logically follows that B_{12} is involved in the enzymatic synthesis of these nucleosides. Thus, in microbes requiring vitamin B_{12}, the metabolic pathway of cobalt can be traced to vitamin B_{12} which plays a role in nucleic acid synthesis and cellular organization (Hutner *et al.*, 1950). Interesting calculations made on Euglena show approximately 4900 molecules of B_{12} synthesized per cell.

BIOCHEMICAL PROSPECTING

Analysis for ecological types of microbes around cobalt deposits should show a predominance of certain species, many of which possess the ability to concentrate cobalt. Microbes concentrating high levels of cobalt in the cell, useful in the exploration for cobalt in mineralized areas, need investigation.

Higher plants have been studied in more detail in this respect. Plants reported to contain more than 3 mg of cobalt/kg tissue are: beech bark, clover, apricots, beets, carrots, radishes, rice straw, and vetch.

The cobalt level in most plants run less than 0.1 mg of cobalt/kg tissue.

An examination of the vegetation and soil of the chromite ore bodies in the Hinokami district in Japan showed that the tree *Clethra barbinerius* accumulated an average cobalt content of 661 ppm to the 5.3 ppm accumulated by other indigenous plants (Yamagota and Murakami, 1958). *Clethra* also accumulates nickel (140 ppm) but not iron. The soils were derived from serpentine rock and contained 65 ppm cobalt, 1005 ppm nickel, and 578 ppm iron.

Warren and Delavault (1957) examined the cobalt content of 14 trees or shrubs in mineralized and nonmineralized areas near Cobalt, Ontario (Canada). Balsam (*Abies balsamea*), white pine (*Pinus strobus*), black spruce (*Picea mariana*), birch (*Betula papyrifera*) and blueberry (*Vaccinium* sp.) contained up to 360 ppm Co^{2+} in mineralized areas. In nonmineralized vicinities, the Co^{2+} decreased to approximately 3.5 ppm.

The requirement and presence of cobalt in plant and animal life demonstrates one phase of the cobalt cycle in nature. The enzymatic reactions by which microbes catalyze weathering, oxidation-reduction, leaching, and metal translocation of cobalt in the lithosphere are yet to be resolved.

REFERENCES

Anderson, A. B. (1945). Activation of jack-bean arginase by cobalt, manganese and iron. *Biochem. J.* **39**, 139–142.

Bass, H., and Zizuma, A. (1955). Assimilation of cobalt by yeasts. *Latvijas PSR Zinatnu Akad Vestis* pp. 87–91.

Bass, H., and Zizuma, A. (1956). Assimilation of cobalt by yeasts. II. Distribution of cobalt in yeast *Saccharomyces cerevisiae*. *Latvijas PSR Zinatnu Akad. Vestis* No. 8 pp. 109–114.

Clarke, G. D. (1958). The effect of cobaltous ions on the formation of toxin and copro-porphyrin by a strain of *Corynebacterium diphtheriae*. *J. Gen. Microbiol.* **18**, 708–719

Drews, F., Olbrich, J. H., and Vogl, J. (1952). Uptake of cobalt and manganese by microorganisms. *Brauerei, Wiss. Beil* **5**, 91–92.

Edlbacher, S., and Baur, H. (1938). The nature of yeast and liver arginase. *Z. Physiol. Chem.* **254**, 275–284.

Ericson, L. E. (1952). Uptake of radioactive cobalt and vitamin B_{12} by some marine algae. *Chem. & Ind.*, *(London)* pp. 829–830.

Erkama, J., and Enari, T. M. (1956). The effect of cobalt on the nitrogen metabolism in yeast. *Suomen Kemistilehti* **29B**, 176–178.

Grossbard, E., and Stranks, D. R. (1959). Translocation of cobalt-60 and cesium-137 by fungi in agar and soil cultures. *Nature* **184**, 310–314.

Hendlin, E. D., and Ruger, M. L. (1950). The effect of cobalt on microbial synthesis of LLD-active substances. *Science* **111**, 541–542.

Hoff-Jorgensen, E. (1949). Difference in growth promoting effect of desoxyribosides and vitamin B_{12} on three strains of lactic acid bacteria. *J. Biol. Chem.* **178**, 525–526.

Hutner, S. H., Provasoli, L., Schatz, A., and Haskins, C. P. (1950). Some approaches to the study of the role of metals in the metabolism of microorganisms. *Proc. Am. Phil. Soc.* **94**, 152–170.

Iswaran, V., and Sundara Rao, W. V. B. (1960). The effect of trace elements on nitrogen fixation by *Azotobacter*. *Proc. Indian Acad. Sci.* **B51**, 103–115.

Katayama, Y. (1961). Receptor site of *Escherichia coli* for the cobaltous ion. *J. Microbiol.* **5**, 27–34.

Kojima, H., and Matsuki, M. (1956). Studies on the influence of cobalt chloride ($CoCl_2$) on the growth of *Actinomycetes* (Part I). *Tohoku J. Agr. Res.* **7**, 175–187.

Marchal, J. G., and Boussard, M. (1957). Polarography of the violet-cobalt pigment of *Pseudomonas cattleyaecolor*. *Bull. Assoc. Diplomas Microbiol. Fac. Pharm. Nancy* **69**, 24–28.

Oginsky, E. L., and Rumbaugh, H. L. (1955). A cobalt-activated bacterial pyrophos-phatase. *J. Bacteriol.* **70**, 92–98.

Patterson, D. S. P. (1959). Influence of Co and Zn ions in the growth and porphyrin production of *Mycobacterium tuberculosis avium*. *Nature.* **185**, 57.

Perlman, D., and O'Brien, E. (1954). Characteristics of cobalt tolerant culture of *Saccharomyces cerevisiae*. *J. Bacteriol.* **68**, 167–170.

Pital, A., Stafseth, H. J., and Lucas, E. H. (1953). Observations on the cobalt enhance-ment of penicillin activity against *Salmonella pullorum*. *Science* **117**, 459–460.

Rosenblum, C., and Woodbury, D. T. (1951). Cobalt-60 labeled vitamin B_{12} of high specific activity. *Science* **113**, 215.

Sawada, Y., Tanaka, K., Hirano, M. Soto, M., Miyamoto, J., and Tanaka, S. (1955). Studies on the cobalt metabolism of microorganisms. Part 2. Absorption of cobalt by *Aspergillus oryzae* and *Bacillus natto*. *J. Chem. Soc. Japan, Pure Chem. Sect.* **76**, 274–277.

Smith, E. L. (1948). Glycylglycine dipeptidases of skeletal muscle and human uterus. *J. Biol. Chem.* **173**, 571–584.

Sutton, J. A., and Corrick, J. D. (1961). Bacteria in mining and metallurgy. Leaching selected ores and minerals. Experiments with *Thiobacillus thiooxidans*. *U.S., Bur. Mines, Rept. Invest.* **5839**, 1–16.

Tsuchiya, T., and Makuta, T. (1951). Occurrence of cobalt in marine product. *Tohoku J. Agr. Res.* **2**, 113–117.

Warren, H. V., and Delavault, R. E. (1957). Biogeochemical prospecting for cobalt. *Trans. Roy. Soc. Can.* **51**, 33–37.

Yamagata, N., and Murakami, Y. (1958). A cobalt-accumulator plant, *Clethra barbinervis*. Sieb. et Zucc. *Nature* **181**, 1808.

Young, R. S. (1956). Cobalt in biology and biochemistry. *Sci. Progr.* (*London*) **44**, 16–37.

Young, R. S. (1960). "Cobalt—Its Chemistry, Metallurgy and Uses," A.C.S. Monograph. Reinhold, New York.

11

URANIUM BIOGEOCHEMISTRY

GEOCHEMISTRY OF URANIUM

The earth's crust is estimated to contain 0.004% uranium as compared to 0.00003% silver. The largest deposits are located at Shinkolobowe in the Congo Republic, Great Bear and Elliott Lakes in Canada, and at Joachimsthal in Czechoslovakia. Uranium minerals are concentrated in pegmatites, hydrothermal veins (main economic deposits), sandstones, and black shales. The deposits vary from 50 to 2000 million years in age.

Uranite (UO_2) is the best known primary mineral. Pitchblende, another black or dark brown mineral of high density, is a microcystalline form of UO_2. Primary minerals, i.e., uranite, pitchblende, brannerite, davidite, and coffinite are generally deposited from molten rock or hot solutions within the earth's crust. The mineral carnotite is of secondary importance as an ore. Carnotite is deposited in areas leached by the action of acid or alkaline ground water and is found in the United States in Utah and Colorado. Vanadium complexes with uranium to form yellow to brown salts. A list of the more common uranium minerals and their typical chemical composition are shown in Table 11.I

Microbial participation in the geochemistry of uranium is related directly to the chemical properties of this element. Uranium is geologically laid down under highly reduced conditions and exists in the quadrivalent state. The accumulation of uranium in Carboniferous and Cretaceous deposits has been attributed to biogenic reactions (Miholic,

TABLE 11.I

MICROBIAL LEACHING OF URANIUM MINERALS

Uranium material[a]	Chemical composition	Predicted degree leaching with microbes[b]
Oxides		
Uranite	UO_2	+
Hydrous oxides		
Gummite	$UO_3 \cdot n\, H_2O$	+
Bacquerelite	$2\, UO_3 \cdot 3\, H_2O$	+
Multiple oxides		
Brannerite	$(U,Ca,Fe,Th,Y)_3Ti_5O_{16}$	+
Davidite	$(Fe,Ce,U)(Ti,Fe,V,Cr)_3(O,OH)_7$	+
Silicates		
Coffinite	$U(SiO_4)_{1-x}(OH)_{4x}$	−
Hydrous silicates		
Uranophane	$CaO \cdot 2\, UO_3 \cdot 2\, SiO_2 \cdot 6\, H_2O$	+ −
Beta-uranotile	$CaO \cdot 2\, UO_3 \cdot 2\, SiO_2 \cdot 6\, H_2O$	+ −
Sklodowskite	$MgO \cdot 2\, UO_3 \cdot 2\, SiO_2 \cdot 6\, H_2O$	+ −
Hydrous phosphates		
Autunite	$CaO \cdot 2\, UO_3 \cdot P_2O_5 \cdot 8\, H_2O$	+
Torbernite	$CuO \cdot 2\, UO_3 \cdot P_2O_5 \cdot 8\, H_2O$	+
Uramphite	$NH_4(UO_2)(PO_4) \cdot 3\, H_2O$	+
Hydrous arsenates		
Zeunerite	$CuO \cdot 2\, UO_3 \cdot As_2O_5 \cdot 8\, H_2O$	+
Carnotite	$K_2O \cdot 2\, UO_3 \cdot V_2O_5 \cdot 3\, H_2O$	+ −
Tyuyamunite	$CaO \cdot 2\, UO_3 \cdot V_2O_5 \cdot 8\, H_2O$	+ −
Hydrous sulfates		
Zippeite	$2\, UO_3 \cdot SO_3 \cdot n\, H_2O$	+
Uranopilite	$6\, UO_3 \cdot SO_3 \cdot n\, H_2O$	+
Johannite	$CuO \cdot 2\, UO_3 \cdot 2\, SO_3 \cdot 7\, H_2O$	+
Carbonates		
Schroeckingerite	$3\, CaO \cdot Na_2O \cdot UO_3 \cdot CO_2$ $\cdot SO_3F \cdot 10\, H_2O$	+
Organic compounds		
Urano-organic complexes hydrocarbons Thucholite (Carburan)		+

[a] From Gittus (1963).

[b] Easy, +; hard, −; variable, + −

1952). Many of the reduced forms are easily oxidized, under a variety of conditions, to the soluble hexavalent state. In the oxidized state, uranium is readily leached by acids and by mild complexing agents such as the carbonate ion. Ferric iron is one of the most effective oxidants for solubilizing uranium. For commercial leaching strong oxidants ordinarily are used, e.g., sodium chlorate, perchloric acid, potassium permanganate, disodium peroxydisulfate, or manganese dioxide. In solubilizing uranium from associated minerals, ferric iron should be present since it appears to mediate the oxidation.

$$2 Fe^{2+} + MnO_2 + 4 H^+ \rightarrow 2 Fe^{3+} + Mn^{2+} + 2 H_2O$$
$$6 Fe^{2+} + ClO_3^- + 6 H^+ \rightarrow 6 Fe^{3+} + Cl^- + 3 H_2O$$

The ferric iron then oxidizes the quadrivalent uranium:

$$UO_2 + 2 Fe^{3+} \rightarrow UO_2^{2+} + 2 Fe^{2+}$$

Ores containing increased amounts of phosphates, fluorides, or arsenates require greater amounts of iron because these anions complex with ferric iron to make it ineffective as an oxidant (Mouret and Pottier, 1961). Precipitates of UO_2HPO_4 and UO_2HAsO_4 do not form at a pH of less than 1.8 (Thunaes et al., 1956). Once it is oxidized uranium is readily leached by acid or carbonate solutions. Addition of sulfuric acid results in the formation of soluble uranyl sulfate $[UO_2(SO_4)_2^{2-}]$, while addition of sodium carbonate gives the soluble tetrasodium uranyl tricarbonate $[Na_4UO_2(CO_3)_3]$.

MICROBIAL RECOVERY OF URANIUM

Both autotrophic and heterotrophic microorganisms produce oxidizing and reducing agents functional in either the solubilization or deposition of uranium. These microbes produce mineral or organic acids, strong oxidants, and, under some conditions, carbon dioxide as by-products.

Controlled in situ bacterial leaching of uranium has been accomplished, and MacGregor (1966) has used biochemical techniques to gradually increase the uranium output of the Stanrock uranium mines in Elliott Lake, Ontario. The technology used is crude and consists in the systematic washing of stopes. These ores contain 10–15% pyrite and small quantities of pyrrhotite, sphalerite, and chalcopyrite. Although the microbes used were not identified, the high yields of sulfuric acid synthesized during washing of the ores indicate the presence of Thiobacillus ferrooxidans and

Ferrobacillus thiooxidans. The acid is formed by the oxidation of the sulfur moiety in pyrite while the oxidation of the iron to ferric sulfate introduces another good leaching agent into the ore.

Natural leaching techniques with no addition of microbes have been utilized in studying Canadian ores (Harrison *et al.*, 1966). Miller *et al.* (1963a,b) crushed ores to $\frac{1}{4}$ in., added pyrite, and conditioned the ore with a leaching solution by adjusting the pH to 4.0 with sulfuric acid. Such conditioning induced growth of the iron-oxidizing autotrophs. An Urgeirica ore containing 0.10% U_3O_8 and 5.0% pyrite leached rapidly upon addition of 1.0% $FeSO_4$.Without ferrous sulfate, leaching was slower. Bica ore containing 0.16% U_3O_8 and less than 1.0% pyrite was not leached unless $FeSO_4$ was added. A Volinhos ore containing calcite and 0.14% U_3O_8 complexed with carbonate. Leaching did not increase upon treatment with 0.5 and 3.0% aqueous solutions of pyrite. The pH of the effluent did not decrease below 6. Substitution of 3.0% $FeSO_4$ for pyrite lowered the pH of the effluent and gave rapid extraction of uranium. Either Cu^{2+} or microbes could be used as catalysts to maintain the iron in the ferric state.

Heap leaching per se was first employed by the Germans and dates back to the 16th century. Current techniques are similar to those first developed. Mashbir (1964) employed the following hydrometallurgical technique to leach uranium ores from the gas hills area in Wyoming. Flow rates of solution through beds averaged 4–5 gal/hr/ft² of "fill area" both during washing and recirculation. The heaps were leached in 3 days, with haulage reduced to up to 99%.

The rate of natural leaching of uranium is affected by the type of deposit, ore depth, particle size, and rate of watering. Heating the ores reduced the rate of leaching which indicated that microbes were involved and in turn destroyed by thermal treatment. Between 70–80% recovery of uranium was obtained from ores from (1) the Dyson's Mine, (2) Northern Territories, Australia, (3) Umtali Mine in Southern Rhodesia, and (4) the Mary Kathleen Mine in Queensland, Australia; this was accomplished by adding either 5–10% pyrite or $FeSO_4$ in the leaching solutions.

In leaching uranium ores, two agents are needed: a good acid or carbonate solubilizing agent and a good oxidant. Sulfuric acid is one of the more common solubilizing agents and ferric iron is a preferred oxidant. The degree of biogeochemical leaching and mobilization of uranium depends upon the biological agent and its capacity for producing acid and oxidant. Deposition of uranium occurs where the reverse conditions exist. Biosynthetically formed chelating chemicals for uranium are not known; however, uranium is trapped in ores containing high concentrations of humus. These humic fractions are usually formed by microbial degradation of plants, etc.

Synthesis of Sulfuric Acid

Several reduced sulfur compounds can be oxidized to yield energy for growth of autotrophic microbes, and sufficient sulfuric acid is produced in the process to leach uranium. Two major morphological groups of chemoautotrophic sulfur microbes oxidize these compounds. The first group consists of large filamentous bacteria and the second group of short rods.

Group 1. Beggiatoa alba is the best example of the filamentous bacteria. This bacterium deposits conspicuous intracellular sulfur globules which may not be entirely sulfur, but rather sulfur dissolved in fat globules. *Beggiatoa alba* oxidizes hydrogen sulfide in two steps as follows (Foster, 1951). Sulfur may accumulate as an intermediate in the first reaction.

$$H_2S + 0.5\ O_2 \rightarrow S + H_2O + \text{Energy}$$

Oxidation of the intracellular sulfur follows:

$$2\ S + 3\ O_2 + 2\ H_2O \rightarrow 2\ H_2SO_4 + \text{Energy}$$

A portion of this energy which is biologically available is coupled to carbon dioxide assimilation.

Group 2. The small rod-shaped bacteria characterizing the second group are capable of deriving their energy from the oxidation of sulfur or sulfur compounds. These bacteria comprise the genus *Thiobacillus*, which has four well-defined species, each characterized by a distinctive metabolic behavior toward sulfur compounds. Important members are classified as follows:

I. **Aerobic Growth**
   A. Obligate autotrophic
      1. *Thiobacillus thioparus*—Optimum growth occurs at pH 7.0.
      2. *Thiobacillus thiooxidans*—Optimum growth occurs between pH 2.0 and 3.5.
   B. Faculative autotrophic
      1. *Thiobacillus novellus*
II. **Anaerobic Growth (in Presence of Nitrate)**
   A. *Thiobacillus dentrificans*

Thiobacillus thioparus occurs in marine and freshwaters, mud, and soil and can be isolated from mineral solutions containing sulfide, thiosulfate, or tetrathionate. Bicarbonate is added to supply the carbon for cell growth. This obligate autotroph deposits globules of elemental sulfur outside the cells. It does not utilize organic matter for growth. Both sulfur and sulfate are produced simultaneously from the oxidation of thiosulfate. The ratio of free sulfur to sulfate sulfur produced is 2 : 3 when calculated on the basis of sulfur content. In the absence of a suitable buffer, the sulfuric acid which accumulates lowers the pH of the medium to about 4.4 which suppresses further microbial growth.

The chemical conversion of thiosulfate by *Thiobacillus thioparus* is represented as follows (Starkey, 1935):

$$5 \, Na_2S_2O_3 + H_2O + 4 \, O_2 \rightarrow 5 \, Na_2SO_4 + H_2SO_4 + 4 \, S + Energy$$

Tetrathionate, trithionate, and sulfite are probable intermediates. The extracellular sulfur can be further oxidized to sulfate, but this process is slow compared to thiosulfate oxidation and is considered less important than that observed during oxidation of sulfur by *Thiobacillus thiooxidans.*

Autotrophs of the *Thiobacillus thiooxidans* type are much more common than other species of *Thiobacillus* and are the most active in oxidizing the thiosulfate formed in nature. Thiosulfate is a predicted intermediate in sulfur oxidation.

Thiobacillus thiooxidans oxidizes elemental sulfur to sulfuric acid. This oxidation is represented as follows (Waksman and Joffe, 1922):

$$S + 1.5 \, O_2 + H_2O \rightarrow H_2SO_4 + Energy$$

Thiosulfate is also oxidized to sulfate without the accumulation of sulfur:

$$Na_2S_2O_3 + H_2O + 2 \, O_2 \rightarrow Na_2SO_4 + H_2SO_4 + Energy$$

Sulfide is also oxidized to sulfate.

Thiobacillus thiooxidans is unique because it produces a higher concentration of acid than any other living microbe. Starting with elemental sulfur it may synthesize up to 1.5 M sulfuric acid. *Thiobacillus thiooxidans* is therefore quite tolerant to acid conditions and is still physiologically active at concentrations of 0.5 M H_2SO_4 (pH 0.3).

Thiobacillus thiooxidans can be isolated by classic enrichment techniques from acid environments in nature such as peat bogs, coal mines, acid soils, and soils treated with sulfur or sulfur-containing fungicides. It is responsible for the breakdown of pipeline sealing mixtures which contain elemental sulfur. This organism is normally cultivated aerobically on a mineral medium consisting of $(NH_4)_2SO_4$, $MgSO_4$, $FeSO_4$, $CaCl_2$, KH_2PO_4, and powdered sulfur. *Thiobacillus thiooxidans* obtains its carbon from carbon dioxide and not from bicarbonate since the latter does not exist in solution at the pH used for growth of this microbe. Carbon dioxide is only sparingly soluble in solutions with such a low pH.

In general, the chemical intermediates involved in the biological conversion of sulfur to sulfate are probably comparable to those observed in other biological oxidations. One can therefore visualize that sulfur is oxidized by successive hydration and dehydrogenation to form sulfate in the same way that aldehyde groups are oxidized heterotrophically by successive hydration and dehydrogenation to form carbon dioxide. The postulated sequence of steps in sulfur oxidation are as follows (Kluyver, 1931);

$$S \xrightarrow[\text{+HOH}]{} S{=}O \xrightarrow[-2H]{+HOH} S{-}OH \xrightarrow[-2H]{+HOH} S{-}OH \xrightarrow[-2H]{+HOH} S{-}OH$$

(Sulfur hydrate)   (Sulfoxylic acid)   (Sulfurous acid)   (Sulfuric acid)

Thiobacillus novellus is a facultative autotroph, i.e., it is capable of growing either autotrophically by oxidation of inorganic sulfur compound and assimilation of CO_2, or heterotrophically by oxidation of organic materials in the absence of inorganic sulfur. It is particularly active in the oxidation of thiosulfate, thus being similar to *Thiobacillus thiooxidans*.

$$Na_2S_2O_3 + H_2O + 2O_2 \rightarrow Na_2SO_4 + H_2SO_4 + \text{Energy}$$

Growth is at a pH near neutrality. Oxidation of thiosulfate is rapid. Intermediate metabolites of elemental sulfur and polythionates have not been observed. When *Thiobacillus novellus* is grown heterotrophically, it retains the ability to oxidize thiosulfate to sulfate.

Thiobacillus denitrificans oxidizes sulfur to sulfate anaerobically, providing nitrate is added as an oxidant. Any calcium carbonate added is neutralized by the acid formed. The oxidation of elemental sulfur and reduction of nitrate yields, respectively, sulfate and free nitrogen:

$$6KNO_3 + 5S + 2H_2O \rightarrow K_2SO_4 + 4KHSO_4 + 3N_2 + \text{Energy}$$

Thiobacillus denitrificans is an obligate anaerobe and depends entirely upon nitrate as an electron acceptor. Various reduced sulfur compounds such as sulfide, thiosulfate, or tetrathionate are oxidized. The energy produced is utilized for CO_2 fixation and other cellular requirements for growth.

Numerous heterotrophic bacteria and fungi oxidize reduced sulfur compounds in a manner analogous to autotrophic oxidation. The fundamental difference is that the heterotrophs cannot utilize the energy derived by these oxidations as their sole source of energy. Microorganisms of this type oxidize thiosulfate to tetrathionate.

$$2\ Na_2S_2O_3 + H_2O + 0.5\ O_2 \rightarrow Na_2S_4O_6 + 2\ NaOH + Energy$$

The alkaline conditions produced during this reaction cause a chemical decomposition of the tetrathionate into tri- and pentathionate. The pentathionate spontaneously decomposes, yielding more tetrathionate and elemental sulfur. Tetrathionate does not accumulate from oxidation of thiosulfate by obligate or faculatively autotrophic sulfur bacteria. The heterotrophic microbes oxidizing sulfur to sulfate are probably not as active in the geochemistry of uranium as the autotrophic forms; however, they do provide an alkaline leach. If organic acids are synthesized they function as well as mineral acids in the leaching of uranium ores.

In leaching studies Menke (1959) has discovered that many organic compounds form complexes and increase the solubility of uranium, viz., oxalate, salicylate, malate, maleate, tartrate, acetate, lactate, pyruvate, glyceryl phosphate, proteinate, acetylacetonate, thioglycollate, thenoyltrifluoroacetone, and sulfosalicylate. With the exception of a few of these compounds all are end products synthesized by diverse microbes found in nature.

Acetate is one of the better-known microbial products. It forms many salts with uranium; the most common follow.

$NaUO_2(CH_3COO)_3$	$UO_2(CH_3COO)_4 \cdot 2\ H_2O$
$KUO_2(CH_3COO)_3$	$RbUO_2(CH_3COO)_3$
$NH_4UO_2(CH_3COO)_3$	$AgUO_2(CH_3COO)_3$
$Mg(UO_2)_2(CH_3COO)_6\ 7\ H_2O$	$PbUO_2(CH_3COO)_4 \cdot 4\ H_2O$
$ZnUO_2(CH_3COO)_4$	$CsUO_2(CH_3COO)_3$

Microbial Synthesis of Chemical Oxidants

Ferric Sulfate

The chemical oxidant of greatest potential to be found in subsurface water is ferric sulfate. It is synthesized from ferrous sulfate by *Thiobacillus ferrooxidans* and *Ferrobacillus ferrooxidans*. These microbes are quite active below pH 3.0 where autooxidation of iron becomes limiting.

$$4 \, FeSO_4 + 2 \, H_2SO_4 + O_2 \rightarrow 2 \, Fe_2(SO_4)_3 + 2 \, H_2O$$

The chemical reaction involved in the microbial dissolution of iron (Sutton and Corrick, 1963) from sulfide minerals is quite simple. Pyrite in the presence of oxygen and water is slowly oxidized to ferrous sulfate and sulfuric acid:

$$2 \, FeS_2 + 7 \, O_2 + 2 \, H_2O \rightarrow 2 \, FeSO_4 + 2 \, H_2SO_4$$

Thiobacillus ferrooxidans, in the presence of oxygen and sulfuric acid, oxidizes available ferrous sulfate to ferric sulfate. Some of the ferric sulfate formed reacts chemically with additional pyrite to form H_2SO_4 and ferrous sulfate:

$$7 \, Fe_2(SO_4)_3 + FeS_2 + 8 \, H_2O \rightarrow 15 \, FeSO_4 + 8 \, H_2SO_4$$

or some of the ferric iron can react with quadrivalent uranium to oxidize it to the soluble hexavalent state:

$$UO_2 + 2 \, Fe^{3+} \rightarrow UO_2^{2+} + 2 \, Fe^{2+}$$

The sulfuric acid produced increases the dissolution. The reduced iron is reoxidized biochemically adding to the continuous nature of the process.

Sulfur synthesized or deposited in any of these reactions is oxidized by *Thiobacillus concretivorus*:

$$2 \, S + 3 \, O_2 + 2 \, H_2O \rightarrow 2 \, H_2SO_4 + \text{Energy}$$

Manganese Dioxide

Manganese dioxide is also a good oxidant. The microbial oxidation of the manganous ion is discussed elsewhere (page 166). Unlike ferric sulfate, manganese dioxide has a greater solubility under alkaline leaching conditions and is one of the preferred oxidants in the carbonate leaching of uranium. Manganese dioxide precipitates under acid conditions. Its function in oxidizing reduced iron has been discussed previously.

Hydrogen Peroxide

A large variety of other oxidants can be used in leaching uranium (Gittus, 1963). Hydrogen peroxide is an effective oxidant but it also complexes with uranium under certain conditions. Numerous aerobic and anaerobic microorganisms synthesize small amounts of H_2O_2, thus providing an oxidant active in the *in situ* leaching of uranium. Normally in microbial oxidations and electron transport, two pairs of electrons and hydrogen ions are transferred to each oxygen molecule with the formation of water (Stainer *et al.*, 1957):

$$O_2 + 4 H \rightarrow 2 H_2O$$

The enzymes active in hydrogen peroxide formation catalyze the transfer of a single pair of electrons and hydrogen ions to the oxygen molecule with the formation of H_2O_2.

$$O_2 + 2 H \rightarrow H_2O_2$$

Although hydrogen peroxide is quite labile, and the amount formed is quite small, its common occurrence as a terminal metabolic product enhances its biogeochemical significance. Among the peroxide-forming enzymes the glucose aerodehydrogenase (glucose oxidase) in *Aspergillus niger* is most typical. It catalyzes the oxidation of glucose to gluconic acid with the uptake of one atom of oxygen. One mole of H_2O_2 is formed per mole of sugar oxidized (Foster, 1949).

$$CH_2OH(CHOH)_4CHO + H_2O + O_2 \rightarrow CH_2OH(CHOH)_4COOH + H_2O_2$$
(Glucose)   (Gluconic acid)

Gluconic acid itself is acidic enough to produce some solubilization of uranium.

Iridophycus flaccidum also possesses a carbohydrate oxidase that oxidizes glucose or galactose to gluconic or galactonic acid, respectively, with the formation of H_2O_2 (Lewin, 1962). Some of the strains of *Mycobacterium* possess an enzyme, lactic acid oxidase, which produces 1.2 micromoles of H_2O_2 and 0.9 micromoles of pyruvate from each 30 micromoles of L-lactate (Cousins, 1956).

L-Amino acid oxidase, an enzyme formed by a number of penicillin-producing fungi, attacks the natural (L-configuration) amino acids according to the general equation:

$$2 \text{ R}-\underset{\underset{\text{NH}_2}{|}}{\text{CH}}-\text{COOH} + O_2 \longrightarrow 2 \text{ R}-\underset{\underset{\text{NH}}{\|}}{\text{C}}-\text{COOH} + H_2O_2$$

$$+ 2 \text{ HOH}$$

(Imino acid)

$$2 \text{ R}-\underset{\underset{\text{O}}{\|}}{\text{C}}-\text{COOH} + 2 \text{ NH}_3$$

Hydrogen peroxide is produced by this reaction (Foster, 1949).

There are numerous microbes which produce H_2O_2. This oxidant is synthesized by both aerobic and anaerobic bacteria and by both pathogenic and free-living microbial forms. Certain salivary bacteria of the *Streptococcus viridans* type produce H_2O_2 (Thompson and Johnson, 1951; Kraus *et al.*, 1957).

Other microbes grown under unusual conditions often show abnormal H_2O_2 production (Table 11.II). For example, cultures of *Streptococcus faecalis* strain B33A, grown anaerobically and transferred to a fresh medium under aerobic conditions, form considerable amounts of free H_2O_2 (Seely and del Rio-Estrada, 1951).

Isoniazid, a compound which quickly permeates the cell of the tubercle bacillus, is oxidized to isonicotinic acid. Isonicotinic acid hinders the action of coenzyme I or II resulting in the formation of abnormally high amounts of H_2O_2 in certain mycobacteria (Thiemer, 1956).

Hydrogen peroxide is not the only peroxide formed by microorganisms. Filtrates of liquid cultures of *Chlorella vulgaris* were shown after a few days storage to possess sufficient peroxides to inhibit growth. *t*-Butyl hydroperoxide, lineolic-oleic acid peroxide, and diethyl peroxide have been identified as microbial products (Scutt, 1964).

An overwhelming number of microbes produce H_2O_2 (Table 11.II). Two gram-negative bacteria, *Alcaligenes faecalis* and *Pseudomonas fluorescens*, produce H_2O_2 (Berger, 1952), as does *Diplococcus pneumoniae* (Aussen *et al.*, 1950). A strain of *Rhizobium meliloti* observed by Cook and Quadling (1962) produced enough H_2O_2 to limit its own growth. These workers also noted that not all Rhizobia strains were equally sensitive to the peroxide.

The organism *Clostridium perfringens* synthesizes a soluble cyanide-insensitive oxidase that catalyzes the reduction of oxygen to H_2O_2 with reduced DPN as the hydrogen donor. Traces of peroxide (2×10^{-7} to 3×10^{-6} M) are formed during the oxidation of reduced diphosphopyridine nucleotide which is catalyzed by the clostridial enzyme (Dolin, 1959; Mallin and Seeley, 1958).

Catalase-negative lactic acid bacteria can grow in the presence of O_2 and accumulate quantities of H_2O_2 during aerobic growth (Stanier *et al.*, 1957). Studies conducted on strains of *Streptococcus, Pediococcus, Aerococcus, Leuconostoc*, and *Lactobacillus* showed that some lactic acid bacteria formed detectable amounts of H_2O_2 while others did not regardless of their preference or requirement for aerobic or anaerobic conditions (Whittenbury, 1964). Anaerobes in general were shown to produce H_2O_2 by Gordon *et al.* (1953).

TABLE 11.II

HYDROGEN PEROXIDE FORMATION BY MICROORGANISMS

Organism	Substrate	Aerobic	Anaerobic	Reference
Algae				
Chlorella vulgaris	—	X		Scutt (1964)
Bacteria				
Aerococcus	Hematin	X		Whittenbury (1964)
Alcaligenes faecalis	—	X		Berger (1952)
Clostridium kluyveri	Butyrate		X	Lieberman and Barker (1954)
Clostridium perfringens	Glucose		X	Mallin and Seeley (1958)
Diplococcus pneumoniae	—	X		Aussen et al. (1950)
Lactic acid bacteria	—	X		Stanier et al. (1957)
Lactobacillus	Hematin	X		Whittenbury (1964)
Leuconostoc	Hematin, glucose	X		Whittenbury (1964)
Mycobacterium phlei	L-Lactate	X		Cousins (1956)
Mycobacterium smegmatis	L-Lactate	X		Cousins (1956)
Mycobacterium stercosis	L-Lactate	X		Cousins (1956)
Pediococcus	Hematin, glucose	X		Whittenbury (1964)
Pseudomonas fluorescens	—	X		Berger (1952)
Rhizobium meliloti	Soft agar	X		Cook and Quadling (1962)
Salivary bacteria	—	X		Kraus et al. (1957)
Salivary streptotocci	—	X		Thompson and Johnson (1951)
Streptococcus	Hematin	X		Whittenbury (1964)
Streptococcus faecalis	—	X		Seely and del Rio-Estrada (1951)
Streptococcus mitis	—	X		Clapper et al. (1950)
Tubercle bacteria	—	X		Thiemer (1956)
Unidentified anerobic bacteria	—		X	Gordon et al. (1953)
Fungi				
Aspergillus niger	Glucose $+ O_2$	X		Foster (1949)
Iridophycus flaccidum	Glucose/galactose	X		Lewin (1962)
Penicillin-producing fungi	L-Amino acids	X		Foster (1949)

Clostridium kluyveri, an obligate anaerobe, oxidizes butyrate and accumulates quantities of H_2O_2 (Lieberman and Barker, 1954). These reports indicate that a wide variety of H_2O_2-producing microorganisms exist (Table 11.II).

The overall significance of microbially produced hydrogen peroxide in geochemistry remains to be assessed. In an evaluation of the biogeochemical significance of H_2O_2, one should consider that probably there are just as many microbes that produce the enzymes catalase and peroxidase which rapidly catalyze the destruction of H_2O_2 as there are H_2O_2-producing microorganisms. Even though H_2O_2 in ground water is short-lived, it must be considered as active in oxidizing many metals.

BIOENERGETICS OF URANIUM OXIDATION

The oxidation of uranium minerals is exothermic, and although the amount of energy produced is small it is adequate to support autotrophic growth:

$$UO_2 + \tfrac{1}{2}O_2 \rightarrow UO_3$$
$$\Delta F_{27°C} = -29.1 \text{ kcal}$$

$$3\,UO_2 + O_2 \rightarrow U_3O_8$$
$$\Delta F_{27°C} = -63.4 \text{ kcal}$$

$$U_3O_8 + \tfrac{1}{2}O_2 \rightarrow 3\,UO_3$$
$$\Delta F_{27°C} = -22.6 \text{ kcal}$$

Enzymes catalyzing this reaction have not yet been described. The energy provided by this reaction is sufficient to support autotrophic growth. A uranium oxidase enzyme will probably also be found in certain heterotrophs.

Polyvalent metals form complexes with a variety of biochemical substances. This is particularly true of uranium in the form of the uranyl ion. It complexes with phosphate and carboxyl groups. Carbohydrate metabolism by yeast is sensitive to uranium (Booy, 1940; Barron *et al.*, 1948). Uranium penetrates cells at 8.5×10^{-11} mol/minute/cm^2 of surface. The permeability constant for uranium is about the same as water but considerably higher than cations such as potassium. The uranyl ions form relatively undissociated complexes with the orthophosphate in cells $(1-2 \times 10^{-2}\ M)$ and carboxyl groups of bicarbonate $(1 \times 10^{-1}\ M)$, organic acids, and proteins. The cytoplasm itself will bind $1 \times 10^{-1}\ M$/liter of uranium.

BIOLEACHING OF URANIUM

With our present knowledge of uranium biochemistry, it can be seen that certain uranium minerals should be more easily attacked by microbial processes than others. Estimates of the amenability of uranium minerals to bio-leaching are given in Table 11.I. Where the minerals listed are found associated with minerals sulfides or organic matter, the uranium should lend itself to recovery through microbial action.

Some microbes are sensitive to uranium (Muntz *et al.*, 1951), with an inhibition of glucose oxidation. In yeast this effect is reversed by the addition of phosphate in sodium bicarbonate. Bacteria differ in their sensitivity to uranium; however, in all instances phosphate, or ortho-phosphate (Rothstein, 1954), can be added to reverse this inhibition.

BIOPROSPECTING FOR URANIUM

Microbiological techniques for prospecting for uranium have not been developed. Botanical methods have received greater emphasis.

Plant physiologists have shown that uranium is a necessary nutrient for plant growth (Cannon, 1954) and is therefore present in all plants. Selenium and, to a lesser extent, sulfide minerals associated with many uranium and vanadium ores are required in large quantities in only certain groups of plants. Indicator plants for sulfide in uranium districts are *Sisymbrium oletissimum*, *Toxicosordion gramineum*, *Eriogonum inflatum*, *Lepidium montanum*, *Stanleya pinnata*, and *Allium acum inatum*; Indicator plants for selenium in uranium districts are *Astragalus cenfertiflorus*, *A. preussi* var. *aritus*, *A. thompsonae*, *A. pattersonii*. *Oryzopsis hymenoides*, and *Aster venusta*. A uranium concentration in any plant exceeding 3.0 ppm may be indicative of a geologically favorable deposit.

BIOGEOCHEMICAL URANIUM CYCLE

The uranium present in uranite is oxidized by ferric ions to UO^{2+} which is soluble in water (see Fig. 11.I). Iron-oxidizing microbes provide a continuous supply of ferric ions which aids in the mobilization of uranium. Other oxidants also help maintain the iron in the ferric state. Two such oxidants are hydrogen peroxide and manganese dioxide. The

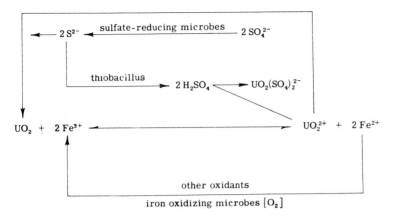

Fig. 11.1. Biogeochemical uranium cycle.

manganese dioxide is especially active in alkaline ores where it is most soluble. Hexavalent uranium is redeposited when it encounters highly reduced conditions and undergoes reduction to the insoluble mineral uranite. These reducing conditions are produced by the sulfate-reducing microbes. Organic matter or humus also chelates uranium and traps this important metal.

The sulfuric acid formed by the sulfur oxidizing bacteria aids in the solubilization of uranium, and once the uranium is in the hexavalent state the soluble uranyl sulfate is formed in an acid leach. Organic acids also increase the rate of leaching although they are too expensive to use the industrial leaching of uranium; however, organic acids synthesized by microbes in the lithosphere contribute to the movement of this metal.

REFERENCES

Aussen, B. S., Kingman Boltjes, T. Y., and Wastmann, B. S. J. (1950). The polarographic determination of the catalase activity and the hydrogen peroxide production of bacteria. *Antonie van Leewuenhock, J. Microbiol. Serol.* **16**, 65–78.

Barron, E. S. G., Muntz, J. A., and Gasvoda, B. (1948). Regulatory mechanisms of cellular respiration. I. The role of cell membranes: Uranium inhibition of cellular respiration. *J. Gen. Physiol.* **32**, 163–178.

Berger, V. (1952). Hydrogen peroxide production by gram negative bacteria. Also a contribution to the diagnosis of hydrogen peroxide forming microorganisms. *Z. Hyg. Infektionskrank.* **134**, 162–172.

Booy, H. L. (1940). The protoplasmic membrane considered as a complex system. *Rec. Trav. Botan. Neerl.* **37**, 1–77.

Cannon, H. L. (1954). Botanical methods of prospecting for uranium. *Mining Eng.* **6**, *Tech. Publ.* 3724L, 217–220.

Clapper, W. E., Grace, H., and Haiby, G. (1950). The peroxide production and peroxidase activity of sulfathiazole-sensitive and resistant strains of *Streptococcus mitis. Prov. Soc. Exptl. Biol. Med.* **87**, 600–602.

Cook, F. D., and Quadling, C. (1962). Peroxide production by *Rhizobium meliloti. Can. J. Microbiol.* **8**, 933–935.

Cousins, F. B. (1956). Lactic acid oxidase of the Mycobacteria. *Biochem. J.* **64**, 297–304.

Dolin, M. I. (1959). Oxidation of reduced diphosphopyridine nucleotide by *Clostridium perfringens.* I. Relation of peroxide to the over-all reaction. *J. Bacteriol.* **77**, 383–392.

Foster, J. W. (1949). "Chemical Activities of Fungi," pp. 460–461. Academic Press, New York.

Foster, J. W. (1951). *In* "Bacterial Physiology," (C. H. Werkman and P. W. Wilson, eds.), pp. 363–373. Academic Press, New York.

Gittus, J. H. (1963). "Metallurgy of the Rarer Metals -8.Uranium," p. 34. Butterworth, London and Washington, D.C.

Gordon, J., Holman, R. A., and McCleod, J. W. (1953). The production of hydrogen peroxide by anaerobic bacteria. *J. Pathol. Bacteriol.* **66**, 527–535.

Harrison, V. F., Gow, W. A., and Hughson, M. R. (1966). Factors influencing the application of bacterial leaching to a Canadian uranium ore. *Dept. Mines Tech. Surveys, Ottawa Canad. Mines Branch Rept. No.* **EMA 66–1**, 1–31.

Kluyver, A. J. (1931). "The Chemical Activities of Microorganisms." Oxford Univ. Press, London and New York.

Kraus, F. W., Nickerson, J. F., Perry, W. L. and Walker, A. P. (1957). Peroxide and peroxidogenic bacteria in human saliva. *J. Bacteriol.* **73**, 727–735.

Lewin, R. A. (1962). "Physiology and Biochemistry of Algae," p. 129. Academic Press, New York.

Lieberman, I., and Barker, H. A. (1954). The production of hydrogen peroxide by an obligate anaerobe, *Clostriduim kluyveri. J. Bacteriol.* **68**, 61–62.

MacGregor, R. A. (1966). Recovery of U_3O_8 by underground leaching. *Can. Mining Met. Bull.* **59**, 583–587.

Mallin, M. L., and Seeley, H. W. (1958). Some relations of hydrogen peroxide to oxygen consumption by *Clostridium perfringens. Arch. Biochem. Biophys.* **73**, 306–314.

Mashbir, D. S. (1964). Heap leaching of low grade uranium ore. *Mining Congr. J.* pp. 50–54.

Menke, J. R. (1959). Method of recovering uranium from underground deposit. U.S. Patent 2,896,930.

Miholic, S. (1952). Radioactivity of waters issuing from Sedimentary Rocks. *Econ. Geol.* **47**, 543–547.

Miller, R. P., Napier, E., and Wells, R. A. (1963a). Natural leaching of uranium ores. *Bull. Inst. Mining Met.* **679**, 271–234.

Miller, R. P., Napier, E., and Wells, R. A. (1963b). Natural leaching of uranium ores. I. Preliminary tests on Portuguese ores. II. Experimental variables. III. Application to specific ores. *Bull. Inst. Mining Met.* **674**, 234–254.

Mouret, R., and Pottier, P. (1961). Capillary leaching of uranium ores. *Energie Nucl.* **3**, 251–256.

Muntz. J. A., Singer, T. R., and Guzman Barron, E. S. (1951). Effect of Uranium on the metabolism of yeast and bacteria. *Natl. Nucl. Energy Ser., Div. IV* **23**, 246–261.

Rothstein, A. (1954). Enzyme systems of the cell surface involved in the uptake of sugars by yeast. *Symp. Soc. Exptl. Biol.* **8**, 165–201.

Scutt, J. E. (1964). Autoinhibitor production by *Chlorella vulgaris*. *Am. J. Botany* **54**, 581–584.

Seely, H. W., and del Rio-Estrada, C. (1951). The role of riboflavin in the formation and disposal of hydrogen peroxide by *Streptococcus faecalis*. *J. Bacteriol.* **62**, 649–656.

Stanier, R. Y., Doudoroff, M., and Adelberg, E. A. (1957). "The Microbial World," pp. 166–167. Prentice-Hall, Englewood Cliffs, New Jersey.

Starkey, R. L. (1935). Products of the oxidation of thiosulfate by bacteria in mineral media. *J. Gen. Physiol.* **18**, 325–349.

Sutton, J. A., and Corrick, J. D. (1963). Mircobial leaching of Copper minerals. *Mining Eng.* **6**, 37–40.

Thiemer, E. K. (1956). Hypothesis on the action of isonicotinic acid hydrazide on the tubercle bacteria. *Naturwissenschaften* **42**, 425.

Thompson, R., and Johnson, R. (1951). The inhibitory action of saliva on the diphtheria bacteria: Hydrogen peroxide, the inhibitory agent produced by salivary Streptococci. *J. Infect. Diseases* **88**, 81–85.

Thunaes, A., Smith, H. W., and Brannen, J. (1956). Uranium separation process. U.S. Patent 2,738,253.

Waksman, S. A., and Joffe, J. S. (1922). Microorganisms concerned in the oxidation of sulfur in the soil. II. *Thiobacillus thiooxidans*, a new sulfur-oxidizing organism isolated from the soil. *J. Bacteriol.* **7**, 239–256.

Whittenbury, R. (1964). Hydrogen peroxide formation and catalase activity in the lactic acid bacteria. *J. Gen. Microbiol.* **35**, 13–26.

12

VANADIUM
BIOGEOCHEMISTRY

Vanadium, a member of the ferrides or iron family, was discovered by Sefstrom in 1830. It is one of the more abundant trace elements and is found in iron meteorites, silicate meteorites, and igneous rocks. Geochemically vanadium is lithophilic with a siderophile tendency. Since it is present in all plants and animals, it is also biophilic.

The geochemical and biochemical behavior of vanadium is controlled by the fact that it possesses three stable oxidation states, i.e., tri-, tetra-, and pentavalent forms. All have a tendency to complex completely with oxygen and partly with sulfur. Vanadium resembles phosphorus and titanium and like these latter two elements is concentrated in basic rocks.

The main minerals of vanadium are vanadinite, carnotite, roscoelite, and, possibly, patronite. Vanadinite is lead chlorovanadate [$Pb_5Cl(VO_4)_3$], a rare mineral of secondary origin found in oxidized lead veins with other lead minerals. Fine crystals are found near Oudjda, Morocco, and Grootfontein, southwest Africa, and in Arizona and New Mexico in the United States.

Carnotite is a hydrated basic vanadate of potassium and uranium [$K_2(UO_2)_2(VO_4)_2 \cdot n\ H_2O$]. The mineral is yellow and does not fluoresce. It is of secondary origin which is usually attributed to the leaching action of water on other uranium and vanadium minerals. Carnotite is concentrated in the plateau region of southwestern Colorado and Utah where it is embedded in sandstone. Tyuyamunite is the calcium analog of the same mineral.

Roscoelite $[K_2V_4Al_2Si_6O_{20}(OH)_4]$ is a vanadium mica. The primary source of vanadium is from ores high in patronite (VS_4) which is a mineral of indefinite composition. It is also found complexed as copper sulfovanadate sulvanite $[Cu_3VS_4]$. The igneous rocks may contain from 150–300 gm vanadium/ton ore, and this concentration may increase in basic igneous rocks. Tetravalent vanadium is found in igneous rocks and it readily replaces Ti^{4+} (atomic rad., 0.64 kX*), Fe^{3+}, and Al^{3+} (atomic rad., 0.57 kX). Pentavalent vanadium is present in sedimentary rocks and is known to replace aluminum from clay minerals.

Titanian magnetites are high in vanadium, containing up to 0.41% vanadium, with V^{4+} replacing some Ti^{4+} in these magnetites. Vanadium is also high in sphene and rutile. In rutile vanadium crystallizes as VO_2. In apatite V^{5+} replaces P^{5+} (atomic rad., 0.35 kX). Only in apatite ores and titaniferous iron is vanadium as concentrated as it is in marine sediments.

Vanadium is used chiefly in the hardening of steel. Vanadium oxide is a good mordant.

VANADIUM IN THE BIOSPHERE

In plants and soils the vanadium content varies from 20–1000 gm/ton (Mitchell, 1944). Coal is high in vanadium and coal ashes may contain 20% or more of this element.

Vanadium promotes nitrogen assimilation by soil microbes. In higher animals it increases the oxidation of lipids. In marine environments the holothurians *Stichopus moebii* and ascidians contain up to 10% vanadium in their blood. Boeri and Ehrenberg (1954) have evidence indicating that hemovanadin is a chromoprotein with reducing properties. The vanadium is probably present in trivalent form as sulfate. The vanadium in crude oil may approach 3.0%; however, the content generally ranges from 0.001 to 0.027% by weight of V_2O_5. Venezuelan oil is very high in vanadium which may make up to 61.8% of the ash; thus it may be the source of a major air pollutant. Sjöberg (1950) estimated that from 180,000 tons of oil burned in Stockholm annually, 65 tons of vanadium pentoxide were

* Where use of atomic or ionic radii are given, they are expressed as a 6-fold coordination of the sodium chloride type to make them comparable (kX and angstroms are approximately the same).

released into the atmosphere. Asphaltic ashes often exceed 43 % vanadium. Porphyrin complexes of vanadium exist in petroleum and in asphaltic and bituminous rocks. The average vanadium content of certain bioliths are shown in Table 12.I (Bader, 1937).

TABLE 12.I

VANADIUM CONTENT OF BIOLITHS

Biolith	Vanadium (lb/ton)
Asphalts	10
Bituminous phosphate	3.2
Coal	1.8
Graphite	0.4

The toxicology and biological significance of vanadium is widely recognized (Hudson, 1964). Its distribution in the earth's crust has been estimated by Bertrand (1950) at 300 ppm. Bertrand found higher plants to average 1.0 ppm vanadium; however, root nodules of leguminous plants averaged 4 ppm and fungi 12 ppm. Bertrand (1941, 1942) considers vanadium a dynamic trace element in microbes since it is required for growth of *Aspergillus niger*. Higher animals contain less than 1.0 ppm in most tissues.

Vanadium markedly increases the growth of the green algae *Scenedesmus obliquus*. It was found to stimulate photosynthesis and could not be replaced by molybdenum or other metals (Arnon and Wessel, 1953). Certain strains of *Azotobacter* (Horner *et al.*, 1942) utilize vanadium for nitrogen fixation; however, molybdenum is much more active. Vanadium in the form of sodium orthovanadate or metavanadate increased nitrogen fixation 5–25 times in eight strains of *A. chroococcum* and one strain of *A. vinelandii*. At 0.00001 ppm there was a detectable increase in the nitrogen fixed. The maximum increase in nitrogen fixation was observed between 0.005 and 0.01 ppm. In vanadium- and molybdenum-deficient media less than 50% of the sugar was consumed. In the presence of tungstate, nitrogen fixation is initiated by molybdenum but not by vanadium (Takahashi and Nason, 1957).

Vanadium activates monamine oxidase in certain deaminating reactions involving tryptamine (Perry *et al.*, 1955). In chickens, 25 ppm of ammonium metavanadate uncouples oxidative phosphorylation in mitochondria isolated from liver (Hathcock *et al.*, 1966). This uncoupling was manifest whether succinate or β-hydroxybutyrate was used, suggesting that all three phosphorylating sites associated with electron transport were uncoupled. The toxic effects of vanadium in animals may involve oxidative phosphorylation. Vanadium also interferes with the synthesis of cholesterol (Azarnoff and Curran, 1957). The interference occurs during the condensation of mevalonic acid with active acetate, prior to the formation of squalene. Ore dust from carnotite is definitely toxic (Wilson *et al.*, 1953) and this can be partly related to its disruption of phosphorylation and cholesterol synthesis.

CHEMISTRY AND BIOCHEMISTRY

The calcium salt of ethylenediaminetetraacetic acid is a good chelating agent for vanadium. Disodium catechol disulfonate also chelates vanadium to form indigo blue complexes.

In examining plants, Ter Meulen (1931) found 3.3 ppm vanadium in the Basidiomycete *Amanita muscaria*. Other Basidiomycetes contain only traces or no vanadium. For plants in general Bertrand (1950) found a mean vanadium content of 0.16 ppm in fresh samples, 1.0 ppm in dry samples, and 7 ppm in ash.

Comparatively, vanadium acts like a transition element between subdivisions A and B of group V of the periodic table. The formation of a pentoxide places it near nitrogen as well as phosphorus.

It forms complex anions similar to those formed by chromium and molybdenum, in the trivalent state it forms such double salts as thiocyanate. Chemically it has bi-, tri-, and pentavalent states which should make it amenable to microbiological attack. On a bioenergetic basis one can expect specific autotrophic bacteria to oxidize the reduced forms and obtain a part or all of their energy for cellular growth and multiplication.

Vanadate, molybdate, selenite, tellurite, and arsenate are all reduced by extracts of *Micrococcus lactilyticus* in the presence of hydrogen (Woolfolk and Whiteley, 1962).

$$H_2VO_4 + \tfrac{1}{2} H_2 \rightarrow VO(OH)_2 + OH^-$$
(Vanadate)

$$Mo_6O_{21} + 3 H_2 \rightarrow 6 MoO_{2.5} + 6 OH^-$$
(Molybdate)

$$HSeO_3 + 2 H_2 \rightarrow Se + 2 H_2O + OH^-$$
(Selenite)

$$HTeO_3 + 2 H_2 \rightarrow Te + 2 H_2O + OH^-$$
(Tellurate)

$$H_2AsO_4 + H_2 \rightarrow HAsO_2 + H_2O + OH^-$$
(Arsenate)

$$UO_2 (OH)_2 + H_2 \rightarrow U(OH)_4$$
(Uranyl
hydroxide)

The reactions shown are associated with the enzyme hydrogenase present in *M. lactilyticus.*

Goren (1966) has coupled the iron-oxidizing bacteria such as *Thiobacillus ferrooxidans*, and *Ferrobacillus thiooxidans* with the oxidation of vanadium and uranium in acidic leaching solutions.

$$2 Fe^{3+} + V^{3+} \rightarrow V^{5+} + 2 Fe^{2+}$$
$$2 Fe^{3+} + U^{4+} \rightarrow U^{6+} + 2 Fe^{2+}$$

The ferric ion acts as an oxidant for these metals and the ferrous ion formed is reoxidized by other iron-oxidizing bacteria. Thus these autotrophs are active in oxidizing vanadium and making it more soluble in aqueous solution for transport in ground water. The interrelationship between ferrous and ferric iron and the oxidation of vanadium are shown

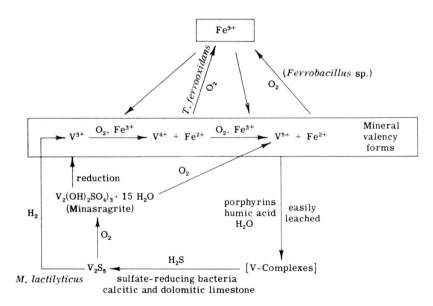

FIG. 12.1. Biogeochemical vanadium cycle.

in Fig. 12.1. *Thiobacillus ferrooxidans* and *Ferrobacillus thiooxidans* catalyze the oxidation of iron which in turn oxidizes vanadium. The sulfate-reducing bacteria generate H_2S and provide reducing conditions for the reduction of pentavalent vanadium and the formation of vanadium sulfides. *Micrococcus lactilyticus* catalyzes the reduction of vanadium sulfide and trivalent vanadium is formed. The sulfide moiety of vanadium sulfide is easily oxidized by microbes, and minasragrite is probably formed in this manner. The biogeochemical cycle for vanadium also points up the fact that chelates synthesized by microbes bind the metal.

REFERENCES

Arnon, D. I., and Wessel, G. (1953). Vanadium as an essential element for green plants. *Nature* **172**, 1039.

Azarnoff, D. L., and Curran, G. L. (1957). Site of vanadium inhibition of cholesterol biosynthesis. *J. Am. Chem. Soc.* **79**, 2968.

Bader, E. (1937). Vanadium in organogenen Sedimenten. I. Die Gründe der Vanadinanreicherung in organogenen Sedimenten. *Zentr. Mineral Geol. A* p. 164.

Bertrand, D. (1941). Le vanadium comme facteur de croissance pour l'*Aspergillus niger*. *Bull. Soc. Chim. Biol.* **23**, 467.

Bertrand, D. (1942). Le vanadium comme élément oligosynergique pour l'*Aspergillus niger*. *Ann. Inst. Pasteur* **68**, 226.

Bertrand, D. (1950). The biogeochemistry of vanadium. No. 2. Survey of contemporary knowledge of biochemistry. *Bull. Am. Museum Nat. Hist.* **94**, 409–455.

Boeri, E., and Ehrenberg, A. (1954). On the nature of vanadium in vanadocyte haemolyzate from ascideans. *Arch. Biochem. Biophys.* **50**, 404.

Goren, M. (1966). Oxidation and recovery of vanadium values from acidic aqueous media. U.S. Patent 3,252,756.

Hathcock, J. N., Hill, C. H., and Tove, S. B. (1966). Uncoupling of oxidative phosphorylation by vanadate. *Can. J. Biochem.* **44**, 983–988.

Horner, C. K., Burk, D., Allison, F. E., and Sherman, M. S. (1942). Nitrogen fixation by *Azotobacter* as influenced by molybdenum and vanadium. *J. Agr. Res.* **65**, 173.

Hudson, T. G. F. (1964). "Vanadium-Toxicology and Biological Significance." Elsevier, Amsterdam.

Mitchell, R. L. (1944). The distribution of trace elements in soils and grasses. *Proc. Nutr. Soc. (Engl. Scot.)* **1**, 183.

Perry, H. M., Teitlebaum, S., and Schwartz, P. L. (1955). Effect of antihypertensive agents on amino acid decarboxylation and amine oxidation. *Federation Proc.* **14**, 113.

Sjöberg, S. G. (1950). Vanadium pentoxide. *Acta Med. Scand. Suppl.* **149**, 238.

Takahashi, H., and Nason, A. (1957). Tungsten as a competitive inhibitor of molybdate in nitrate assimilation in N_2 fixation by *Azotobacter*. *Biochim. Biophys. Acta* **23** 433.

Ter Meulen, H. (1931). Sur la répartition du molybdène dans la nature. *Rec. Trav. Chim.* **50**, 491–504.

Wilson, H. B., Stockinger, H. E., and Sylvester, G. E. (1953). Acute toxicity of carnotite ore dust. *Arch. Ind. Hyg. Occupational Med.* **7**, 301.

Woolfolk, C. A., and Whiteley, H. R. (1962). Reduction of inorganic compounds with molecular hydrogen by *Micrococcus lactilyticus*. I. *J. Bacteriol.* **84**, 647–657.

13

ALUMINUM AND MICROBES

Aluminum is a metal with many interesting properties. It is strong, light in weight, resistant to atmospheric corrosion, and a conductor of electricity. The number of applications of its use exceed several thousand.

Bauxite, which is the primary source of aluminum, is a mixture of hydrous aluminum oxides of indefinite composition. Some bauxites resemble gibbsite [$Al(OH)_3$] but most are composed of mixtures of gibbsite, boehmite [$AlO(OH)$], and diaspore ($HAlO_2$) and usually contain iron. The finely grained amorphous constituent in bauxite is cliachite [$Al(OH)_3$].

Bauxite has a supergene origin, having been produced by prolonged weathering and leaching of silica from aluminum rocks in tropical and subtropical climates. It originates as a colloidal precipitate which may remain in place as a primary derivative product from the rock or transported to form deposits such as the lateritic soils, largely hydrous aluminum and ferric oxides, of the tropics. This deposition can also occur in temperate zones where drainage is poor as in the terra rossa soils around the Mediterranean Sea which are similar in composition to laterites.

Calcareous bauxites are formed from argillaceous limestones. The clay material in limestone is converted to bauxite, and the calcium carbonate is removed as bicarbonate.

Bauxite does not refer to a mineral species but is a lithic or rock name. The name was derived from Baux, France, where bauxite occurs in high quantity.

The principal sources of aluminum are in Surinam, Jamaica, and British Guiana with large deposits found also in Indonesia, Russia, and Hungary. The United States is the world's third largest producer with its main deposits located in Arkansas, Georgia, and Alabama. Notably all aluminum ores of commercial importance are of sedimentary and continental origin.

Aluminum is always present in river water but only in small quantities. The aluminum content of seawater is also low. This is caused by the rapid precipitation of the weakly basic hydroxides of aluminum. Soluble aluminum salts in weathering residues hydrolyze and are physically removed as solid products. They are redeposited almost quantitatively in aluminum-enriched sediments.

Aluminum is the most abundant metal in igneous rocks. Compared to all the elements it is exceeded only by oxygen and silicon in the accessible lithosphere. Hutchinson (1943) reports an average aluminum lithosphere content of 8.13%. If overall occurrence is considered, then iron and magnesium are more abundant than aluminum. Analysis of meteorites show a mean content of 1.79% Al. Aluminum is concentrated in the lithosphere (soil crust) and is practically absent from the deeper geochemical shells of the earth. It is an oxyphilic element and no known sulfide minerals have been identified. Early crystallates from magma contain only insignificant amounts of aluminum. Aluminum is not found in elemental form, and the occurrence of unhydrated oxides, trisodium aluminum fluoride, and cryolite is insignificant.

The feldspars, which are the most important constituents of igneous rocks, are aluminosilicates. They contain univalent and bivalent metals and possess the bulk of the lithospheric aluminum. The aluminum content of the important feldspars is given in the following table.

Mineral	Aluminum (%)
Orthoclase and adularia, $K[AlSi_3O_8]$	9.69
Microcline and sanidine $(K,Na)AlSi_3O_8$	9.69
Albite, $Na(AlSi_3O_8)$	10.29
Anorthite, $Ca(Al_2Si_2O_8)$	19.40

In the weathering of feldspar, aluminum and the silicates go into ionic solution. Aluminum remains dissolved in both acidic (pH 4.0) and basic (pH 9.0) solutions. A shift in the concentration of the solution or a change in pH toward neutrality will cause aluminum hydroxide to precipitate. Aluminum and silica may react to form crystalline clay minerals. Clay minerals have an aluminosilicate structure, with Al^{3+} as the most important cation. The aluminum is nearly always located externally to the silicon-oxygen framework as a cation with coordination number 6.

The micas contain aluminum as a major constituent. Biotite varies from 10–20% Al_2O_3. Muscovite contains even greater amounts of alumina, which, in some cases, exceeds 30%. At high temperatures, sillimanite ($Al_2O_3SiO_2$) and mullite (3 Al_2O_3 2 SiO_2), two less-known silicates, and aluminum, are formed.

In soils aluminum is complexed in a variety of forms (Mattson, 1941). Aluminosilicate rock, feldspar, and mica fragments of varying composition are common. These fragments vary from ultramicroscopic to macroscopic in size. Alumino silicates are complexed with clays and soil colloids derived from weathering and decomposition. Colloidal and undispersed hydroxides are formed as heterogenous gel and sol complexes with hydrated ferric oxide, hydrated silica, and organic compounds. Aluminum phosphate in leached soils provides an important source of phosphate for biological activity. Some of the aluminum also stays in true solution in soil water.

In bauxite ores *F. ferrooxidans*, *T. ferrooxidans*, and other iron-oxidizing microbes will oxidize the ferrous iron to $Fe(OH)_3$ which may be leached away by ground water to, in effect, increase the Al_2O_3 content of the soil.

BIOGEOCHEMISTRY OF ALUMINUM

Plants contain higher quantities of aluminum than animals. The tree *Orites excelsa* accumulates large quantities of aluminum succinate in trunk cavities. Ashed samples of *Lycopodium alpinum* yield up to 33% Al_2O_3. Aluminum is also concentrated in coal ash where it is complexed with organic acids.

Three thermal algae, of the indigo blue type, were found to contain aluminum when analyzed spectrographically (Umemoto and Mifune, 1953). These were *Mastigoculadus laminosus*, *Phormidium ambiguum*, and *Oscillatoria cortiana*. One freshwater form, *Rhizocolonium hieroglyphicum*, also contained aluminum. Only algae found in hot springs possessed significant quantities of Ge. All four of these algae also contained Cu, Zn, Pb, Mn, and Mo.

Marine algae contain 0.5% aluminum based on dry weight or up to 3–4% based upon ash (Vinogradov, 1935). These results are probably high and most algae do not contain more than 0.1% Al in ash. Literature on the aluminum content of lichens is reviewed by Hutchinson (1943). *Cetraria islandica*, *Lecanora esculenta*, *Parmelia scruposa*, and *Stereocaulon denudatum* have been analyzed.

In the growth of algae both *Chlorella* and *Crucigena* are stimulated by the addition of 0.02–0.2 mg Al/liter of medium (Young, 1935). Aluminum has also been found as a constituent of contemporary and fossil red algae (Maslov, 1962). Diatoms decompose clay minerals utilizing the silica and liberating the aluminum.

Microbial requirements for aluminum appear to be limited to a few organisms. Al^{3+} may be substituted for Fe^{3+} to activate histidine carboxylase activity in *Lactobacillus* sp. (Guirard and Snell, 1954); pyrodoxal phosphate was also required for decarboxylase activity.

DIRECT AND INDIRECT MICROBIAL CORROSION OF ALUMINUM

Aluminum—as an ubiquitous element—finds microbial enemies in strange places: in airplane parts, jet fuel lines, and fuel storage tanks; tap water and seawater; in natural and artificial media; in aerobic and anaerobic environments. With the increased use of aluminum as a prime fabrication material, the problem of its corrosion has become critical. Microbes often cause tremendous damage to metal parts in relatively short periods of time. Trainer aircraft shipped to Africa in unseasoned, poorly ventilated wooden crates were found to be discolored in some areas and covered with white deposits which upon close examination showed pitting. The principal agent active in pitting this aluminum was the ascomycete *Ceratostomello*. These crates were in transit only two months (Copenhagen, 1950).

Microbial contamination of jet fuel and aviation gasoline is an important factor in the corrosion of aluminum aircraft. Bacterial and fungal growth are present in most jet fuel storage facilities and have been found in aircraft wing tanks. Beneath the mats or slimes formed by these microbes, corrosion of the aluminum and degradation of the coating material takes place which weakens the structure (Churchill, 1963). Some of the more predominant cultures isolated from jet fuel-water samples are *Pseudomonas aeruginosa*, *Aerobacter aerogenes*, *Desulfovibrio desulfuricans*, *Penicillium lutem, Aspergillus flavus, Sphaerotilus natans, Hormodendrum hordei* or *Cladosporium* spp., *Clostridium* spp., *Spicaria* spp., *Bacillus* spp., *Fusarium* spp., and *Micrococcus* spp. Environmental changes, i.e., acid production, CO_2 production, surface tension changes, or decreased available O_2, attributed to microbial activity are probably responsible for the corrosion. Acid production is the most likely causative agent. Since mixed populations are present in the corrosive loci, a symbiotic relationship between aerobic and anaerobic species may develop as a result of changes in oxygen concentration. This creates conditions which are more conducive to corrosion. The addition of nitrate or phosphate to media often inhibits corrosion of aluminum (Blanchard and Goucher, 1964).

Rogers and Kaplan (1964) surveyed microbial contamination of fuel distributor systems and found *Pseudomonas* spp. to be the most common contaminants with *Bacillus* spp. and *Aerobacter aerogenes* also present. However, no direct correlation was made between the microbial populations and any corrosion of the systems or aircraft.

Miller *et al.* (1964) showed that microbes can metabolize JP-4 jet fuel and penetrate the tank coating leaving the metal unprotected and creating small galvanic cells which induce corrosion.

Laboratory studies made to reproduce the conditions of aluminum corrosion in a Bushnell-Haas medium and JP-4 jet fuel revealed that $Fe(OH)_3$ and Ca^{2+} ions stimulate corrosion in sterile medium while nitrate and phosphate ions inhibit this reaction (Blanchard and Goucher, 1964). Media containing potassium nitrate from 0.02–1.2 gm/liter and a proper balance of corrosion stimulating and inhibiting ions have been devised. Aluminum alloys 7075 and 2024 were placed in uninoculated media and media inoculated with selected mixed or pure cultures of *Pseudomonas* growing on fuel. The corrosion induced by either mixed or pure cultures was comparable.

Extensive corrosion of aluminum alloy 7075 was observed in inoculated flasks after 97 days at every level of nitrate tested. Aluminum alloy 7075 in the inoculated media which contained the greatest amount of nitrate was the last to corrode. When corrosion began, it progressed more rapidly at lower levels of nitrate. In uninoculated medium, aluminum 7075 was corroded only where the nitrate ion was absent. Aluminum alloy 2024 subjected to the same tests showed signs of corrosion in inoculated media containing no nitrate or high concentrations of nitrate. In uninoculated controls essentially no corrosion occurred. Corrosion of both alloys took place with and without the presence of nitrate in an inoculated proteinaceous medium.

It was suggested that microbes utilize nitrate and phosphate more rapidly than calcium or iron thus creating a medium which is chemically corrosive.

Hedrick *et al.* (1964) conducted tests on aluminum corrosion using aluminum 7178-T651 in four types of media containing JP-4 jet fuel. *Pseudomonas aeruginosa* GD/FW B-3, *Desulfovibrio desulfuricans* Prince ASD, *Cladosporium resinae* QMC 7998, *Aspergillus niger* GD/FW FI, and mixed microflora (predominantly pseudomonads from contaminated fuels) were used as inocula both in pure and artificially mixed populations. After 90 days, corrosion was most extensive in inoculated deionized water-fuel medium, with the deepest pitting occurring at the water phase or the water-fuel interface. Weight losses of the aluminum strips varied from 0.015 and 0.574%. The heaviest corrosion was observed in naturally and artificially mixed cultures. Bacterial cultures were more destructive than fungi. Aerated media were more conducive to corrosion than nonaerated.

Dugan and Lundgren (1963) found that *Ferrobacillus ferrooxidans* corroded aluminum sheeting 3–16 hours in 9K medium which contained ferrous sulfate. They proposed the general reaction:

$$3\ Fe^{3+} + Al \rightarrow 3\ Fe^{2+} + Al^{3+}$$

and postulated that *Ferrobacillus ferrooxidans* accelerates or catalyzes this reaction. No corrosion occurred in the absence of iron.

Corrosion of aluminum can be viewed as a series of electrochemical reactions. Important reactions taking place at the anode and cathode are as follows.

Anodic Reaction:

$$\tfrac{1}{2} O_2 + 2\ H_2O + Al \leftrightarrow Al(OH)_3 + H^+ + e^-$$
$$4\ OH^- + Al \leftrightarrow AlO_2^- + 2\ H_2O + 3\ e^-$$
$$3\ OH^- + Al \leftrightarrow Al(OH)_3 + 3\ e^-$$
$$Al \leftrightarrow Al^{3+} + 3\ e^-$$
$$HSiO_3^- + 2\ Al + 2\ H_2O \leftrightarrow Al_2SiO_5 + 5\ H^+ + 6\ e^-$$

Cathodic Reactions:

$$Al(OH)_3 \leftrightarrow \tfrac{3}{4} O_2 + Al^{3+} + 1\tfrac{1}{2}\ H_2O + 3\ e^-$$
$$AlO_2^- \leftrightarrow O_2 + Al^{3+} + 4\ e^-$$
$$Al(OH)_3 + 3\ Cl^- \leftrightarrow 3\ HOCl + Al^{3+} + 6\ e^+$$

As corrosion proceeds the electrodes become polarized and tend to seek an equipotential unless influenced by catalysts which cause depolarization.

Free hydrogen-oxidizing bacteria which produce the enzyme hydrogenase increase the corrosion of aluminum by furthering cathodic depolarization.

Free hydrogen-oxidizing microbes indirectly cause corrosion of aluminum by cathodic depolarization. Where a trivalent metal (M) such as aluminum is involved, metallic hydroxide, cathodic hydrogen, and electrons are usually formed (von Wolzogen Kühr and van der Vlugt, 1953).

$$M + 2 H_2O \rightarrow M(OH)_3 + H^+ + e^-$$

The depolarization reaction follows with oxygen acting as a hydrogen acceptor with the formation of H_2O and the release of energy:

$$2 H + O \rightarrow H_2O + 6.9 \text{ kcal}$$

Removal of the hydrogen through microbial metabolism shifts the equilibrium to the right in the first reaction causing increased rates of corrosion.

Von Wolzogen Kühr and van der Vlugt (1953) showed that hydrogen oxidizers were present in tap water by running the water over plates of iron, zinc, and aluminum. The tubercles of corrosion which formed on the anodic surfaces were collected for gas utilization studies. Microorganisms present in the corrosive areas reduced an oxyhydrogen gas mixture at an average rate of 9 cc/day for several weeks. The maximum daily gas reduction was 77 cc.

In contrast, Baudon (1958), in a review of microbial corrosion phenomena, restated the view that the bacteria most commonly found in drinking water have almost no effect on aluminum except to cause a slight depolarization.

In natural flowing seawater, bacterial colonies, especially *Bacillus mycoides*, form calcite-magnesium tubercles resembling coral reefs on aluminum plates (Kalinenko, 1959). *Leptothrix* species were also active in this respect.

Colonies of bacteria apparently intensify the electrochemical process of corrosion and the base of the colony per se is the focal point for the energetic destruction of the metal. A 2.0-mm thick sheet of aluminum foil was laced with holes after 48 hours. Colonies forming on aluminum gave a distinct acid reaction in contrast to the alkaline seawater. The hollow tubercles emit hydrogen gas, possibly due to electrolysis of the seawater which releases hydrogen and deposits calcium and magnesium carbonates. These structures may act as energy reactors utilizing electric current flowing from the metal plate for synthesis of bacterial protein.

TABLE 13.I

MICROBIAL CORROSION OF ALUMINUM

Medium	Microbe	Type of aluminum	Type of corrosion	Reference
	Ceratostomella sp.	Alclad	Discoloration; pitting	Copenhagen (1950)
9K Medium	Ferrobacillus ferrooxidans	6-mil Sheeting	Pitting; holing	Dugan and Lundgren (1963)
Tap water	Hydrogen oxidizers	Plates	Tubercles	von Wolzogen Kühr and van der Vlugt (1953)
Seawater	Bacillus mycoides Leptothrix spp.	2-mm Foil	Tubercles; pitting; holing	Kalinenko (1959)
Nutrient medium containing 10% glucose	Aspergillus niger Aspergillus amstelodami Penicillium cyclopium Penicillium brevi-compactum Paecilomyces varioti	0.420-mm Oxidized wire	Disintegration; loss of breaking strength; decrease in diameter	Al'bitskaya and Shaposhnikova (1960)

Fuel/medium	Organisms	Application/Alloy	Corrosion	Reference
Jet fuel	*Pseudomonas aeruginosa* *Aerobacter aerogenes* *Desulfovibrio desulfuricans* *Penicillium luteum* *Aspergillus flavus* *Sphaerotilus natans* *Hormodendrum hordei* or *Cladosporium* spp.	Aircraft and fuel storage tanks	Pitting	Churchill (1963)
JP-4 Jet fuel Bushnell-Haas and JP-4 fuel	*Clostridium* spp. *Spicaria* spp. *Bacillus* spp. *Fusarium* spp. *Micrococcus* spp. Not identified Pseudomonads	7075-T6 7075 Alloy 2024 Alloy	Galvanic Pitting; holing; discoloration	Miller *et al.* (1964) Blanchard and Goucher (1964)
Bushnell-Haas fuel, deionized water-fuel, Bushnell-Haas cystine-fuel, sea-water medium-fuel	*Pseudomonas aeruginosa* *Desulfovibrio desulfuricans* *Cladosporium resinae* *Aspergillus niger* Mixed microflora	7158-T651 Alloy	Pitting; intergranular blister; exfoliation; grain fragmentation	Hedrick *et al.* (1963)

Conducting research on the effects of molds on materials and manufactured products, Al'bitskaya and Shaposhnikova (1960) found the breaking strength of oxidized aluminum wire (0.420 mm diameter) was weakened by acids produced by *Aspergillus niger*, *A. amstelodami*, *Penicillium cyclopium*, *P. brevi-compactum*, and *Paecilomyces varioti* after 28 days of exposure to actively growing cultures. The diameter of the wire was decreased to an average of 0.355–0.365 mm.

Thus, corrosion of aluminum metals by microorganisms is not an isolated uncommon occurrence. The complexities and broad scope are becoming more apparent as more investigators look into this problem (Table 13.I).

ALUMINUM AS A STIMULANT TO OR INHIBITOR OF MICROBIAL GROWTH

Aluminum may function as a trace element to increase growth of some microbes. Bertrand (1963) reported that 5, 10, 20, 50, and 100 μg of aluminum/liter of medium increased the dry cell weight of *Aspergillus niger* (Table 13.II). The maximum increase of 1.625 gm of cells was obtained with 50 μg of aluminum/liter of medium. Controls with no aluminum yielded a cell harvest of 1.375 gm/liter. For some cultures of *A. niger* and *Penicillium glaucum*, 0.01 % Al is required for optimal growth.

Soil dilutions plated on Thornton's agar with the addition of 0, 50, 100, 200, and 400 ppm of aluminum showed that aluminum inhibits bacterial growth at all levels (Sulochana, 1952). These results differ from those obtained with fungi. Actinomycetes were aided slightly by the addition of aluminum, with the greatest increase in growth occurring at 200 ppm. Fungal growth was increased threefold over control growth when 200 ppm of aluminum was added to the medium. Lithium was also tested and 100–200 ppm stimulated the growth of actinomycetes and increased the fungal population fivefold. In order of relative importance Li, Mn, Mo, Co, and B all increased the fungal population to a greater extent than did aluminum.

Matsuda and Nagata (1958), while testing soil, found that the number of bacteria decreased with the addition of 10 ppm of aluminum while the fungi present were not affected. Concentrations of aluminum greater than 25 ppm were toxic to *Penicillium* but not to *Aspergillus niger*. Aluminum chelated with ethylenediaminetetraacetate was toxic to *Torula* and *Acetobacter* at more than 10 ppm and concentrations in excess of 25–50 ppm were toxic to *Saccharomyces* and *Pseudomonas*.

TABLE 13.II

ALUMINUM AND MICROBIAL GROWTH

Medium	Microbe	Effect of aluminum on microbe	Amount of aluminum	Reference
Original Thornton's agar	*Aspergillus niger*	Increased growth	5–100 ppm	Bertrand (1963)
	Actinomycetes	Slightly increased growth	> 50 ppm	Sulochana (1952)
	Bacteria	Inhibited growth	> 50 ppm	
	Fungi	Greatly increased growth	> 50 ppm	
	Bacteria	Inhibited growth	> 10 ppm	Matsuda and Nagata (1958)
	Actinomyces	No change	> 10 ppm	
	Penicillium spp.	Toxic	> 25 ppm	
	Aspergillus niger	No change	> 25 ppm	
	Torula spp.	Toxic	> 10 ppm	
	Acetobacter spp.	Toxic	> 10 ppm	
	Saccharomyces spp.	Toxic	> 25–50 ppm	
	Pseudomonas spp.	Toxic	> 25–50 ppm	
Nutrient broth	*Staphylococcus albus*	Inhibited	> 6% Al_2O_3	Bünemann *et al.* (1963)
	Escherichia coli	No change	10% Al_2O_3	
	Mycobacterium tuberculosis	No change	100/ml	Kostrzenski and Poklerska-Pobratyn (1959)
Peptone-free nutrient agar	*Staphylococcus aureus*	Inhibited growth	$15 \times 35 \times 1$-mm	Berger and Einstmann (1959)
	Pseudomonas aeruginosa	Inhibited growth	Lamellae	
	Escherichia coli	Inhibited growth		

Aluminum oxide, at concentrations if 0.4, 6.0, and 10.0%, was examined for its effect on *Staphylococcus albus* and *Escherichia coli* (Bünemann *et al.*, 1963). *Staphylococcus albus* was inhibited at 6.0 and 10.0% levels, but no adverse effect was observed at 0.4%. *Escherichia coli* was not affected by these three levels of Al_2O_3. Aluminum concentrations of 100 mg/ml failed to inhibit the growth of BCG bacteria (*Mycobacterium tuberculosis*) (Kostrzenski and Poklerska-Pobratyn, 1959).

Aluminum and anodized aluminum were tested for bactericidal effects on *Staphylococcus aureus*, *Pseudomonas aeruginosa*, and *Escherichia coli* using diffusion, suspension, and germ-carrier techniques (Berger and Einstmann, 1959). Diffusion tests showed no inhibition of growth. Suspention tests showed that the presence of aluminum stopped the growth of *Staphylococcus aureus* within 4 hours, *Pseudomonas aeruginosa* within 6 hours, and *Escherichia coli* within 8 hours. Anodized aluminum had no appreciable effect on these bacteria. In germ-carrier tests neither aluminum nor anodized aluminum affected the growth of *Staphylococcus aureus*. Aluminum was toxic to both *Escherichia coli* and *Pseudomonas aeruginosa* in 24 hours in germ-carrier tests. Anodized aluminum had no effect on *Pseudomonas aeruginosa* and was only slight bactericidal to *Escherichia coli* in 24 hours.

Yeast extract (Difco) analyzed spectrographically contains an average of 3.1 μg of aluminum/gm dry weight (Grant and Pramer, 1962). This knowledge is important where aluminum-sensitive microbes are cultured, since toxicity of aluminum in parts per million has been reported (Sulochana, 1952; Matsuda and Nagata, 1958).

ALUMINUM EARTH MINERALS AND MICROBES

Aluminosilicates, which represent the most numerous group of earth crust minerals, are reported to be decomposed by soil bacteria (Aleksandrov and Zak, 1950). Some aluminosilicates such as mica, feldspar, etc., contain 13–20% potassium in insoluble form. Bacterial decomposition of these minerals is accompanied by release of potassium and other alkaline earth metals into the soil. Magnesium and iron are released more slowly, and aluminum and silicon appear during the latter stages of degradation. Two bacteria possessing this ability are *Bacillus mucilaginosus* var. *silicius* and *Bacillus megatherium* De Bary.

A mucoid bacillus isolated from aluminosilicates by Novorossova *et al.* (1947) was able to decompose microcline $K(AlSi_3O_3)$ and kaolinite

$(Al_2O_3 \cdot 2\,SiO_2 \cdot 2\,H_2O)$. Asparagine, ammonium phosphate, and sugar were required in the medium. After 10 days incubation, 1.0% SiO_2 and 0.5% Al_2O_3 were leached from kaolinite, and 0.7% SiO_2 and 0.9% Al_2O_3 were leached from microcline.

Karavaiko (1962) theorizes that bauxite deposits (essentially $Al_2O_3 \cdot 2\,H_2O$) were formed or at least concentrated by the action of sulfur-oxidizing bacteria, and particularly by *Thiobacillus thiooxidans* and *T. ferrooxidans*. Through biological oxidation of sulfur compounds to sulfuric acid these microbes contribute to sulfuric acid weathering of aluminosilicates. This weathering is a prime factor in the mobilization of alumina. Free aluminum accumulates in the lower, more alkaline areas of swamps where it is absorbed by peat. In bed-rock areas sulfuric acid leaches aluminum from the weathering zones. However, as these acidic waters reach the surface, aluminum may be precipitated as an hydroxide.

The formation of halloysite beds in Indiana as "kaolin" deposits $(Al_2O_3 \cdot 2\,SiO_2 \cdot 2\,H_2O)$ may have been due to the action of sulfate-reducing bacteria (Bucher, 1921). Bucher's scheme is summarized as follows:

1. Oxidation of iron sulfide in shale:

$$FeS_2 + 3\tfrac{1}{2}\,O_2 \rightarrow FeSO_4 + H_2SO_4 + H_2O$$

2. Action of sulfuric acid on shale:

$$x\,Al_2O_3 \cdot y\,SiO_2 \cdot z\,H_2O + 3x\,H_2SO_4 \rightarrow x\,Al_2(SO_4)_3 + y\,SiO_2(aq.)$$

3. Ferrous sulfate, aluminum sulfate, and colloidal silica in solution in sulfuric acid.
4. Reduction of sulfuric acid by bacterial action to form hydrogen sulfide:

$$H_2SO_4 + 4\,H_2 \rightarrow H_2S + 4\,H_2O$$

5. Possible reduction of aluminum sulfate to the sulfite form by bacteria:

$$Al_2(SO_4)_3 + 3\,H_2 \rightarrow Al_2(SO_3)_3 + 3\,H_2O$$

6. Removal of sulfuric acid to precipitate silica out of solution.
7. Hydrolysis and separation of aluminum hydroxide from aluminum sulfite:

$$Al_2(SO_3)_3 + 3\,H_2O \rightarrow Al_2O_3(aq.) + 3\,H_2SO_3$$

Thus, microbial activity is partly responsible for the precipitation of fine particles of silicia and alumina, in the formation of halloysite ($Al_2O_3 \cdot 2 SiO_2 \cdot H_2O$) beds.

The iron-oxidizing bacteria such as *Leptothrix ochracea* precipitates Al_2O_3 in its filaments; filaments may contain 14.7% Fe_2O_3 and up to 33.3% Al_2O_3. *Crenothrix manganifera* is reported to deposit 33.9% Mn_2O_3 in its sheath (Jackson, 1901). The mechanism of deposition of these minerals is not known.

REFERENCES

Al'bitskaya, O. N., and Shaposhnikova, N. A. (1960). The effects of molds on the corrosion of metals. *Mikrobiologiya* **29**, 725–730.

Aleksandrov, G., and Zak, G. A. (1950) Bacteria which decompose aluminosilicates (silicate bacteria). *Mikrobiologiya* **19**, 97–104.

Baudon, L. (1958). The part played by microorganisms in certain corrosion phenomena. *Ind. Chim. Belge* **23**, 983–990.

Berger, von U., and Einstmann, E. (1959). Oligodynamic action of a few newer structural metals. *Arch. Hyg. Bakteriol.* **143**, 634–640.

Bertrand, D. (1963). Aluminum—A dynamic trace element for *Aspergillus niger. Compt. Rend.* **257**, 3057–3059.

Blanchard, G. C., and Goucher, C. R. (1964). The corrosion of aluminum by microbial cultures. *Develop. Ind. Microbiol.* **6**, 95–104.

Bucher, W. (1921). Logan's explanation of the origin of Indiana's "Kaolin." *Econ. Geol.* **16**, 481–492.

Bünemann, von G., Klosterkotter, W., and Ritzerfeld, W. (1963). Effects of inorganic dusts on growth of microorganisms. *Arch. Hyg. Bakteriol.* **147**, 58–65.

Churchill, A. R. (1963). Microbial fuel tank corrosion: Mechanisms and contributory factors. *Mater. Protect.* **2**, 18–23.

Copenhagen, W. J. (1950). The pathology of metals. Corrosion of steel and alclad parts by a fungus. *Metal Ind. (London)* **77**, 137.

Dugan, P., and Lundgren, D. G. (1963). Acid production by *Ferrobacillus ferrooxidans* and its relation to water pollution. *Develop. Ind. Microbiol.* **5**, 250–257.

Grant, C. L., and Pramer, D. (1962). Minor element composition of yeast extract. *J. Bacteriol.* **82**, 869–870.

Guirard, B. M., and Snell, E. E. (1954). Pyridoxal phosphate and metal ions as cofactors for histidine decarboxylase. *J. Am. Chem. Soc.* **76**, 4745–4746.

Hedrick, H. G., Miller, C. E., Halkias, J. E., and Hildebrand, J. E. (1963). Selection of a microbiological corrosion system for studying effects on structural aluminum alloys. *Appl. Microbiol.* **12**, 197–300.

Hutchinson, G. E. (1943). The biogeochemistry of aluminum and of certain related elements. *Quart. Rev. Biol.* **18**, 1–262.

Jackson, D. D. (1901). A new species of *Crenothrix (C. manganifers) Trans. Am. Microscop. Soc.* **23**, 31–40.

Kalinenko, V. O. (1959). Bacterial colonies on metallic plates in seawater. *Mikrobiologiya* **28**, 750–756.

Karavaiko, G. I. (1962). Some points concerning the study of biological sulfuric acid weathering. Geologic activity of microorganisms. *Trans. Inst. Microbiol. (USSR) (English transl.)* **9**, 111–112.

Kostrzenski, W., and Poklerska-Pobratyn, H. (1959). Growth of BCG bacteria in trace element-supplemented mediums. *Gruzlica* **27**, 39–45.

Maslov, V. P. (1962). Stony red algae as concentrators of chemical elements. *Akad. Nauk. SSSR* **53**, 142–148.

Matsuda, K., and Nagata, T. (1958). Effects of aluminum concentration on growth of microorganisms. *Nippon Dojo-Hiryogaku Zasshi* **28**, 405–408.

Mattson, S. (1941). Laws of soil colloidal behavior. XXIII. The constitution of the pedosphere and soil classification. *Soil Sci.*, **54**, 407–426.

Miller, R. N., Herron, W. C., Krigens, A. G., Cameron, J. L., and Terry, B. M. (1964). Research program shows microorganisms cause corrosion in aircraft fuel tanks. *Mater. Protection.* **3**, 60–67.

Novorossova, L. E., Remezov, N. P., and Sushkina, N. N. (1947). Breakdown of alumino-silicates by soil bacteria. *Dokl. Akad. Nauk. SSSR* **58**, 655–658.

Rogers, M. R., and Kaplan, A. M. (1964). A survey of the microbiological contamination in a military fuel distribution system. *Develop. Ind. Microbiol.* **6**, 80–94.

Sulochana, C. B. (1952). The effect of microelements on the occurrence of bacteria, actionmycetes and fungi in soils. *Proc. Indian. Acad. Sci.* **B36**, 19–33.

Umemoto, S., and Mifune, M. (1953). Chemical studies on thermal algae (1) on inorganic constituents (1). *Okayama Daigaku Onsen Keukyusho Hokoku* pp. 11–14.

Vinogradov, A. (1935). The elementary chemical composition of marine organisms. Part I. *Trav. Lab. Biogeochim. Acad. URSS Sci.* **3**, 67–278.

von Wolzogen Kühr, C. A. H., and van der Vlugt, L. S. (1953). Aerobic and anaerobic iron corrosion in water mains. *J. Am. Water Works Assoc.* **45**, 33–46.

Young, R. S. (1935). Certain rarer elements in soils and fertilizers and their role in plant growth. *Cornell Univ., Agri. Expt. Sta. Mem.* **174**, 1–70.

14

BIOGEOCHEMISTRY
OF SILICON

GEOCHEMISTRY

In the upper lithosphere, oxygen is the first and silicon the second most abundant element. Silicon is to the mineral world what carbon is to the organic. Silicon and carbon are similar in that both combine with oxygen to form a variety of structures. The framework of the silicate minerals consists of a basic unit of a silicon-oxygen tetrahedron (SiO_4). Silicon is an essential cationic constituent in most major minerals. The radius of Si^{4+} is 0.34 kX. Its spatial demands are extremely small compared to the larger oxygen anions. Silicon is highest in granitic minerals. The repulsive force between charged Si^{4+} ions increases with the number of oxygen ions shared by the tetrahedron. Two neighboring tetrahedra (SiO_4) rarely share more than one oxygen ion, e.g., olivine $(Mg,Fe)_2(SiO_4)$. Of the silicates olivine, which belongs to the orthosilicates, is the poorest in silica. Feldspars, the major constituents of igneous rock, contain most of the silica found in the upper lithosphere. These silicates are classified as tectosilicates and crystallize at lower temperatures than do olivines which are nesosilicates.

In nature the greatest solubilization of silica and dissolution of silicate minerals should occur where proteolysis is taking place. Basic amines released during proteolysis should be good eluting agents. Ionic silica solutions are best prepared under alkaline conditions. The higher the pH the greater the solubility. When silica becomes concentrated, amorphous gels precipitate. Organic materials may help stabilize silica hydrosols by forming protective colloids. Microbial decomposition of the organic phase and CO_2 production will destroy the organic material, lower the pH, and precipitate the silica. Magnesium salts accelerate the solution and weathering of silica. Clay minerals are formed during weathering. Surprisingly enough, the silicon becomes enriched in such argillaceous sediments because of the loss of water. Sandstone sediments high in quartz may contain as much as 90% SiO_2 compared to 59% in igneous rocks. Of the resistates quartz has no cleavage to promote its disintegration. Inland and ocean waters contain silicon in true colloidal solution. River water may contain 5.5% Si, whereas this decreases 10,000 times in ocean water (2–5 ppm). Silicon is removed from the sea by the formation of silicious sediments by plant and animal action.

The primary oxidation state is Si^{4+}. Silicon dioxide forms several acids. The orthoforms of the di-, tri-, and tetraacids are $H_4Si_2O_7$, $H_8Si_3O_{10}$, and $H_{10}Si_4O_{13}$, respectively. Many dehydrated forms such as metamonosilicic acid are known. In aqueous solution these acids present complicated equilibria and although the difference between the free energies of the various polyacids are small, it is likely that it is sufficient to support microbial growth. Manufacturers of colloidal solutions of silica gel encounter microbial growth so frequently in these solutions that preservatives must be added. Just what chemical reactions are involved in microbial growth in these solutions are not known.

The bonds between atoms in solid silicon are so strong and the activation energies for reactions involving the free element are so high that the oxidation-reduction couples of silicon are not thermodynamically reversible at low temperature. Silicon itself is not stable and is rarely, if ever, found as such in nature. It disproportionates with water to form a gaseous silica tetrahydride and silica dioxide. Note that these hydrides are similar to gaseous methane. Since SiO_2 is one of the predominant forms found in nature, one assumes that it is stable and that the thermodynamics greatly favor its formation.

The silicate minerals are classified into five groups according to the type linkage of the silicon oxygen tetrahedra: nesosilicates, sorosilicates, inosilicates, phyllosilicates, and tectosilicates (Rankama and Sahama, 1960). The more common silicate minerals are listed in Table 14.I.

TABLE 14.I

COMMON SILICATE MINERALS

Mineral	Formula
Forsterite	$Mg_2(SiO_4)$
Grossularite	$Ca_3Al_2(SiO_4)_3$
Andalusite	$Al_2(O)(SiO_4)$
Hemimorphite	$Zn_4[(OH)_2(Si_2O_7)H_2O]$
Wallastonite	$Ca_3(Si_3O_9)$
Benitoite	$BaTi(Si_3O_9)$
Cordierite	$Mg_2Al_3(AlSi_5O_{18})$
Sillimanite	$Al(AlSiO_5)$
Pyroxenes (mineral group)	
Amphiboles (mineral group)	
Talc	$Mg_3[(OH)_2(Si_4O_{10})]$
Mica (mineral group)	
Kaolinite	$Al_4[(OH)_8SiO_{10})]$
Feldspars, e.g., orthoclase, microcline, sanidine, abite, anorthite	
Feldspathoids (mineral group)	
Zeolites (mineral group)	
Moissanite	SiC

The complex humic acid complexes derived by microbial plant decomposition are reputed to be strong agents in degrading rocks. Contrary to early suggestions, they do not necessarily solubilize silica unless in alkaline solutions. Oxidation of humic acids by microbes to yield carbon dioxide or carbonic acid will normally cause precipitation of any colloidal silica.

The high concentration of leachable silica in granitic rocks may have some evolutionary significance. When early life appeared sedimentary deposits were limiting and igneous rocks predominated. Fresh waters high in silica supplied a natural enrichment favoring development of diatoms, radiolarians, foraminifera, siliceous sponges, etc. Silicon rather than carbonate or phosphate served as the primary substrate for synthesis of solid supporting structures. The complexities involved during this evolutionary development, and the present-day importance of biochemistry in explaining the transition of silicon through its many and varied states, is not appreciated.

MICROBIAL CHEMISTRY OF SILICON

Bacteria

Webley *et al.* (1960) prepared insoluble silicates of calcium, zinc, and magnesium. These silicates were ball milled and incorporated into molten agar, and the degradation of the silicates by soil bacteria studied. Wollastonite [Ca(SiO$_3$)], apophyllite [Ca$_4$K(Si$_4$O$_{10}$) F · 8 H$_2$O], and olivine [(Mg,Fe)$_2$SiO$_4$] were ground and also evaluated. A culture of *Pseudomonas* was found to destroy the crystalline structure of the minerals. Products formed during growth of *Pseudomonas* were 2-ketogluconic acid, calcium, zinc, and magnesium.

Although the silica in silicate minerals may not be attacked directly, many silicates are solubilized through microbial action (Webley *et al.*, 1963). Rock specimens selected from areas in northeast Scotland heavily colonized with lichens, bacteria, fungi, and plants contain a high percentage of microbes which solubilize silicates (Table 14.II). Of 149 cultures of fungi tested, the percent dissolution of Ca-silicate, Mg-silicate, and Zn-silicate average, respectively, 94, 76, and 96%. Fungi were more active than bacteria and actinomycetes in solubilizing silicates. Bacteria and actinomycetes isolated from soils give an average solubilization of 83–87% of Ca-silicate. The bacteria were more active than actinomycetes in solubilizing wollastonite and Mg-silicate.

Addition of Na$_2$SiO$_3$ to a glucose-peptone mineral salt medium permitted limited growth of *Bacillus subtilis* and *Escherichia coli* (Heinen,

TABLE 14.II

INCIDENCE OF SILICATE-DISSOLVING BACTERIA, ACTINOMYCETES, AND FUNGI
ISOLATED FROM ROCK SEQUENCES AND WEATHERED STONE

Organism	Total isolates	Percent of mineral dissolved			
		Ca-silicate	Wollastonite	Mg-silicate	Zn-silicate
Bacteria	265	83	57	65	NT
Actinomycetes	39	87	38	46	NT
Fungi	149	94	NT^a	76	96

^{a} NT = not tested. (After Webley *et al.* 1963.)

1960). *Serratia marcescens* would not grow on this medium. Uptake of silica was highest between pH 6.6–7.5 at a concentration of 10^{-3} M SiO_2. In decreasing order of preference, silica uptake was observed from Na_2SiO_3, silica gel, and quartz.

Bassalik (1914a) carefully isolated and identified species of *Pseudomonas Bacillus, Bacterium, Vibrio, Micrococcus, Sarcina,* and *Streptothrix* from the intestines of the earthworm *Lumbricus terrestris* which decomposed aluminosilicates. Organic bouillon-gelatin or urea-gelatin type media supplemented with minerals and silica were used in the isolation. These bacteria corroded polished marble and penetrated the lamellae of mica. Feldspar was added and provided an adequate supply of minerals for growth. Orthoclase was degraded. *Bacillus extorquens* was one of the most active bacteria. Organic acid-producing *Clostridium pasteurianum* partially solubilized apatite which, although not a silicate, is difficult to solubilize. Of twelve silicates tested *Bacillus extorquens* gave the highest degradation of magnesium mica; this was followed by nepheline, potash-mica, leucite, and olivine (Bassalik, 1914b). *Bacillus extorquens* grows over the grainy surface of mica encasing any grainy particles and breaking them into finer laminae. With other bacteria leucite and nepheline were easy to degrade and orthoclase the most difficult.

Common soil bacteria such as *Azotobacter* utilize only a small amount of ^{31}Si from media (Keeler and Varner, 1958). Germanium did not effect silica uptake although it inhibited *Azotobacter* growth. Silica had no effect on the uptake of ^{99}Mo. When cells grown on a ^{31}Si-medium were sonically disrupted and centrifuged, 85% of the ^{31}Si was located in the supernatant and 5.0% in the sedimentable fraction.

Some soil microbes carry out an exchange reaction in which Si is absorbed and phosphate excreted from the cell (Heinen, 1963a,b,c). Under controlled conditions 3 hours were required to adapt cells. The highest exchange rate was obtained on addition of 1.2 mg/ml of glucose. Silica uptake was inhibited by cadmium chloride but not by dinitrophenol. Silica is lost from cells in media high in either phosphate or carbonate.

The potassium complexed with aluminosilicates is probably not assimilated directly by higher plants; if it is, assimilation is slow. Although reports are varied, the silicate bacteria are purported to decompose aluminosilicates, mobilizing potassium for plant utilization. In decomposing aluminosilicates, the alkali earth metals appear first in solution. Potassium, when present, solubilizes next, and magnesium and iron follow. The last elements to appear in solution are silicon and aluminum (Aleksandrov and Zak, 1950). The most active bacteria isolated on silicate agar medium were *Bacillus mucilaginosus* var. *silicius* and *Bacillus megaterium.* The cultures were isolated from the Asiatic and black soil of the Kuyhyshev region. *Bacillus mucilaginosus* grew best without addition of nitrogen to the medium. Low-nitrogen media induced the formation of large mucoid colonies which spread over the surface of the experimental medium.

Bacteria per se degrade aluminosilicates and use the potassium for nutritional purposes (Tesic and Todorovic, 1959). Atmospheric nitrogen is believed to be fixed since only trace quantities of nitrogen were present in the cultivation medium. Potassium is also released into the surrounding soil. Phosphorus was solubilized from phosphorites and granite powder. Assimilable silicon was reported as the most important element for growth of these bacteria at pH 2.0–3.0.

Fungi

Potassium is obtained from clay minerals by *Aspergillus niger*. Not all clays possess sufficient potassium to support growth. Biotite and muscovite supply the greatest amounts of potassium, whereas green sand and microcline provide the least to fungi (Eno and Reuszer, 1955).

Species of *Spicaria, Penicillium, Botrytis, Cephalosporium, Fusarium, Hormodendron, Mucor, Trichoderma,* and *Aspergillus niger* decompose silicate-containing saponite and vermiculite (Henderson and Duff, 1963). Dissolution, observed by the formation of clear zones around fungal colonies, was attributed to citric and oxalic acids which caused active release of metallic ions and silica from rocks.

BIOGEOPROSPECTING

Prospecting for silicates is not a problem, and it is doubtful that bioprospecting for this element will ever be important. However, Wlodnimierz (1963) presents an interesting discussion on using the Si/Ca ratio in plants and animals as an index to ontogenetic age and phylogenetic position. Silica itself is an important evolutionary factor in skeletization, starting with the Infusoria or protozoans and going back through the phylogenetic series.

Silica is absorbed and deposited in most plants. It is believed to be essential for the growth of barley, sunflower, and beets. As shown in bacteria culture, it may be involved in phosphorus metabolism. Silica deficiencies in plants have been correlated with increased susceptibility to plant disease (Lanning and Linko, 1961). For example, of our varieties of sorghum tested, *Spur feterita* and Pink kafir whose content of silica always tested higher than Atlas and Dwarf, were more resistant to smut, Milo disease, and chinch bugs.

Aleksandrov (1949) increased the yield of corn 67% and wheat 56% by inoculating soils with a suspension of silicate bacteria. Straw formation was not as high. The amount of potassium assimilated by these plants upon fertilization with silicate bacteria increased 5–30 fold. The decomposition of aluminosilicates by silicate bacteria was continuous and gave a steady supply of potassium.

REFERENCES

Aleksandrov, V. G. (1949). The role of silicate bacteria in the mobilization of potassium in the soil and the increase in yield of spring wheat and corn. *Dokl. Vses. Akad. Sel'skohoz. Nauk*. **14**, 12–19.

Aleksandrov, G., and Zak, G. A. (1950). Bacteria which decompose aluminosilicates (silicate bacteria). *Mikrobiologiya* **16**, 97–104.

Bassalik, K. (1914a). Silicate decomposition by soil bacteria. I. Activity of earthworms in regard to soil bacteria. *Ze. Gaerungsphysiol*. **2**, 1–30.

Bassalik, K. (1941b). Silicate decomposition by soil bacteria and yeast. II. *Ze. Gaerungsphysiol*. **3**, 15–42.

Eno, C. F., and Reuszer, H. W. (1955). Potassium availability from biotite, muscovite, greensand, and microcline as determined by growth of *Aspergillus niger*. *Soil Sci*. **80**, 199–209.

Heinen, W. (1960). Silicon metabolism in microorganisms. I. Silicon uptake by bacteria. *Arch. Mikrobiol*. **37**, 199–120.

Heinen, W. (1963a). Silicon metabolism in microorganisms. III. Influence of different anions on bacterial Si metabolism. *Arch. Mikrobiol*. **45**, 145–161.

Heinen, W. (1963b). Silicon metabolism in microorganisms. IV. Effect of organic compounds, especially glucose, on silicon metabolism in bacteria. *Arch. Mikrobiol*. **45**, 162–171.

Heinen, W. (1963c). Silicon metabolism in microorganisms. V. Mobility of incorporated silicic acid. *Arch. Mikrobiol*. **45**, 172–178.

Henderson, M. E. K., and Duff, R. B. (1963). The release of metallic and silicate ions from minerals, rocks and soils by fungal activity. *J. Soil Sci*. **17**, 236–246.

Keeler, R. F., and Varner, J. E. (1958). Silicate in the metabolism of *Azotobacter vinelandii Nature* **184**, 127.

Lanning, F. C., and Linko, Y-Y. (1961). Absorption and deposition of silica by four varieties of sorghum. *Agr. Food Chem*. **9**, 463–465.

Rankama, K., and Sahama, T. G. (1960). "Geochemistry." Univ. of Chicago Press, Chicago, Illinois.

Tesic, Z. P., and Todorovic, M. S. (1959). Specific properties of silicate bacteria. *Zemljuste Bilkja* **7**, 233–238.

Webley, D. M., Duff, R. B., and Mitchell, W. A. (1960). A plate method for studying the breakdown of synthetic and natural silicates by soil bacteria. *Nature* **188**, 766–767.

Webley, D. M., Henderson, M. E. K., and Taylor, I. F. (1963). The microbiology of rocks and weathered stones. *J. Soil Sci*. **14**, 102–112.

Wlodzimierz. (1963). The silicon component in living cells and organisms. *Kosmos (Warsaw)* **A12**, 497–504.

15

MICROBIAL CORROSION AND OXIDATION OF METALLIC IRON

In its broadest concept, corrosion is an electrochemical reaction. It is caused by any kind of heterogeneity at different points on a metal surface. Once a surface makes contact with an electrolyte, however weak, different heterogenous points show a variance in potential. They develop an electromotive force (emf) and form small local piles or corrosion cells. The more reactive metal acts as the anode, and its atoms pass into the ionic state; at the same time, hydrogen ions are driven back to the less reactive metal which acts as the cathode. When hydrogen forms at the cathode, the corrosion cell becomes polarized. The surface seeks an equipotential state and as this state is reached, metallic dissolution of the anode follows and continues until the emf decreases and corrosion stops. Outside forces, of course, can intervene and it is these intervening forces which sustain or increase corrosion.

For the corrosion of a solid metal, there must be water or moisture in contact with the metal and the water must contain ions to act as an electrolyte (Fig. 15.I). To form a corrosion cell, an electrical conductor

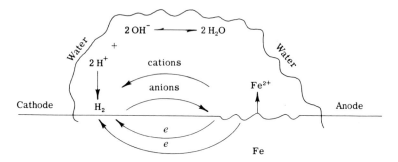

FIG. 15.1. Sketch of corrosion process.

is required and this is usually the metal itself. Water is corrosive when it dissolves the metal to yield positive ions or furnishes constituents which react with the metal interface. The corrosion process is oxidative and it releases electrons which, as described previously, flow back toward the metal. This latter reaction is reductive and results in electron capture by the metal. Thus, oxidation occurs at the anode while reduction occurs at the cathode. These reactions may only be molecules apart.

The reactions at the electrode are called half-cell or single electrode reactions. Reactions at the anode require oxygen or some other oxidant, e.g., Cl_2, NO_3, NO_2, etc.

	$\Delta F°25$	$E°25\ (volts)$
$\frac{1}{2} O_2 + 2 H_2O + Fe \rightharpoonup Fe(OH)_3 + H^+ + e^-$		2.37
$2 OH^- + Fe \rightharpoonup Fe(OH)_2 + 2 e^-$	-40.0 kcal	0.877
$CO_3^{2-} + Fe \rightharpoonup FeCO_3 + 2 e^-$		0.755
$HSiO_3^- + Fe \rightharpoonup FeSiO_3 + H^+ + 2 e^-$		0.567
$Fe \rightharpoonup Fe^{2+} + 2 e^-$	-20.3 kcal	0.440

Common cathodic reactions involved are:

H_2(hydrogen polarization) $\rightharpoonup 2 H^+ + 2 e^-$	0.00
$Fe(OH)_3 \rightharpoonup \frac{3}{4} O_2 + Fe^{2+} + \frac{3}{2} H_2O + 2 e^-$	-1.386
$3 Cl^- + Fe(OH)_3 \rightharpoonup 3 HOCl + Fe^{2+} + 5 e^-$	-1.58
$H_2O \rightharpoonup \frac{1}{2} O_2 + 2 H^+ + 2 e^-$	-1.272

Lead, copper, zinc, aluminum, tin, nickel, and chromium metals undergo similar type reactions. The general reaction would be:

$$\text{Metal} \underset{\text{Oxidation}}{\overset{\text{Reduction}}{\rightleftharpoons}} \text{Cations} + \text{Electrons}$$

Metals can be made more resistant or more passive to corrosion. Passivity results when an oxide film forms on the metal surface. Chemical passivators are nitric acid and sodium nitrate. For passivators to be effective, they must be present in concentration of 100–1000 ppm. It should be mentioned that concrete, cement mortar, and asbestos cement are also subject to corrosion. This corrosion is primarily by leaching.

In corrosion microbes contribute in many ways:

1. Sulfur oxidizers form sulfuric acids.
2. Heterotrophs form organic acids.
3. They change the E_h or electrode potential (basis of oxygen differential cell).
4. They create microgalvanic cells.
5. They depolarize surfaces by oxidizing hydrogen.
6. Sulfate reducers produce H_2S which is itself corrosive.

Von Wolzogen Kühr and van der Vlugt (1934, 1953) proposed a theory for the anaerobic corrosion of iron by bacteria. The reaction involved may be presented schematically. Species of the genus *Desulfovibrio* produce the enzyme hydrogenase which removes hydrogen from the surface of steel (Fig. 15.2).

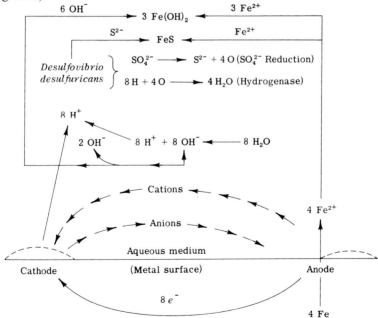

FIG. 15.2. Microbial corrosion of metallic iron.

The anaerobic microorganisms producing hydrogenase are most active. In the process, sulfate, which acts in place of oxygen as a terminal electron acceptor, is reduced. The electrons removed from the cathode as a result of hydrogen utilization permit more iron to be dissolved which increases the corrosion at the anode. The hydrogen sulfide and hydroxyl ions formed combine with the ferrous ions to form secondary reactions products at the anode.

One of the key mechanisms in the above theory is the removal of hydrogen or electrons at the cathode and the dissolution of ferrous iron at the anode. Iverson (1966) using cells of *Desulfovibrio* at the cathode and no cells at the anode, substituted the dye, benzyl viologen, for sulfate as an electron acceptor. The formation of hydrogen sulfide was avoided and the reduction of the dye was followed by the formation of a violet color. With the use of 1010 steel coupons (density 7.85) on an iron agar medium and quite specific conditions, the formation of ferric ion was demonstrated by the addition of potassium ferricyanide. A heavy concentration of ferrous ions was found at the anode (at which there were no cells) as compared to the cathode which was in contact with the cells.

Three reactions which take up electrons or act as "electron sinks" were proposed:

1. The combination of electrons with residual oxygen and water to form OH^- ions, which react with Fe^{2+} ions and more oxygen to form an "edge" effect [in (c), the carbon dioxide usually contained in water reacts with the OH^- ions]:
   (a) $2 e^- + H_2O \rightarrow H_2 + 2 OH^-$ (acid electrolytes)
   (b) $2 e^- + 0.5 O_2 + H_2O \rightarrow 2 OH^-$ (neutral salt solutions)
   (c) $2 e^- + 0.5 O_2 + H_2O + 2 CO_2 \rightarrow 2 HCO_3^-$
2. The direct uptake of electrons by oxidized benzyl viologen to give the violet reduced form of the dye.
3. The reduction of H^+ ions at the iron surface to form molecular or atomic hydrogen which is then removed

$$2 H^+ + 2 e^- \rightarrow H_2$$

(released electrons reduce benzyl viologen).

Two electrodes, with a surface area of 1.1 cm^2, encased in Lucite and placed on the agar surface, were used with a sensitive Hewlett-Packard vacuum tube voltmeter to measure electron flow. No current flow was observed without microbes. Over a 9-hour test period in the presence of microbes, a sustained current density of 1.0 $\mu A/cm^2$ was observed. This gave a calculated corrosion rate of 2.5 mg/dm^2/day (mdd) or about 0.00046 inches/year (ipy).

$$mdd = ipy \times 696 \times density$$

Aluminum was also cathodically depolarized. The more noble metals such as tin, zinc, and lead were resistant.

Booth and Tiller (1960) have shown that the enzyme hydrogenase is directly associated with the ability to bring about cathodic depolarization; no cathodic depolarization was observed with *Desulfovibrio orientis*, a microbe which does not produce hydrogenase. The same phenomena was demonstrated with hydrogenase- and nonhydrogenase-producing strains of *Clostridium nigrificans* which also reduce sulfate. Both strains of clostridia caused anodic polarization due to the production of partially protected sulfide films. Cathodic polarization was observed only when hydrogenase was present.

These studies support a view that the ability of anaerobic microorganisms to utilize hydrogen is a more important factor in microbiological corrosion than sulfate reduction per se.

The primary internal corrosion reaction of pipes and other iron materials is

$$Fe + 2\,H_2O \leftrightarrow Fe(OH)_2 + 2\,H^+ + 2\,e^-$$

If the hydrogen and ferrous iron are continually removed, equilibrium of the reaction is to the right. Under aerobic conditions, oxygen reacts in this system to form ferric hydroxide and H_2O. The chemical oxidation of ferrous iron and removal of hydrogen is slow, especially if oxygen is limiting. Corrosion is increased by biochemical oxidation. Both the iron bacteria and oxyhydrogen bacteria catalyze reactions which cause corrosion. The oxidation of hydrogen to water by microbes containing hydrogenase occurs at the cathodic iron surface and the oxidation of ferrous to ferric iron is catalyzed by the iron bacteria in the anodic area (von Wolzogen Kühr and van der Vlugt, 1953). In localized spots, iron rust, bound by iron bacteria and strengthened by incrustations of calcium carbonate and manganese dioxide, form resistant masses and induce tubercle formation.

Tubercles are a visual manifestation of anodic action at the internal iron surface. Whereas the initial oxidation of iron was aerobic and supported an iron-oxidizing microbial population, gradually anaerobic conditions are established and the sulfate-reducing microbes become predominant.

Schematically, iron corrosion can be represented as follows:

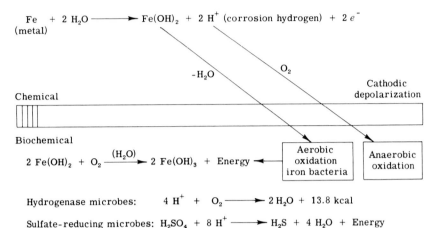

$$\begin{array}{l} \text{Fe} + 2\,H_2O \longrightarrow Fe(OH)_2 + 2\,H^+ \text{ (corrosion hydrogen)} + 2\,e^- \\ \text{(metal)} \end{array}$$

Chemical

Biochemical

$$2\,Fe(OH)_2 + O_2 \xrightarrow{\;(H_2O)\;} 2\,Fe(OH)_3 + Energy$$

Aerobic oxidation iron bacteria

Anaerobic oxidation

Cathodic depolarization

$-H_2O$ O_2

Hydrogenase microbes: $4\,H^+ + O_2 \longrightarrow 2\,H_2O + 13.8\ \text{kcal}$

Sulfate-reducing microbes: $H_2SO_4 + 8\,H^+ \longrightarrow H_2S + 4\,H_2O + Energy$

The initial corrosion reaction remains in heterogenic equilibrium and follows the "principle of labile equilibrium" of van't Hoff and Le Chatelier which states that if both reaction products are removed the reaction will proceed to the right. Aerobic and anaerobic microbes containing hydrogenase are implicated in the removal of corrosion hydrogen. The actual removal of this hydrogen is called depolarization. When hydrogen is removed from the iron surface, the iron goes into solution as ferrous ions, and electrons are liberated according to the anodic reaction:

$$Fe \rightarrow Fe^{2+} + 2\,e^-$$

The electrons from the anode flow to the cathode where they are accepted by hydrogen ions to form hydrogen gas.

$$2\,H^+ + 2\,e^- \rightarrow H_2$$

Numerous bacteria containing the enzyme hydrogenase are active in catalyzing the oxidation of hydrogen (Table 15.I).

Desulfovibrio desulfuricans

$$4\,H_2 + SO_4^{2-} \rightarrow S^{2-} + 4\,H_2O + Energy$$

Clostridium aceticum

$$4\,H_2 + 2\,CO_2 \rightharpoonup CH_3COO^- + H_2O + H_3O + Energy$$

Hydrogenomonas facilis

$$2\,H_2 + O_2 \rightarrow 2\,H_2O + Energy$$

Micrococcus denitrificans

$$4 H_2 + NO_3^- \rightarrow NH_3 + 2 H_2O + OH^- + \text{Energy}$$

Methanobacterium omelianskii

$$4 H_2 + CO_2 \rightarrow CH_4 + 2 H_2O + \text{Energy}$$

It should be reemphasized that iron corrosion should be more specifically related to hydrogenase activity (Tiller and Booth, 1962) than to sulfate reduction.

Other chemical reactions may occur following tubercle formation. For example, ferric compounds are good hydrogen acceptors.

$$Fe(OH)_3 + H \rightarrow Fe(OH)_2 + H_2O$$

If bicarbonate is present in water, the $Fe(OH)_2$ readily changes to soluble iron carbonate:

$$2 Fe(OH)_2 + 2 HCO_3^- \rightarrow Fe(HCO_3)_2 + 2 OH^-$$

When sulfate reduction itself occurs in the tubercle, the ferric ion may be reduced.

$$2 Fe(OH)_3 + 3 H_2S \rightarrow 2 FeS + S + 6 H_2O$$

Some pyrite also forms.

$$FeS + S \rightarrow FeS_2$$

By analysis, the tubercles are shown to be hydrated ferrous-ferric oxide mixed with ferric hydroxide, along with FeS, H_2S, FeS_2, and, possibly, free sulfur.

TABLE 15.I

HYDROGENASE-CONTAINING
BACTERIA[a]

Azotobacter vinelandii
Aerobacter aerogenes
Aerobacter cloacae
Clostridium aceticum
Clostridium sporogenes
Clostridium welchii
Desulfovibrio desulfuricans
Escherichia coli
Hydrogenomonas facilis
Methanobacterium omelianskii
Proteus vulgaris

[a] After Sharpley (1961).

CORROSION THEORY

If two metals of different composition are exposed to an electrolyte and connected externally by an electrical conductor, galvanic current flows from one metal to the other. It is important to know that galvanic current also flows between electrodes of identical composition if they are exposed to different electrolytes connected by a salt bridge. Under these conditions the anodic member of the circuit will suffer a loss of weight. The amount of loss is directly related to the galvanic current according to Faraday's Law. In corrosion electrochemistry this relation is conveniently expressed by:

$$W_a = \frac{I\,t}{F}\,Ja$$

where W_a = weight loss at the anode (grams), I = galvanic current (amp), t = time (seconds) of current flow, F = Faraday (96,500 coulombs), Ja = gram-equivalent weight of anodic metal.

It is not always satisfactory to study weight loss of galvanic corrosion as it is related to current and Faraday's Law. In some experiments it is impossible to duplicate the natural circuit. Under these conditions investigations are based upon galvanic current caused by polarization at one or both electrodes. Wesley and Brown's (1948) approach to this situation is of fundamental importance.

The potentials used are oxidation potentials. The relationship between *potential*, *current*, and *resistance* in galvanic cells is expressed by:

$$E_a - E_c = I(R + R_i)$$

where E_a = polarized potential of the anode, E_c = polarized potential of the cathode, I = galvanic current flow through cell, R = resistance of the metallic portion of the circuit (external resistance), and R_i = resistance of the electrolytic portion of the circuit (internal resistance).

Emf measurements made with a potentiometer are called open circuit potentials because, at the balance point, opposing voltages are equal and no current flows in the circuit being measured. The open circuit potential at the cathode (E_c') and anode (E_a') are related to their polarized potential by the following expressions:

$$E_c = E_c' + f_c(I/A_c)$$
$$E_a = E_a' - f_a(I/A_a)$$

where f_c and f_a = polarizing functions of cathode and anode, respectively, and A_c and A_a = areas of cathode and anode, respectively.

Polarization is defined as the change in the potential of an electrode resulting from current flow, i.e., the potential of the anode in the anodic direction as the galvanic current increases. This is illustrated by polarization curves. The electrodes in a galvanic circuit cannot be polarized indefinitely. The theoretical maximum galvanic current is obtained by plotting I against E_c and E_a. The intersection of the cathodic and anodic polarization curves shows the maximum current.

Under neutral or acid electrolytic conditions one of the common causes of polarization is the electrodeposition of hydrogen upon the cathode. Hydrogen is adsorbed in the atomic state and cannot be evolved unless the cathode potential is equal to or greater than its equilibrium potential plus its hydrogen overvoltage. Adsorbed hydrogen on the metallic surface and hydrogen ions in the electrolyte form the basic components of a gas electrode, a situation analogous to the hydrogen electrode. Such electrodes produce an emf in opposition to that of the cell with a resulting decrease in galvanic current. This essentially describes the mechanism of cathodic polarization by adsorbed hydrogen.

Cultures using hydrogen, especially those of the sulfate-reducing bacteria, have been shown to use hydrogen which has been adsorbed on cathodic surfaces more readily.

In spite of the fact that the experimental methods used are quite precise, corrosion testing must be approached statistically. For the corrosion engineer, the conditions used should duplicate those encountered in the field. In laboratory tests duplicating conditions found in nature, the galvanic currents obtained have been too low to be regarded as the cause of serious corrosion (Bradley, 1954).

Rozenberg and Ulanovskii (1960) studied the growth of B. megaterium, P. fluorescens, and B. mycoides on cathodic surfaces in waters of the Black Sea. Current densities were varied up to 0.25 mA/cm². Optimal growth of these bacteria occurred on the cathodic surface between 0.08–2.0 mA/cm². Growth decreased tenfold at 2.5 mA/cm².

OXYGEN DIFFERENTIAL CELLS

One of the simplest types of corrosion is produced during growth of the aerobic microbes in a particular environment. The area in which bacteria are most active has less oxygen than adjacent areas supporting little or no bacterial activity. The difference in oxygen concentration between two such areas produces an oxygen differential cell. The oxygen differential cell will persist in an area as long as bacteria use oxygen faster than new oxygen can diffuse into and stabilize the system.

Oxidation-reduction potentials have been investigated in two primary areas by microbiologists; first, in reactions within the cell and second, in oxidation-reduction potential changes in the medium during cellular growth. The oxidation-reduction potential of intracellular and enzyme-substrate systems are determined by actual measurement or by thermodynamic considerations. The E_o value is derived from the Gibbs free energy equation.

$$E_o = \frac{RT}{nF} \ln \frac{[\text{Oxidized compounds}]}{[\text{Reduced compounds}]}$$

The value of E_o represents the potential with respect to a normal hydrogen reference electrode when the concentration of the particular oxidized and reduced compounds are equal. From this it has been established that the electrical potentials encountered within living bacteria are within the same range as those involved in metal oxidations of interest to the corrosion engineer.

The measurement of electrical potentials of the medium of growing bacteria have shown they produce reduced conditions. Potentials reported vary from $+100$ to -400 mV (Hewitt, 1950).

The iron- and sulfur-oxidizing bacteria which synthesize sulfuric acid actually produce one of the most common and corrosive solvents known to man. Hydrogen sulfide, itself produced by sulfate-reducing bacteria, dissociates in water to give a weak acid. Carbon dioxide, synthesized as an end product of carbohydrate oxidation, hydrolyzes to give carbonic acid. Nitrous and nitric acids produced by nitrifying bacteria are corrosive. The organic acids formed in nature by microbial degradation and oxidation of organic matter yield a large and comprehensive list of mono-, di-, and tricarboxylic acids, most of which are corrosive and increase the solubilization of cations.

BACTERIA ASSIMILATION OF IONIC IRON

Ehrenberg in 1836 observed that microbes were active in the formation of ferruginous deposits in waters containing soluble iron. These ferruginous bacteria are filamentous forms which accumulate iron in slimy sheaths.

They have been studied by many workers, but they need a thorough investigation to clarify their biochemistry and taxonomy. Water containing 0.2–0.3 ppm iron will almost always contain ferruginous bacteria. Many of these bacteria require organic matter for growth, and iron does not appear to be a vital metabolite. Iron ions are assimilated, and, in the presence of oxygen, the iron oxide formed gives the filaments a brown color. Increased oxidation yields insoluble ferric hydroxide.

$$4 \text{ Fe}^{2+} + 3 \text{ O}_2 + 6 \text{ H}_2\text{O} \rightarrow 4 \text{ Fe(OH)}_3$$

Some of these microbes assimilate manganese as well as iron. The more important microbes involved in precipitating or solubilizing iron are shown in Table 15.II (Silverman and Ehrlich, 1964).

TABLE 15.II

MICROBES WHICH PRECIPITATE OR
SOLUBILIZE IRON

Microbe	Nutrition
Clonothrix[a]	Heterotrophic
Clostridium nigrificans	Heterotrophic
Crenothrix[a]	Heterotrophic
Desulfovibrio	Heterotrophic
Ferribacterium[a]	Heterotrophic
Ferrobacillus	Autotrophic
Gallionella	Autotrophic
Leptothrix[a]	Heterotrophic, autotrophic
Micromonospora	Heterotrophic
Naumaniella	Heterotrophic
Ochrobium	Heterotrophic
Siderobacter	Heterotrophic
Siderocapsa	(?)
Sideromonas[a]	Autotrophic, heterotrophic
Sideronema[a]	Heterotrophic
Siderosphaera	Heterotrophic
Siderophacus	Heterotrophic
Sphaerotilus	Heterotrophic
Thiobacillus ferrooxidans	Autotrophic
Toxothrix	Heterotrophic
Bacillus[a]	Heterotrophic

[a] After Silverman and Ehrlich (1964).

The filamentous forms of iron bacteria are not as active in the initial phases of biochemical corrosion of iron metals as the other forms discussed. They normally do not attack the metal directly. Most are regarded as a nuisance and are found plugging or fouling water pipes, e.g., *Leptothrix ochracea, Sphaerotilus natans, Crenothrix polyspora,* etc. These forms do aid in corrosion by oxidizing ferrous to ferric iron.

$$4 \, FeCO_3 + O_2 + 6 \, H_2O \rightarrow 4 \, Fe(OH)_3 + 4 \, CO_2$$

In oil-field water injection systems *Desulfovibrio, Pseudomonas, Sphaerotilus, Bacillus,* and *Clostridium* are the predominant microbes. They are implicated in hydrogen-sulfide production, metal corrosion, slime formation, hydrocarbon oxidation, and plugging. For a comprehensive investigation of the microbes involved in these processes see Carlson *et al.* (1961) and Allred (1954).

Algae are very active in generating the oxygen present in ground water. They are also known to synthesize potent organic oxides which should aid in iron oxidation (Fig. 15.3). The involvement of manganese in iron metabolism by microbes is quite well known. Bromfield (1956) has observed the oxidation of the manganous ion to the manganic state by *Chromobacterium* and *Corynebacterium,* with manganese dioxide formation. The manganic ion should be a good oxidant for ferrous iron.

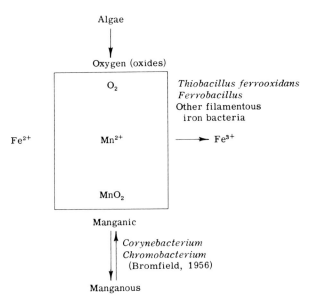

FIG. 15.3. An algae-bacteria-manganese triad in the oxidation of iron.

DETECTION OF MICROBIAL CORROSION

There are no known reliable methods for measuring the corrosive effect of bacteria *in situ*. The number of bacteria present can be determined indirectly by removing material from the corroded area and using one of several plating methods. Sulfate-reducing bacteria are most commonly sought and are the best indicators of microbial corrosion. Well casing profiles, where the positive or negative potentials are taken at 10-ft intervals along the surface of the pipe, have not been found useful in detecting localized corrosion caused by bacteria.

CONTROL OF MICROBIAL CORROSION

The most practical approach to the control of microbial corrosion is to control the conditions which initiate corrosion. Pipes when possible should be located in well-drained alkaline soils. Soils low in electrolytes should be selected, if possible. Since sulfate reducers require organic matter, areas low in organic matter are ideal. Pipes should be coated with resistant material and cathodic protection devices are advantageous. Bactericides are used. Some of the important ones are chlorinated phenols, quarternary ammonium compounds, amines, and surface-active agents. Chlordihexidine is effective but too expensive. The following table lists bactericides in common use.

Bactericide	Concentration (ppm)
Panacide (dichlorodihydroxydiphenyl methane)	5–10
Citrimide (trimethyl cetylammonium bromide)	—
Laurylpentachlorphenate[a]	—
Roccal (benzylalkonium chloride)	2
Copper sulfate	5
Sodium tetrachlorophenate	5–10
Methyl violet	10

[a] Toxic for humans.

Some sulfate reducers produce protective sulfide films on mild (low-carbon) steel plates. Thick coatings of bitumin or coal-tar pitch reinforced with asbestos or glass fiber are good for pipes. Tapes of pressure-sensitive polyvinyl chloride are also used with a 55% overwrap.

One of the best-known methods of corrosion protection is that of cathodic protection. Best results are obtained when used in conjunction with protective coatings. The pipe or structure to be protected is made into the cathode of an electrolytic cell. This is done by connecting it to remotely situated bars of a more electronegative metal (anodes) and by imposing a direct current between the structure and the expendable anodes to which the corrosion is transferred. The alkalinity produced in areas adjacent to the cathodic structure affords additional protection. An alkaline condition is not conducive to sulfate reduction because it aids in the formation of an iron sulfide protective layer at the anodic surface. Sulfides themselves have a depolarizing action.

Experience has shown that the principal source of corrosion in natural underground water is the presence of free carbon dioxide. Carbon dioxide is eliminated from many water supplies by addition of lime to bring the pH up to 8.2–8.4.

REFERENCES

Allred, R. C. (1954). The role of microorganisms in oil field water flooding operations—bacterial control in the North Burbank Unit Water Flood, Osage County Oklahoma, *Producers Monthly* pp. 18–22.

Booth, G. H., and Tiller, A. K. (1960). Polarization studies of mild steel in cultures of sulfate reducing bacteria. *Trans. Faraday Soc.* **56**, 1689–1696.

Bradley, W. G. (1954). A study of galvanic corrosion in marine pseudo-sediments. *Texas Agr. Mech. Rept. Coll.* **24-A**.

Bromfield, S. M. (1956). The oxidation of manganese by *Chromobacterium* and *Corynebacterium*. *Australian J. Biol. Sci.* **9**, 238–252.

Carlson, V., Bennett, E. O., and Rowe, J. A. (1961). Microbiol flora in a number of oil field water-injection systems. *Soc. Petrol. Engrs. J.* **1**, 71–80.

Hewitt, L. F. (1950). "Oxidation-Reduction Potentials in Bacteriology and Biochemistry," 6th ed. Livingstone, Edinburgh and London.

Iverson, W. P. (1966). Direct evidence for the cathodic depolarization theory of bacterial corrosion. *Science* **151**, 986–988.

Rozenberg, L. A., and Ulanovskii, I. B. (1960). Growth of bacteria during cathode polarization of steel in sea water. *Mikrobiologyia* **29**, 721–724.

Sharpley, J. M. (1961). Microbiological corrosion in water floods. *Corrosion* **17**, 386–390.

Silverman, M. P., and Ehrlich, H. L. (1964). Microbial formation and degradation of minerals. *Advan. Appl. Microbiol.* **6**, 153–206.

Tiller, A. K., and Booth, G. H. (1962). Polarization studies of mild steel in cultures of sulfate reducing bacteria. *Trans. Faraday Soc.* **58**, 110–115.

von Wolzogen Kühr, C. A. H., and van der Vlugt, L. S. (1934). The graphitization of cast iron as an electrobiochemical process in anaerobic soils. *Water, Den Haag* **16**, 147–165.

von Wolzogen Kühr, C. A. H., and van der Vlugt, L. S. (1953). Aerobic and anaerobic iron corrosion in water mains. *J. Am. Water Works* **45**, 33–46.

Wesley, W. A., and Brown, R. H. (1948). *In* "Corrosion and Anti-corrosives" (H. H. Uhlig, ed.), 1188pp. Wiley, New York.

16

CALCIUM

GEOCHEMISTRY

Calcium and magnesium are two of the major electropositive constituents found in the earth's crust. In meteorites the weight of magnesium
exceeds calcium by a factor of about nine. In the upper lithosphere the
weight of calcium is about twice that of magnesium. This is especially
true of igneous rocks where magnesium is eliminated by gravitational
settling of the early crystallizates of magnesium silicates. Lighter minerals
are pushed toward the surface. In an analysis by weight, calcium exceeds
silicon in content in both the lithosphere and interstellar space (Goldschmidt,
1954).

Even though sodium is univalent and calcium is divalent, these two
metals have crystallochemical properties in common because of the
similarity of their ionic radii (1.06 Å for calcium and 0.98 Å for sodium);
this is also true of barium and potassium.

The calcium minerals associated with the main igneous rocks are
shown in Table 16.I.

Plagioclase minerals are more subject to dissolution than the amphi-
boles. Solubilization usually requires bicarbonate ion to be present.
Finely precipitated carbonate accumulates quickly as evidenced by the
marl deposits in the lakes of Northern Scotland and Norway. In freshwater environments the calcareous shells of molluscs, such as the bivalves
and gastropods, contribute to the carbonate sediments.

TABLE 16.I

CALCIUM MINERALS ASSOCIATED WITH THE MAIN IGNEOUS ROCKS

Igneous rocks	Type
Calcic monoclinic pyroxenes	$CaMg(SiO_3)_2$
Plagioclase feldspars	
Anorthite	$CaAl_2Si_2O_8$
Albite	$NaAlSi_3O_8$
Monoclinic Amphiboles	$Mg_5Ca_2Si_8O_{22}(OH,F)_2$
Apatite	$Ca_5(PO_4)_3F$
	$Ca_5(PO_4)_3Cl$
Nelpheline syenites, granites, and grandiorites	
Sphene	$CaTiSiO_5$
Cancrinite	$3\ NaAlSiO_4 \cdot CaCO_3$
Calcite	$CaCo_3$
Gabbro and basalt	
Akermanite	$Ca_2MgSi_2O_7$
Gehlenite	$Ca_2Al_2SiO_7$
Monticellite	$CaMgSiO_4$
Perovskite	$CaTiO_3$

The bulk of calcium is in the ocean. The oceans have their own calcium cycle in which chemical and biochemical processes are involved in deposition and solubilization. This cycle involves a balance between dissolved calcium ions, carbonic acid, calcium bicarbonate, calcium carbonate, and the availability of oxygen and carbon dioxide. Note should be made that the surface waters in the lower tropical latitudes of the Atlantic are supersaturated with calcium carbonate which is conducive to the biosynthesis of calcium carbonate shells of plankton-forming globigerina ooze. In shallow northern waters calcareous algae such as *Lithothamnium* are found.

In bivalves the calcite crystals reach several inches in length. Sea urchins of the genus *Cidaris* found in subtropical and tropical regions form large spines, with each spine a single calcite crystal. The optical axis of these crystals coincide with the length of the spine. Protoplasm is accommodated in uniformly arranged channels within these single calcite crystals. The mechanism by which these living systems control calcite deposition is not known.

Under evaporative conditions calcium is precipitated from seawater as gypsum ($CaSO_4 \cdot 2\ H_2O$ or $CaSO_4$), polyhalite ($K_2SO_4 \cdot 2\ CaSO_4$

$MgSO_4 \cdot 2 H_2O$), and tachhydrite ($CaCl_2 \cdot 2 MgCl_2 \cdot 12 H_2O$). Even with the deposition of these minerals from the ocean the continents are gradually being leached of calcium. At present the resources of limestone, marl, and dolomite are in excess of the agricultural demand.

Calcium in Metamorphic Rock

In volcanic rock, burnt lime or CaO is found, since heating $CaCO_3$ releases carbon dioxide to the atmosphere. Calcium oxide is not encountered in older geological formations because it is too reactive. Calcium carbonate and magnesium oxide are formed in the thermal decomposition of dolomite:

$$CaMg(CO_3)_2 \leftrightarrow CaCO_3 + MgO + CO_2$$
(Dolomite) Δ

The MgO (periclase) rehydrates to form $Mg(OH)_2$ (brucite). Thermal transformation of calcite limestone and dolomite with the silica in quartz yields silicates of wollastonite and fosterite, respectively.

$$CaCO_3 + SiO_2 \leftrightarrow CaSiO_3 + CO_2$$
(Calcite) (Quartz) Δ (Wollastonite)

$$2 CaMg(CO_3)_2 + SiO_2 \leftrightarrow Mg_2SiO_4 + 2 CaCO_3 + 2 CO_2$$
(Dolomite)   (Quartz) Δ (Fosterite)   (Calcite)

Microbes are not involved in wollastonite and fosterite synthesis.

BIOCHEMISTRY OF CALCIUM

Algae

Biochemical processes are active in beach sandstone formation in the Red Sea (Nesteroff, 1954). Microscopic observation of samples shows calcareous detritus (pseudo-ooliths) surrounded by a thin brown layer of material. Cementation between grains is by fine needles of aragonite. Incrustation of amorphous $CaCO_3$ is formed on all grains. There is a sub-layer of organic material. The aragonite needles formed are perpendicular to the sand grains. Their existence is temporary as they later appear as calcite rhombohedra. This latter type of crystal has no organic sub-layer. This transformation occurs in the intertidal zone. It would appear that amorphous $CaCO_3$ and aragonite are formed through microbial activity. Similar carbonate deposition is observed on the western edge of the Great Bahama Bank (Seibold, 1962). The algae may settle out to form sedimentary deposits or undergo degradation to release sedimentary materials, as has been observed by Lowenstan (1955) in west Atlantic waters.

Algae are quite active in the deposition of calcium (Wolf, 1962). Walter-Levy and Strauss (1961) collected 35 species of Corallinaceae, 3 species of Squamiaceae, and 1 species each of Ceramiaceae, Dietyotaceae, Udoteaceae, and Dasycladaceae. Cultures were taken from various geographical locations. The calcium carbonate deposited occurred in two forms. These were aragonite and a type of calcite containing some magnesium at the nodes of the crystal lattice. For cultures forming Mg^{2+} the Mg/Ca ratio increased with the temperature of the sea. Algae containing the most Mg^{2+} grew in the warmest seawater.

Of the red algae, *Glaxaura*, *Laurencia*, and *Liagora* deposit small quantities of aragonite and lesser amounts of magnesium carbonate (Maslov, 1962). The ash residue of Corallinaceae contains up to 73% carbonates of calcite and $MgCO_3$. This may be attributed to the high amounts of pectin around the cell walls of these algae which selectively adsorbs calcium. The $MgCO_3$ in Corallinaceae varied from 4–36%. Deposition of $MgCO_3$ increases and $CaCO_3$ decreases in warmer seas. For a comprehensive analysis of algae for mineral content, Vinogradov's classic work (1953) should be consulted. The $MgCO_3$ content of various Corallinaceae varied from 11.8–20.6%. In analyzing red algae, Maslov (1962) found the following amounts of calcium and magnesium as carbonates. Although old and young parts of algae were examined no

Algal species	Percent		
	$CaCO_3$	$MgCO_3$	$CaSO_4$
Lithophyllum pachydermum	73.6	25.0	1.24
Goniolithon strictum	75.4	23.0	1.22

definite conclusions could be made as to whether there is a shift in metal selectivity and deposition between the old and young parts by the same algal species.

In Vinogradov's review of the elemental content of stony red algae, phosphorus, sulfur, chloride, and aluminum were found in small quantities. The aluminum in *Lithophyllum proboscideum* was 0.89%, whereas, in

L. kaiseri and *Porolithon oniodes* it was 0.29 and 0.22% respectively. *Jania rubens* contained 0.3% arsenic and 0.023% manganese.

Iodine, boron, radium, and fluoride are present in much reduced amounts. Maslov (1962) has also completed a spectral analyses of fossil tertiary Ukranian red algae.

The brown algae *Ascophyllum nodosum* has been dried, ground to 5–50 mesh, and extracted with 1.0 N HCl for 30 hours at 20°C (Wassermann, 1948). Cells treated in this manner were tested for their ability to adsorb calcium. These algal tissue behaved like a cation-exchange material.

Bacteria

Krasilnikov (1949) determined bacteria counts for calciferous bacteria and actinomycetes in soils showing $CaCO_3$ rock deposits. The calciferous bacteria numbered 0.2–1.75 million/gm ore (MGO) while the actinomycetes numbered 1.3–8.5 MGO. Biogenic deposits of $CaCO_3$ were found on granite, basalt, and tuff in several areas of Armenia. Calciferous bacteria are quite active in the rhizosphere of plants. Over a 2-month test period with soils inoculated with these bacteria, the number of aggregates doubled. The greatest increase was in the 0.25–0.5-mm aggregates (Tulaikova, 1952).

Up to 17% of the calcium in solution is precipitated as calcium carbonate by *Bacterium precipitatum* in 5–6 days (Rozenberg, 1950). Over 250 mg of $CaCO_3$/liter of medium was recovered. The carbon dioxide produced tends to keep $CaCO_3$ in solution by the formation of bicarbonate.

Proteus vulgaris, *Bacillus mycoides*, *Bacterium calcis*, and *Bacterium sewanense* form $CaCO_3$ in their slimes. As crystals form they may completely enclose the bacterial cell (Isachenko, 1948). Several cultures of this type have been isolated from hot springs or from lakes similar to Lake Sevang in Armenia. Amorphous carbonate is deposited in the slime secretion of the bacteria. Oolites or sphaerolites are formed. Travertine deposits are possibly formed by biolithogenesis.

A drill core at Koblenz Aargau Canton contained biotite gneiss at a depth of 160 meters (Neher and Rohrer, 1958). The sample possessed idiomorphic crystals of dolomite surrounded by a dark border of viable bacteria. These bacteria were active in precipitating dolomite from solutions containing the elements present in gneiss.

Marine pseudomonads (Greenfield, 1963) are also active in concentrating calcium and magnesium. Viable and nonviable cells absorb considerable quantities of Ca^{2+} and Mg^{2+}. If sufficient carbonate is present aragonite crystals are formed. The Ca^{2+}-rich surface of the cell acts as a nucleus for precipitating $CaCO_3$. Acid solution of such crystals shows that bacteria occupy a central position in each crystal.

Other Microbes

Lime-loving molds have been observed in calcareous sands of Australia and on the surface of shells of *Foraminifera, Ostracoda*, and the spicules of calcareous sponges. The fungal forms in these incrustations were placed in the Phycomycetes, a few being allied with the Cladochytriaceae and the Myxomycetes (Porter and Zebrowski, 1937).

LIMESTONE MILIEU

Baas-Becking *et al.* (1956) has examined the limestone milieu in many environments. In the Jenolean caves, west of the Blue Mountains in New South Wales, samples were taken from two underground rivers, stalactite drips, pools in the caves (often blue with so-called cave-milk containing suspended calcium and iron salts), and the Jenolean River below the caves. The E_h distribution was unimodal at $+375$ mV with limits of $+455$ as to $+100$ mV and a pH varying from 5.9 to 9.0. The total base was highest in the stalactite drip. Water was slightly chalybeate, i.e., impregnated with iron salts.

Baas-Becking believes that cell walls of *Sphagnum* and other cells are capable of cation exchange. In one study conducted, rainwater near a moor contained 21 mg NaCl/liter. On *Sphagnum* walls the following reaction occurred:

$$NaCl + HX \leftrightarrow HCl + NaX$$
$$2\,NaCl + HCO_3 \leftrightarrow HCl + Na_2CO_3$$

Under some conditions where Na_2CO_3 is removed or deposited, the acid produced decreases the pH. Cation exchange takes place in waters with high electrolyte contents, but sometimes these effects cannot be measured because of the high buffering capacity of the water. This is especially true where photosynthesis is occurring. Carbonates are not assimilated during photosynthesis unless acid is produced. Therefore, during photosynthesis CO_2 is obtained by microbes from the bicarbonate ion:

$$2\,HCO_3^- \leftrightarrow H_2O + CO_2 + CO_3^{2-}$$

Thus the pH increases, offsetting the decrease which is observed during carbonate deposition.

At Trona, a mineral typical of alkaline brines has a composition of $NaHCO_3$, Na_2CO_3, n H_2O. This desert brine often is teeming with *Dunaliella* spp. which fixes CO_2 and converts bicarbonate into carbonate increasing the pH.

Bicarbonates are probably formed by straight chemical reactions in terrestrial or marine environments, however microbes are probably active in supplying the reactants, e.g., ammonia from denitrification, carbon dioxide from organic oxidations, and amines from proteolysis. Under temperatures of 14–28°C, alkaline conditions, and pressures of 15 psi or greater bicarbonates may be synthesized:

$$H_3 + CO_2 + KCl + H_2O \rightarrow NH_4Cl + KHCO_3$$
$$\text{Amine} + CO_2 + KCl + H_2O \rightarrow \text{Amine-HCl} + KHCO_3$$

Mono-, di-, and triethylamines react readily under conditions in the second equation above. The conversion with triethylamine gives a 90%

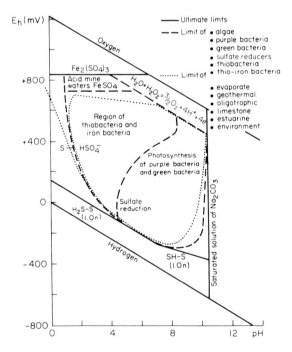

FIG. 16.1. The limits of the natural aqueous environment. (From Baas-Becking *et al.*, 1956.)

recovery of potassium bicarbonate. This is the basis of the Solvay process for producing potassium carbonate.

The relationship of E_h to the pH encountered in natural environments has been developed by Baas-Becking (Fig. 16.1). The thiobacteria and iron bacteria produce a low pH; however, they function in a medium E_h range (~ 300–700 mV), whereas, the sulfate reducer and photosynthetic microbes are functional at a higher pH with the E_h varying from -200 to $+500$ mV.

Several bacteria are active in the deposition of apatite [$Ca_5(PO_4)_3Cl$] in the stomachs of animals (Rizzo et al., 1962). Bacteria involved are *Streptococcus salivaras*, *Actinomyces isralii*, *Actinomyces naeslundii*, and *Bactereonema matruchotii* (*Leptothrix bucialis*). Of course, other bacteria synthesizing organic acids are known to dissolve apatite (Sperber, 1958).

CALCIUM CYCLE

Calcium is removed from the aqueous milieu (Fig. 16.2) by a variety of microbes and biological systems. The red algae, e.g., *Glaxaura*, *Laurencia*, and *Liagora* are quite active. Almost all species of Corallinaceae deposit $CaCO_3$, sometimes with globigerina ooze. Bacteria are less active however. *Bacterium precipitatum*, *Proteus vulgaris*, *Bacillus mycoides*, *Bacterium calcis*, *Bacterium sewanese*, and marine pseudomonads all deposit $CaCO_3$ in their slimes. Fungi have not been adequately investigated. Freshwater molluscs, gastropods, and sea urchins make a large contribution to carbonate sediments. Apatite deposition in nature requires further investigation. Microbes have been found which deposit this mineral in the stomach of animals, but such habitats are not going to add to deposits of apatite geologically; however analogous habitats of this type may occur in nature.

The solubilization of carbonate minerals by microbial activity can in general be related directly to organic acid or mineral acid synthesis which will yield a mobile cation and bicarbonate.

The biosynthesis of gypsum by microbes needs investigation. Certainly numerous autotrophs synthesize sulfuric acid. If the calcium ion increases in such systems and the pH were to increase $CaSO_4 \cdot 2\,H_2O$ deposition should follow. The solubility of gypsum in water at $25°C$ is less than 0.25 gm/100 gm water which is quite low. The calcium in gypsum is returned to the aqueous milieu through sulfate reduction.

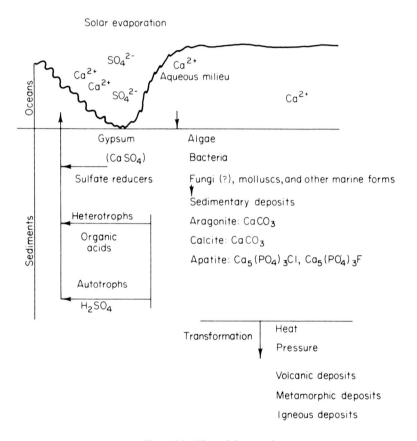

FIG. 16.2. The calcium cycle.

REFERENCES

Baas-Becking, L. G. M., Wood, E. J. F., and Kaplan, I. R. (1956). Biological processes in the estuarine environment. *Biochemistry* pp. 96–102.

Goldschmidt, V. M. (1954). *In* "Geochemistry," (A. Muir, ed.), 730 pp. Oxford (Clarendon) Press, London and New York.

Greenfield. L. J. (1963). Metabolism and concentration of calcium and magnesium and precipitation of calcium carbonate by a marine bacterium. *Ann. N. Y. Acad. Sci.* **109**, 23–45.

Isachenko, B. L. (1948). Biogenesis of calcium carbonate. *Mikrobiologiya* **17**, 118–125.

Krasilnikov, N. A. (1949). Microbiological factors in depositing calcium carbonate on rocks. *Mikrobiologiya* **18**, 128–131.

Lowenstan, H. A. (1955). Aragonite needles secreted by algae and some sedimentary implications. *J. Sediment Petrol.* **25**, 270–272.

Maslov, V. P. (1962). Stony red algae as concentrators of chemical elements. *Tr. Geol. Inst., Akad. Nauk, SSSR* **53**, 142–148.

Neher, J., and Rohrer, E. (1958). Dolomite formation with bacterial action. *Ecolgae Geol. Helv.* **51**, 213–215.

Nesteroff, M. W. (1954). On the formation of beach sandstone or beach rock in the red sea. *Compt. Rend.* **238**, 2547–2548.

Porter, C. L., and Zebrowski, G. (1937). Lime-loving molds from Australian sands. *Mycologia* **29**, 252–257.

Rizzo, A. A., Martin, G. R., Scott, D. B., and Mergenhagen, S. E. (1962). Mineralization of bacteria. *Science* **135**, 439–441.

Rozenberg, L. A. (1950). Physiochemical conditions for bacterial precipitation of calcium. *Mikrobiologiya* **19**, 410–417.

Seibold, E. (1962). Study of carbonate precipitation and solution at the western edge of the Great Bahama Bank. *Sedimentology* **1**, 50–74.

Sperber, J. I. (1958). Solution of apatite by soil microorganisms producing organic acids. *Australian J. Agr. Res.* **9**, 782–787.

Tulaikova, K. P. (1952). The separation of calcium carbonate crystals by bacteria of the rhizosphere. *Agrobiologiya* **4**, 43–51.

Vinogradov, A. P. (1953). "The Elementary Chemical Composition of Marine Organisms." Sears Found. Marine Res. Memoir II. Yale Univ. Press, New Haven, Connecticut.

Walter-Levy, L., and Strauss, R. (1961). The mineral constituents of calcareous algae. *Colloq. Intern. Centre Natl. Rech. Sci. (Paris)* **103**, 39–50.

Wassermann, A. (1948). Adsorption of calcium ions by an acid-extracted brown algae under continuous flow conditions. *Nature* **161**, 562.

Wolf, K. H. (1962). The importance of calcareous algae in limestone genesis and sedimentation. *Neues Jahrb. Geol. Palaeontol., Monatsh.* pp. 245–261.

17

SODIUM CHLORIDE, SALTS, AND HALOPHILES

GEOCHEMISTRY

Considerable amounts of sodium occur in combination with other elements. In magmatic rocks sodium averages 2.83%; in the silicate phase of meteorites this decreases to 0.77%. Sodium is concentrated in the upper lithosphere in rocks of light residual magmas. It is also present in the upper atmosphere, arising from oceans or meteoric dust. The ionic radius of the sodium ion is 0.98 Å and is close to that of calcium which is 1.06 Å. This radius differs considerably from lithium (0.70 Å) and potassium (1.33 Å) which limits isomorphous substitution of these metals for sodium at hydrothermal temperatures. At magmatic temperatures mutual solubility of sodium and potassium is possible.

In igneous rocks sodium is found in abite [$Na(AlSi_3O_8)$] and aegirine [$NaFe(SiO_3)_2$]. A sodium zeolite, analcite ($NaAl\,Si_2O_6 \cdot H_2O$), occurs in rocks containing nepheline. Argillaceous sediments, shales, phyllites, and mica schists increase their sodium content to form feldspathic rocks.

Sodium is easily moved by ground water through leaching action and accumulates in ocean water. This is not true of potassium, rubidium, and cesium which are more strongly adsorbed by clay minerals. The primary minerals containing sodium which may be easily leached are the feldspathoids and the plageoclases of labradorite and bytownite. Microbes are very active in the mobilization of sodium in soil and nature.

In arid climates with suitable topography salts leached from soils accumulate in lakes which evaporate to form alkali soils, saltspurs, etc. Sodium chloride, sodium sulfate, sodium borate, sodium nitrate, sodium chlorate, and sodium-calcium carbonate are deposited under these conditions. In marine sediments sodium is found in small elastic grains of albite. Glauconite sediments, the essential component of green sands, are low in sodium and high in potassium. Sodium is concentrated in rock salt and polyhalite from marine environments during evaporation, as described by J. H. van't Hoff in 1900 for the deposits in Central Germany.

HALOPHILISM

In the deposition of salt in lacustrine environments, halophilic micro-organisms catalyze a variety of reactions. To understand their involvement with salts an examination must be made of the metabolic characteristics of these microbes.

Most microbiologists are prone to regard halophilic environments as abnormal and nonhalophilic environments as normal. Considering the

TABLE 17.I

TOLERANCE TO CONCENTRATION OF SALT[a]

Halophilism	Degree of salt tolerance	Microbe	Reference
I. Intolerant	High	Coliforms	Shewan (1942)
		Pseudomonads	Shewan (1942)
	Moderate	*Staphylococcus*	Scott (1953)
	Slight	*Urobacillus*	Hof (1935)
II. Facultative	Low	*Urobacterium*	Hof (1935)
		Bacillus spp.	Shewan (1942)
	Moderate	*Urobacterium*	Hof (1935)
	High	Micrococci	Shewan (1942)
III. Obligate	Low	*Vibrio costicolus*	Christian (1956)
	Moderate	Colorless halophiles	Shewan (1942)
		Bacterioides halosmophilus	Shewan (1942)
	High	Red halophiles[b] (*Halobacterium halobium*)	

[a] From Ingram (1957).

[b] Some of the red halophiles require at least 15% NaCl for growth and often grow in saturated brine solutions.

vast oceans and the relative ratio of land to water it would appear that the reverse is true. Halophilism may be (or has been) the norm with non-halophilism developing as a consequence of biological development in the lithosphere. This would explain why halophiles can be isolated from almost all ecological terrestrial habitats. As with almost all biological phenomena there exists a continuum between those cultures which possess no tolerance to salt and those which tolerate saturated solutions (Table 17.I). Many of the coliforms and pseudomonads appear to be quite intolerant of halophilic environments. Facultative halophiles are found in the genera *Urobacterium*, *Bacillus*, and *Micrococcus*. Obligate halophiles include *Vibrio costicolus*, *Bacterioides halosmophilus*, and *Halobacterium halobium* as well as certain yeasts and other microbes.

OSMOTIC PRESSURE

The concentration of a dissolved substance on one side of a membrane may be interpreted in terms of the osmotic pressure it produces. Theoretically it is preferable to discuss concentration of the solute in terms of molality (moles/1000 gm of solvent) rather than molarity which changes with temperature. Differences between molarity and molality are not serious at the low concentrations encountered in normal biological systems. However, at higher concentrations the molality will often exceed the molarity two- to threefold at a given temperature.

Frequently osmotic pressure of concentrated solutions is calculated from vapor pressure data by means of the equation

$$P = -\frac{1}{v} RT \ln \frac{P}{P_0}$$

where P = osmotic pressure, R = gas constant, T = absolute temperature, P_0 and P = vapor pressure of water and solution, and v = partial molal volume of water.

The ratio of P/P_0 represents the aqueous activity of water. This ratio varies quite markedly for different solutes. Of those solutes tested (Table 17.II), a wide range of P/P_0 values or what is termed aqueous activity can be obtained with either $CaCl_2$ or glycerol. The aqueous activity decreases from 0.98 to 0.65 as the molality of $CaCl_2$ increases from 0.42 to 3.80. All the other solutes tested require much larger molal concentration to give smaller changes in P/P_0 values. The principles involved in relating

TABLE 17.II

AQUEOUS ACTIVITY OF WATER WITH SOLUTES[a]

P/P_0	Solute molarities (25°C)						
	NaCl	KCl	CaCl$_2$	Na$_2$SO$_4$	Glucose	Sucrose	Glycerol
0.98	0.61	0.61	0.42	0.58	1.12	1.03	1.11
0.95	1.48	1.55	0.92	1.50	2.65	2.31	2.76
0.90	2.83	3.08	1.58	2.94	5.10	4.11	5.57
0.85	4.03	4.60	2.12	4.08	—	5.98	8.47
0.80	5.15	—	2.58	—	—	—	11.5
0.75	—	—	3.00	—	—	—	14.8
0.70	—	—	3.40	—	—	—	18.3
0.65	—	—	3.80	—	—	—	22.0

[a] From Ingram (1957).

P/P_0 to osmotic pressure have been reviewed by Scott (1957). Interpretation in terms of osmotic pressure implies that the cell wall boundary components behave as an ideal semipermeable membrane. This is probably not true for any bacterium and especially for halophiles where large concentrations of solutes enter the cell and specific ions are selected.

Bacteria growing in saturated NaCl solutions with a P/P_0 value of 0.75 would have to withstand an osmotic pressure of some 200 atm. For a spherical microbe of 1μ radius, this corresponds to a cell wall tension of about 10^4 dynes/cm. If the cell wall is 10 mμ thick the tensile strength is of the order of 10^{10} dynes/cm^2 (Salton, 1955). This is about one-hundredth of that observed with cellulose (Preston, 1952). More information is needed for a comparison of tensile strength of the cell walls of non-halophilic and halophilic microbes.

MECHANISMS OF CATION MOVEMENT ACROSS MICROBIAL MEMBRANES

The number of mechanisms proposed for ion transport are many. Ussing (1957) covers the older theories in detail and the subject is still not settled. Hypotheses listed were:

1. Pinocytosis. Uptake of minute droplet of environmental fluids.
2. Fluid circuit mechanism (Ingrahm *et al.*, 1938). Net transfer, e.g., of salts, is brought about by the one-way transport of salt solution through relatively large pores; transport in the opposite direction is of more or less pure water through narrow pores.
3. Simple membrane carrier transport (Osterhout, 1936). The ion transported forms a complex with a membrane constituent. By metabolic processes the concentration at one boundary is maintained lower than at the other, setting up a diffusion gradient.
4. Electron-linked carrier transport (Conway, 1953). It is assumed that an essential step in the transfer is an electron transfer. The path of electron flow differs from the metal ion flow.
5. Propelled carrier transport (Danielli, 1964; Goldacre, 1952). The ion complex is drawn through the membrane by a contraction of a contractile fiber.

In living tissues ionic constituents are transferred across membranes by simple diffusion, thermal agitation, facilitated diffusion, and active transfer.

Danielli (1964) lists at least five possible mechanisms in molecular transport:

1. A diffusion shuttle carrier molecule.
2. A propelled shuttle (with a contractile protein).
3. A rotating molecular carrier of same diameter as plasma membrane.
4. A rotating carrier with a molecular segment.
5. An expanded lattice which collapses asymmetrically.

In all these mechanisms there must be an absorption center which is specific for a limited range of molecules. Forces available for adsorption are van der Waals, electrostatic, and hydrogen bonding. The specificity of these forces must reside in an organized unit.

This organization can be supplied through contractile proteins as envisioned by Goldacre (1952) in his classic theories. Since native proteins consist of lamellae composed of one or more polypeptide chains, Danielli depicts a lipoid component interspersed with protein (polypeptide) lamellae possessing a nonpolar and a polar structure (Fig. 17.1).

The plasma membrane of most cells appears to consist basically of a lipoid layer of 50 Å thickness with protein layers on either side. The plasma membrane contains a monolayer of protein ($\sim 41\%$ by weight) and a monolayer of lipid ($\sim 23\%$ by weight). Active enzymes may be located within 50 Å of the external medium. Components identified

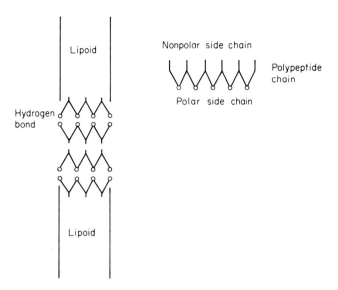

FIG. 17.1. Unit organization for cellular molecular transport. (After Danielli, 1964.)

with plasma membrane are cytochrome, succinic dehydrogenase, lactic dehydrogenase, malic dehydrogenase and phosphatase, α-glycerophosphate dehydrogenase, glucose-6-phosphatase, and glucose-6-phosphate dehydrogenase.

Movement of molecules across membranes is limited by a hydrogen-bonding formation between the water and the molecules concerned and the viscous structural resistance of the membrane.

Divalent cations even in minute traces (1 part in 10^8) have a remarkable effect upon the physiological properties of ionized fatty acid monolayers and change the strong repulsion between molecules to a strong attraction. The divalent cations chemically bind the carboxyl head-groups highly polarizing their charges and, thus, vertically eliminating the zeta potential and the Gaucy-Chapman layer.

The osmotic diffusion pump has been proposed by Franck and Mayer (1947). A connecting link is believed to exist between internal and external solutions in terms of a "solution compartment" bounded on either side by a membrane. The connecting link contains two solutes A and B in which B is an n-fold polymer of A. The diffusion coefficient of A is less than n-fold greater than the diffusion coefficient of B. Thus the same quantity of material can diffuse more readily in the B-form than in the

A-form. The conversion of nA to B and vice versa does not take place within the connecting link but a reaction is catalyzed

$$nA \leftrightarrow B$$

and an equilibrium is attained at the membrane, binding a portion of the outside solution. The supply of some other material at the membrane binding the inside solution dissociates B to nA. Energetic reactions favor the production of more nA than would be formed under normal equilibrium conditions. A is continuously cycled by diffusion from the inside membrane wall to the outside wall where it is converted catalytically to form B. This latter compound diffuses back to the inside to dissociate to form more A. The concentration gradient of B will be less than A. A supply of free energy is required to maintain the dynamic equilibrium by which solute is transferred.

The concepts of solvent drag in which water carries the ion in question across the membrane and the Redox pump of Lund do not appear to be as important as originally postulated.

MICROBIAL MEMBRANES AND RELATED STRUCTURES

Microbial membranes possess many special properties. The microbe itself is small ranging from 1–20 μ in size. These small cells possess all the metabolic activities required for survival and multiplication. Cells may contain from 300 to several thousand enzymes with average molecular weights of around 10^5. The cell cytoplasm of microbes is surrounded by a plasma membrane (or Cytoplasmic membrane).

Microbe	Plasma membrane[a]
Bacillus megaterium	25–50 Å
Spirillum serpens	60 Å
Bacillus cereus	25–50 Å
Escherichia coli	+
Rhodospirillum rubrum	+
Chlorobium limicola	+
Bacillus subtilis	+
Thiovulum majus	+

[a] Plasma membrane demonstrated, +.

Thiovulum majus possesses a quite interesting cytoplasmic membrane. The microbe itself is 18 μ in diameter, and the cytoplasmic membrane has small penetrating invaginations reminiscent of pinocytes observed in mammalian cells.

Chemical analyses of cytoplasmic membranes are difficult. Microbial protoplasts may be prepared by treating cells with lysozyme. This enzyme strips off the cell wall. If isotonic conditions are not used the cell vesicles burst and leave "ghost cells" and small granules. These "ghost cells" may be recovered and their chemical and enzymatic compositions studied. The cytoplasmic membrane is about 10 % by weight of the cell.

Membranes function:

1. To create a barrier between internal medium of cell and external environment.
2. To establish a contact between internal and external solutions media.
3. To control the movement of ions required for both cellular growth and disposal of waste products.

Passive permeability of microbial membranes have been studied by six methods.

1. Microscopic observation of cell plasmolysis in hypertonic solutions.
2. Measurement of light-scattering dependence upon shrinkage or swelling of cells of different solute concentration.
3. Measurement of volume change of cells at different solute concentrations.
4. Measurement of the rate of lysis of protoplasts in isotonic solutions.
5. Measurement of net influx of solutes into the cell by analyses of the cell.
6. Measurement of net influx of solutes into the cell by analysis of the external medium.

In *Escherichia coli* the permeability of the plasma membrane to sucrose, lactose, D-mannose, D-sorbose, D-sorbital, L-rhamnose, L-arabinose, and D-xylose were too low to be measured. The time for half equilibration for erythritol and pentaerythritol was 15 minutes, for D-ribose, 5 minutes, whereas glycerol required less than 1 minute. Similar results have been obtained for *Sarcina lutea* and three other microbes.

The significance of the lipoid phase of the plasma membrane is not known. It definitely appears to be an obstacle to free diffusion and, along with the configuration of structural molecules, helps control penetration.

The cytoplasmic membrane of animal and plant cells is around 100–200 Å thickness. This membrane branches throughout the cell forming an endoplasmic reticulum extending from the cell membrane to the nucleus. The endoplasmic reticulum in turn possesses two membranes, the ectoplast and the tonoplast (70–100 Å).

Physiological studies of microbial or higher plant and animal cells do not show that the entire cell encased within the cell wall is behind an osmotic barrier. Some cytoplasm is always available to externally applied electrolytes or nonelectrolytes. The areas freely available for solute penetration are called "free space." This includes what is known as Donnan free space; and these areas are characterized by high levels of physically bound water.

Since many scientists regard cells as possessing relatively impermeable membranes many reactions are interpreted as taking place at the cell surface.

Rothstein (1954) has developed a newer concept which is referred to as the compartmentalization of the cell surface. He suggests that the cell surface should be regarded as multimembranous and compartmentalized.

Yeast cells are relatively impermeable to divalent cations (Ca^{2+},Mg^{2+}). The cytoplasmic concentration of K^+ and H^+ are about 2×10^{-1} and 1×10^{-6} M. If cells rich in Na^+ are placed in a medium containing K^+, cells will exchange Na^+ for K^+.

TOLERANCE TO DIFFERENT SALTS

Some halophiles have a specific requirement for NaCl, while others can substitute other salts. *Micrococcus denitrificans* failed to grow on fifteen salts other than NaCl (Brown and Gibbons, 1955). The red halophilic microbes, in general, have a high specificity for NaCl. The anaerobic halophile isolated by Baumgartner (1937) grew in NaCl-saturated media, but it was found that chlorides of other cations could be substituted for NaCl. The amount of growth at the optimum osmotic pressure diminished and followed the Hofmeister series: $Na^+>K^+>Li^{2+}>Mg^{2+}>Ca^{2+}$. Similar results are exhibited by *Sporendonema epizoum* (Vaisey, 1954).

In examining the "continuum" from nonhalophilism to halophilism a dependency upon NaCl is seen, facultative halophiles being somewhat independent of the NaCl requirement and the obligate halophiles always requiring NaCl. In those halophiles tolerating divalent cations, the optimal molal concentrations for Mg^{2+} and Ca^{2+} is one-tenth or less that required for NaCl. This is true for both *Dunaliella viridis* and *Halobacterium halobium* (Baas-Becking, 1931; Brown and Gibbons, 1955). The effect of higher valency metals such as ferric iron, uranium, and vanadium on halophiles has not been evaluated.

The physical and chemical conditions of any habitat will affect the halophilic microbes in it. In general a microbe will tolerate a maximal salinic environment at its optimum temperature and pH for growth. With respect to oxygen metabolism the microbes requiring high concentrations of NaCl are usually strong aerobes, the notable exception being the culture of *Bacteroides* isolated by Baumgartner (1937).

Problems are encountered in staining halophiles. Fixation with acetic acid is preferred to heat fixing.

One remarkable characteristic of strict halophiles is their strong tendency to form pink or red pigments. The higher the salt concentration, the greater the pigmentation. Oxygen is required for pigment production but this is not unusual considering the general aerobic nature of halophiles. Glycerol increases pigment production and carotenoids have been identified. No involvement of these pigments in electron transport has been demonstrated. It has been suggested that halophiles derive their resistance to sunlight from their pigments (Horowitz-Wlassowa, 1931).

Cell morphology often changes as the ionic strength of the growth medium increases. *Pseudonomas salinaria* changes from a coccoid to a long rod-shaped cell (Harrison and Kennedy, 1922). Such variance in morphology with ionic strength is correlated with a specific culture than with a general phenomena associated with all microbial cells. The surface structure of the extreme halophile, *Halobacterium cutirubrum*, undergoes definite morphological changes when it is placed in solutions with progressively lower ionic strength. In 4.5 M NaCl at pH 7.0 it is a rod-shaped bacterium. As the pH is decreased to 4.0 cocci are formed with some pleomorphism developing (Kushner and Bayley, 1962). Replacement of the sodium chloride with 1.0% acetic acid gives club-shaped rods showing some granulation. A further decrease in ionic strength of the acetic acid gives circular amorphous cells.

Irreparable damage to the cell occurs when its rod shape is lost. In water, cells of *H. cutirubrum* undergo plasmolysis and dissolve. Several salts added in place of NaCl protect the structural integrity of *H. cutirubrum*. At concentrations of 3.5–4.0 M, $MgCl_2$, $CaCl_2$, $Na_2S_2O_3$, and sodium acetate protect cellular integrity but only NaCl supports growth (Boring *et al.*, 1963).

METABOLIC ACTIVITIES OF HALOPHILES

In at least some halophiles carbohydrate metabolism decreases and the metabolism of amino acid and utilization of protein hydrolysates increases (Katznelson and Robinson, 1955; Smithies *et al.*, 1955). Growth of halophiles is slow which may be associated with amino acid utilization. Alkaline conditions often favor growth. Magnesium (0.1 *M*) is definitely required for growth of halophiles (Brown and Gibbons, 1955), whereas requirements for potassium and, especially, iron do not differ from other microbes. The lipoprotein cell walls of halophiles are higher in nitrogen and lower in carbohydrate than the cell walls of other microbes. A polymer or slime layer of nucleic acid may surround cells of *Vibrio costicolus* and *Micrococcus halodenitrificans*; this layer can be removed with deoxyribonuclease (Smithies and Gibbons, 1955). The polymer produced increases the viscosity of the medium. This polymer may affect survival mechanisms of halophiles in natural environments, because when it is removed cells are often more resistant to desiccation. Thermal temperature requirements for growth of halophiles often increase at higher salt concentrations. *Pseudomonas salinaria* grows over a temperature range from 5–42°C. The optima for growth of *Sporendonema epizoum* is 25°C.

The natural sedimentation of halophilic cells in lacustrine environments is slow. This is particularly true of *Halobacterium halobium* which contains gas vacuoles. Vacuoles can be eliminated by increasing the pressure. When the cells are then centrifuged, the specific gravity is found to be 1.2. They are not highly hydrated.

Nitrification, particularly in the initial stages, is not inhibited by high salt concentrations. Thus, nitrates are formed under halophilic conditions. Rubentschik (1929) observed the conversion of ammonia to nitrite by halophiles at concentrations of 15% NaCl. A culture of *Nitrosomonas* was isolated which showed optimal growth at 4.0% NaCl. The oxidation of nitrite to nitrate was not demonstrated.

The rate of denitrification is inhibited by sodium chloride. Hof (1935) found that at 30% NaCl 10 days was required to give the same amount of gas formed in 3 days at a concentration of 3% NaCl.

Sulfate reducers are active at high salt concentrations. *Spirillum desulfuricans* has been isolated from the Black Sea (Issatschenko, 1929). Sulfate reduction occurs in solutions containing up to 20% NaCl. Saslawsky (1928) concluded that 30% salt was the upper limit.

Several autotrophs and facultative autotrophs are obligate halophiles while others are halo-tolerant. These include *Thiobacillus* sp., *Beggiatoa*, and *Gallionella*. Heterotrophs which break down cellulose, chitin, fat, and proteins are functional at high salt concentrations. In testing 14 genera of microbes representing widely varying groups, Hof (1935) observed that most grew at salt concentrations up to 6%. A culture of *Clostridium acetobutylicum* was the most sensitive and did not grow at 3% NaCl.

A compilation of bacteria, fungi, and algae found in halophilic environments is shown in Table 17.III. Several species of *Cytophaga* which are active in cellulose breakdown tolerate high levels of salt. Cultures of *Lactobacillus* and *Pediococcus* isolated from cucumber brines (Etchells

TABLE 17.III

SALT-TOLERATING BACTERIA, FUNGI, AND ALGAE

Microbe	Salt tolerance characteristics	Pertinent references
Actinomyces	Liman	Hof (1935)
Bacteria		
Bacterioides halosmophilus (anaerobe)		
Bacterium trapanicum (Grissee near Java)		Hof (1935)
Chromobacterium marismortui	0.5–30%	
Cytophaga krzemieniewskae		
Cytophaga diffluens		
Cytophaga sensitiva		
Halobacterium cutirubrum	>4.0	
Halobacterium salinarium	28–32	
Halobacterium halobium	>12% NaCl	
Halobacterium marismortui	>18%	
Halobacterium trapanicum	>18%	
Lactobacillus cucumeris fermentatis	9% NaCl	Hof (1935)
Lactobacillus plantarum	5.6% NaCl	Etchells *et al.*, (1964)
Lactobacillus brevis	5.6% NaCl	Etchells *et al* (1964)
Leucothrix mucor (surface granules)		
Micrococcus morrhuae (Sambhar Lake, India)		Hof (1935)
Pediococcus halophilus	12% NaCl	Hof (1935)
Pediococcus cerevisiae (cucumber)	5.6% NaCl	Etchells *et al.* (1964)
Pseudomonas beijerinckii (salt beans)		
Rhodospirillum halophilum (Sands Springs, Nevada)		Bass-Becking (1928)
Sarcina literalis	16–32%	

(Continued)

TABLE 17.II—(*Continued*)

Microbe	Salt tolerance characteristics	Pertinent references
Urobacillus hesmogenes (Chadji bey Liman, Odessa)	19% Liman	Rubentschik (1926a,b)
Urobacterium sp.	24% NaCl	Hof (1935)
Thiobacillus sp.	12%	Hof (1935)
Vibrio halonitrificans	1–23%	Hof (1935)
Fungi		
Candida tropicalis		
Candida mycoderma		
Debaryomyces nicotianeae		
Debaryomyces membranaefaciens		
Debaryomyces hansenii		
Endomycopsis ohmeri		
Hansenula subpelliculosa		
Pichia membranaefaciens		
Saccharomyces rosei		
Saccharomyces rouxii		
Zalerion eistla		
Algae		
Brachiomonas submarina		
Brachiomonas pulsifera		
Carteria chuii		
Chlamydomonas sp.		
Chlorella sp.		
Dunaliella salina		
Dunaliella euchlora		
Microcoleus chthonoplastes		
Monochrysis sp.		
Platymonas suburdiformis		
Porphyridium sp.		
Nannochloris otomus		
Stichococcus basillaris		

et al., 1964) are very active in brines containing 6.0% NaCl. A culture of *Urobacillus* tolerates 19% liman salt (Rubentschik, 1926a,b) but has lower tolerances for other salts, as follows: NaCl, 13%; KCl, 14%; $MgCl_2$, 9%; $CaCl_2$, 6%. Optimal growth for the halobacteria is around 15% NaCl and many will grow in saturated solutions. *Thiobacillus* active in sulfuric acid synthesis will grow in solutions of 12% NaCl (Hof, 1935).

Most of the fungi listed in Table 17.III are yeasts. They are encountered in various pickling operations (Etchells *et al.*, 1964).

The list of algae in Table 17.III is in no way comprehensive, and those interested in a full listing should consult the classic works of Vinogradov (1953).

SALT CONCENTRATION BY MICROBES

Earlier research has indicated that the internal concentration of salts in cells is the same as the external. In *Micrococcus halodenitrificans* the internal concentration of salt increases from 1.5% in cells grown at 2–4% NaCl to 5.0% in cells grown in 8% salt (Table 17.IV). The internal salt

TABLE 17.IV

INTERNAL AND EXTERNAL CONCENTRATION OF SALT IN HALOPHILES

Halophile	Percent or molarity		References
	External salt	Internal salt	
Micrococcus			
halodenitrificans	2–4%	1.5%	Gibbons and Baxter (1955)
	8.0%	5.0%	
Sarcina littoralis	18%	10%	Gibbons and Baxter (1955)
Bacillus premilus	10.8%	8.9%	Yamada and Shiio (1953)
Vibrio costicolus	1.0 M NaCl	0.3 M NaCl	Baxter and Gibbons (1954)
	1 0 M KCl	0.3 M KCl	
	1 M KCl	0.9 M KCl	

concentration did not increase with higher concentrations of salt (Gibbons and Baxter, 1955). With *Sarcina littoralis* the internal salt concentration increased linearly from 10–19% when the external salt concentration was increased from 18–25%. The potassium ion is somehow involved in the ability of halophilic cells to handle NaCl. In comparing less strict halophiles like *Vibrio costicolus* with *Pseudomonas salinaria*, the less salt tolerant cultures appear to expend energy to keep salt out, whereas strict halophiles tolerate high internal concentrations of salt and produce enzymes resistant to such salts (Baxter and Gibbons, 1954; Egami, 1955), In the strict halophile NaCl is required for growth.

In a comparison of nonhalophilic and halophilic microbes it is found that all concentrate potassium over the amount present in the medium (Christian and Waltho, 1962). The intracellular concentration of Na^+ increase from ~ 100 mM/kg cell water in nonhalophiles, to ~ 500 in moderate halophiles, to a high of 3170 (Table 17.V). There is only a slight difference between the intracellular concentration of chloride ion between nonhalophiles and moderate halophiles, but the chloride level increases tremendously in halophilic cultures of *S. morrhuae* and *H. salinarium.* The phosphate content remained about the same in all those cultures examined.

TABLE 17.V

CELLULAR SOLUTE CONCENTRATION OF NONHALOPHILIC AND HALOPHILIC BACTERIA[a]

| Bacteria | Solute concentration[b] | | | |
	Na^+	K^+	Cl^-	PO_4^{3-}
Nonhalophiles[a]				
Staphlococcus aureus	98	680	8	83
Salmonella oranienburg	131	239	<5	46
Moderate Halophiles[c]				
Micrococcus halodenitrificans	311	474	55	64
Vibrio costicolus	684	221	139	71
Halophiles[c]				
Halobacterium salinarium	1370	4570	3610	91
Sarcina morrhuae	3170	2030	3660	46

[a] After Christian and Waltho (1962).
[b] Meq. or mmoles/kg cell water.
[c] Nonhalophiles grown on 0.15 *M* NaCl and 0.025 *M* KCl; moderate halophiles grown on 1.0 *M* NaCl and 0.004 *M* KCl; halophiles grown on 4.0 *M* NaCl and 0.032 *M* KCl.

Salt tolerance has been correlated with the ability of a microbe to concentrate intracellular potassium. The salmonellas which are intolerant to high salt concentration contain little potassium (Christian and Scott, 1953; Christian, 1958), whereas the staphylococci which are salt tolerant are rich in potassium. In examining 32 strains of nonhalophilic bacteria the ability to tolerate salt was definitely correlated with the ability to accumulate potassium within the cell (Christian and Waltho, 1961). One coccus grew at a sodium chloride concentration of 4.0 *M* which is unusual for a nonhalophile and is comparable to that of *Vibrio costicolus*.

Extreme halophiles such as *Ps. salinaria* and *Micrococcus halodenitrificans* have an intracellular salt concentration of 0.5 *M*. Cytochrome oxidase is active in maintaining this level against a concentration gradient (Baxter and Gibbons, 1956). In *Ps. salinaria* isocitric, succinic, malic, and lactic dehydrogenases were most active at high concentrations of sodium or potassium chloride. In *M. halodenitrificans* lactic dehydrogenase and cytochrome oxidase were most active at high salt concentrations.

The transport of sodium and potassium across membranes is also linked in mammalian cells. The cell is unable to pump sodium out unless extracellular potassium is available to be pumped into the cell (Harris and Maizels, 1951). In human erythrocytes cholinesterase inhibitors decrease the rate of potassium uptake whereas choline acetylase inhibitors increase the rate of loss of potassium from the cell. Adenosine triphosphatase in the membranes in human erythrocytes is a part of the system active in the transport of sodium outward and potassium inward (Post *et al.*, 1960).

Marine bacteria have a higher requirement for Na^+, K^+, and Mg^{++} ions than terrestrial forms (Tomlinson and MacLeod, 1957). Out of 96 isolates made from Atlantic coastal waters off Florida all required sodium chloride for good growth. In most cultures K^+ and Mg^{2+} also improved carbohydrate metabolism (Tyler *et al.*, 1960).

Nonsporulating bacteria normally die when the media they are grown upon gradually dries up. However, Dombrowski (1966) observed that addition of salt until the medium is supersaturated followed by a slow drying until all the salt is crystallized out will give crystals containing entrapped viable bacteria. Some bacteria can be revived after seven years of storage. In an extension of this study to actual ores of salt, bacteria were isolated from ores of the following ages:

Origin of salt	Age of deposit ($\times 10^6$ years)
Germany	200
Canada	260
New York (Siluric Salts)	400
Irkutsk (Lower Cambrian)	500–600

These data are supported by actual experience with crude solar salt which is found to be heavily contaminated with microbes. Mined salts are much less contaminated.

The algae *Nitella* concentrates a number of ions over the concentration found in pond water (Hoagland and Davis, 1923).

Ion	Cell (mg/gm)	Pond water (mg/gm)	Concentration factor
K	2120	0.0	—
Na	230	5	46
Ca	410	31	13
Mg	430	41	10
Cl	3220	32	100
SO$_4$	800	31	36
PO$_4$	350	0.4	870
NO$_3$	—	0.5	—

Blinks and Nielsen (1940) found potassium concentrated in *Hydrodictyon* over 4000 times the external concentration. On cultivation of algae cultures of *Valonia macrophysa*, *Valonia ventricosa*, and *Halicystis osterhoutii* under conditions with an external sodium ion concentration of 0.50 *M*, the latter culture contained a cellular concentration of potassium of 0.56 *M*; cellular concentrations in the other cultures were less by at least a factor of one-fifth.

Three fungi have been consistently isolated from the Dovey Salt Marshes in England (Elliott, 1930). These were *Torula allii*, *Penicillium hyphomycetis*, and *Fusarium oxysporum*. These cultures normally are found in soils, but evidently they can adapt to a salinic environment. Other fungi withstanding salinic environments are *Phoma* and *Curvularia* (Ritchie, 1959).

Dunaliella has been isolated from dry salt samples stored for 7 years (Hof, 1935).

REFERENCES

Baas-Becking, L. G. M. (1928). An organism living in concentrated brine. *Tijdschr. Ned. Dierk. Ver.* **1**, 1.

Baas-Becking, L. G. M. (1931). Salt effects on swarmers of *Dunaliella viridis* Teod. *J. Gen. Physiol.* **14**, 765.

Baumgartner, J. G. (1937). The salt limits and thermal stability of a new species of anaerobic halophile. *Food Res.* **2**, 321.

Baxter, R. M., and Gibbons, N. E. (1954). The glycerol dehydrogenases of *Pseudomonas salinaria*, *Vibrio costicolus* and *Escherichia coli* in relation to bacterial halophilism. *Can. J. Biochem. Physiol.* **32**, 206.

Baxter, R. M., and Gibbons, N. E. (1956). Effects of sodium and potassium chloride on certain enzymes of *Micrococcus halodenitrificans* and *Pseudomonas salinaria*. *Can. J. Microbiol.* **2**, 599–606.

Blinks, L., and Nielsen, J. (1940). The cell sap of hydrodicyton. *J. Gen. Physiol.* **23**, 551–559.

Boring, J., Kushner, D. J., and Gibbons, N. E. (1962). Specificity of the salt requirement of *Halobacterium cutirubrum*. *Can. J. Microbiol.* **9**, 143–154.

Boring, J., Kushner, D. J., and Gibbons, N. E. (1963). Specificity of the salt requirement of *Halobacterium cutirubrum*. *Can. J. Microbiol.* **9**, 143–154.

Brown, H. J., and Gibbons, N. E. (1955). The effect of magnesium, potassium and iron on the growth and cell morphology of halophilic bacteria. *Can. J. Microbiol.* **1**, 486.

Christian, J. H. B. (1956). The physiological basis of salt tolerance in halophilic bacteria. Dissertation, Cambridge University, Cambridge.

Christian, J. H. B. (1958). The effects of washing treatments on the composition of *Salmonella oranienburg*. *Australian J. Biol. Sci.* **11**, 538.

Christian, J. H. B., and Scott, W. J. (1953). Water relations of Salmonellae at 30°C. *Australian J. Biol. Sci.* **6**, 565–573.

Christian, J. H. B., and Waltho, J. A. (1961). The sodium and potassium content of non-halophilic bacteria in relation to salt tolerance. *J. Gen. Microbiol.* **25**, 97–102.

Christian, J. H. B., and Waltho, J. A. (1962). Solute concentrations within cells of halophilic and non-halophilic bacteria. *Biochim. Biophys. Acta* **65**, 506–508.

Conway, E. J. (1953). "The Biochemistry of Gastric Acid Secretion." Thomas, Springfield, Illinois.

Danielli, J. F. (1964). Morphological and molecular aspects of active transport. *Symp. Soc. Exptl. Biol.* **8**, 502–516.

Dombrowski, H. J. (1966). Reviving bacteria from salts. *New Scientist* p. 89.

Egami, F. (1955). Recherches biochimiques sur les bactéries halotolérantes at halophiles. *Bull. Soc. Chim. Biol.* **37**, 207.

Elliott, J. S. B. (1930). The soil fungi of the Dovey Salt Marshes. *Ann. Appl. Biol.* **17**, 284–305.

Etchells, J. L., Costilow, R. N., Anderson, T. E., and Bell, T. A. (1964). Pure culture fermentation of brined cucumbers. *Appl. Microbiol.* **12**, 523–535.

Franck, J., and Mayer, J. E. (1947). An osmotic diffusion pump. *Arch. Biochem. Biophys.* **14**, 297–313.

Gibbons, N. E., and Baxter, R. M. (1955). The relation between salt concentration and enzyme activity in halophilic bacteria. *Proc. 6th Intern. Congr. Microbiol.*, Rome, 1953 Vol. 1, p. 210. Staderini, Rome.

Goldacre, R. J. (1952). The folding and unfolding of protein molecules as a basis of osmotic work. *Intern. Rev. Cytol.* **1**, 135–164.

Harris, E. J., and Maizels, M. (1951). The permeability of human red cells to sodium. *J. Physiol. (London)*, **113**, 506–524.

Harrison, F. C., and Kennedy, M. E. (1922). The red discoloration of cured codfish. *Trans. Roy. Can. Inst.* **16**, 101.

Henneberg, W. (1926). "Handbuch der Gärungsbakteriologie," 2nd ed. Parey, Berlin.

Hoagland, D. R., and Davis, A. R. (1923). A composition of cell sap of the plant in relation to the absorption of ions. *J. Gen. Physiol.* **5**, 629–646.

Hof, T. (1935). Investigations concerning bacterial life in strong brines. *Rec. Trav. Botan. Neerl.* **32**, 92–173.

Horowitz-Wlassowa, L. M. (1931). Über die Rotfärbung gesalzener Därme (der rote Hund). *Zentr. Bakteriol., Parasitenk., Abt. II* **85**, 12.

Ingrahm, R. C., Peters, H. C., and Vesscher, M. B. (1938). On the movement of materials across living membranes against concentration gradients. *J. Phys. Chem.* **42**, 141–150.

Ingram, M. (1957). Microorganisms resisting high concentrations of sugar or salts. *Symp. Soc. Gen. Microbiol.* **7**, 90.

Issatschenko, B. L. (1929). *Spirillum desulfuricans* from the Black Sea. *Proc. 1st Intern. Congr. Plant. Scl., Ithaca,* 1926 p. 211. Banta Publ., Menasha, Wisconsin.

Katznelson, H., and Robinson, J. (1955). Observation on the respiratory activity of certain obligately halophilic bacteria with high salt requirements. *J. Bacteriol.* **71**, 244–249.

Kushner, D. J., and Bayley, S. T. (1962). The effect of pH on surface structure and morphology on the extreme halophile, *Halobacterium cutirubrum. Can. J. Microbiol.* **9**, 53–63.

Osterhout, W. J. V. (1936). The absorption of electrolytes in large plant cells. *Botan. Rev.* **2**, 283–315.

Post, R. L., Merritt, C. R., Kinsoloving, C. R., and Albright, C. D. (1960). Membrane adenosine triphosphatase as a participant in the active transport of sodium and potassium in the human erythrocyte. *J. Biol. Chem.* **235**, 1796–1802.

Preston, R. D. (1952). "The Molecular Architecture of Plant Cell Walls," p. 82. Chapman & Hall, London.

Ritchie, D. (1959). The effect of salinity and temperature on marine and other fungi from various climates. *Bull. Torrey Botan. Club* **86**, 367–373.

Rothstein, A. (1954). Enzyme system of the cell surface involved in the uptake of sugars by yeast. *Symp. Soc. Exptl. Biol.* **8**, 165–201.

Rubentschik, L. (1926a). Über die Einwirkung von Salzen auf die Levenstaligkeit der Urobakterien. *Zentr. Bakteriol., Parasitenk., Abt. II* **67**, 167–236.

Rubentschik, L. (1926b). Zur Entwicklungs Geschichte einiger Urobacterien mit beson deser besucksichtigung der gleichzeitigen ein Wirk ungen von NaCl und von $(NH_4)_2CO_3$ auf dieselben. *Zentr. Bakteriol., Parasitenk, Abt. II* **68**, 161–179.

Rubentschik, L. (1929). Studies on the bios question. *Zentr. Bakteriol., Parasitenk., Abt. II* **77**, 1–17.

Ruinen, J. (1933). Salt flagellates. II. Distribution of the salt flagellates. *Arch. Protistenk.* **90**, 210–258.

Salton, M. R. J. (1955). Bacterial Cell walls. *Symp. Soc. Gen. Microbiol.* **6**, p. 81.

Saslawsky, A. S. (1928). Zur frage der Wirkung hoher Salzkonzentrationen auf die biochenischen Proze im Limanschlamm. *Zentr. Bakteriol., Parasitenk., Abt. II* **73**, 18–28.

Scott, W. J. (1953). Water relations of *Staphlococcus aureus* at 30°C. *Australian J. Biol. Sci.* **6**, 549.

Scott, W. J. (1957). The water relations of food spoilage microorganisms. *Advan. Food Res.* **7**, 83–127.

Shewan, J. M. (1942). Some bacteriological aspects of fish preservation. *Chem. & Ind. (London)* **20**, 312–314.

Smithies, W. R., and Gibbons, N. E. (1955). The deoxyribose nucleic acid slime layer of some halophilic bacteria. *Can. J. Microbiol.* **1**, 614–621.

Smithies, W. R., Gibbons, N. E., and Baylcy, S. T. (1955). The chemical composition of the cell and cell wall of some halophilic bacteria. *Can. J. Microbiol.* **1**, 605–613.

Tomlinson, N., and MacLeod, R. A. (1957). Nutrition and metabolism of marine bacteria. IV. The participation of Na^+, K^+, and Mg^{++} salts in the oxidation of exogeneous substrates by a marine bacterium. *Can. J. Microbiol.* **3**, 627–638.

Tyler, M. E., Bielling, M. C., and Pratt, D. B. (1960). Mineral requirements and other characters of selected marine bacteria. *J. Gen. Microbiol.* **23**, 153–161.

Ussing, H. H. (1957). General principles and theories of membrane transport. *In* "Metabolic Aspects of Transport Across Cell Membranes" (Q. R. Murphy, ed.). Univ. of Wisconsin Press, Madison, Wisconsin.

Vaisey, E. B. (1954). Osmophilism of *Sporendonema epizoum*. *J. Fisheries Res. Board, Can.* **11**, 901–903.

Vinogradov, A. P. (1953). "The Elementary Chemical Composition of Marine Organisms," Sears Found. Marine Res., Memoir II. Yale Univ. Press, New Haven, Connecticut.

Yamada, T., and Shiio, I. (1953). Effects of salt concentrations on the respiration of a halo-tolerant bacterium. *J. Biochem.* (*Tokyo*) **40**, 327–337.

18

BIOGEOCHEMICAL TOPICS

Some biochemical topics have not been adequately studied to merit more than a brief discussion. Four topics in this category are: (1) coal, (2) biogeochemical prospecting, (3) bioelectricity, and (4) chelates synthesized by microbes.

COAL

Until recently coal has been our chief mineral fuel and major source of energy. Natural gas, hydrocarbons, and uranium are now competing with coal. The coalfields in Pittsburgh, Pennsylvania, are the most valuable single mineral deposit of this type in the world. This field has yielded income in excess of 6.5 billion dollars, and the total deposit is estimated at 20,000 million tons.

Complex biological processes are involved in coal deposition, since coals are a compact mass of carbonized plant debris. As a rule they are embedded between sandstones and shales. Many coalfields have a clay base. Coals are classified according to rank which depends upon their content of fixed carbon, combustible hydrocarbons and other volatiles, moisture, ash, and sulfur. Lignites are the lowest ranked coals and when first mined, they may contain 30–40% moisture. Upon drying they are subject to spontaneous ignition. Bituminous coals are of medium rank. They do not slack or crumble like lignites upon drying, and they generally have a prismatic jointing perpendicular to their banding. The highest ranking coal is anthracite, which is distinguishable physically from other ranks by the presence of conchoidal fracture and the absence of cross joining. As the rank increases the fixed carbon increases and the water and volatile matter decrease.

Coals were formed in forested coastal swamps. These swamps were stagnant, and anaerobic conditions existed which permitted the accumulation of organic carbon. Such swamps gradually disappeared leaving large accumulations of plant debris. The nature of this environment is derived from the botanical character of the coal and the flora which have been identified in it. Microbes were very active in the depositional stages. In the very early stages of development the amount of available plant residue far exceeded the amount of oxygen present in the swamps so that the aerobic microbes depleted any oxygen present by oxidizing a small part of the great abundance of plant residue. The anaerobic bacteria continued to degrade plant residues and, because of an imbalance in the excess plant debris and a limitation in the mineral nutrients required for optimal microbial activity, only partial degradation of the plant residue occurred. Also, many toxic microbial and plant residues accumulated to inhibit further microbial activity. Sulfate reducers are active in such habitats, and mineral sulfides such as pyrite (FeS_2) readily precipitate.

With the tremendous known reserves of coal, new uses for coal are being sought. Possible exploitation of coal as a filter aid (Table 18.I) in sewage removal (Office of Coal Research and Development—Report 12) and the catalytic cracking of coal to gasoline and natural gas appear to be feasible.

TABLE 18.I

COMPONENTS REMOVED FROM SEWAGE BY COAL[a]

Components	Treatment	
	Primary	Secondary
Suspended matter	35–60%	80–95%
Oxygen-demanding substances	15–35%	70–90%
Phosphates		
Insoluble	0	20–65
Soluble	0	0–20
Detergents	0–5%	5–30

[a] Rand Development Corporation Report: Office of Coal Research and Development—Report 12.

Sixty percent of the suspended matter and up to 35% of oxygen-utilizing material were removed from sewage by using coal adsorbants (Table 18.I). Phosphate removal in a primary sewage treatment system has not been successful, but both insoluble and soluble phosphate decrease in sewage upon secondary treatment. Detergent reduction varies from 0–30%. In testing samples of sewage for the removal of nitrogen, erratic results are observed. The range of nitrogen removal varies from 0–52%.

Removal of bacteria has been tested with *Escherichia coli*. Definite adsorption occurred in the early stages but the coal appears to saturate quickly, and further removal of bacteria ceases. These tests are preliminary and many other microbial systems will have to be evaluated. Tests with autotrophic bacterial cultures (*Pseudomonas*) capable of metabolizing the mineral and organic components in coal were not performed. Such microbes may give entirely different results.

The rate of removal of pyrite from coal depends on particle size and the type of coal. *Ferrobacillus ferrooxidans* can leach 80% or more of the pyrite from some bituminous coals in 3–4 days. The desulfurization of subbituminous coals and lignites is slower. This is attributed to the acid-neutralizing capacity of lower rank coals. Addition of ferrous sulfate increases the rate of pyrite removal from all coals. *Thiobacillus ferrooxidans* is active but *T. thiooxidans* is not (Leathen *et al.*, 1953). There have been conflicting reports as to the pyrite-oxidizing capacity of this last microbe.

Mycobacterium ranae and *Nocardia coeliaca* change the rheological properties of asphalts (Traxler *et al.*, 1965). Sterile bentonite-asphalt emulsions were prepared and inoculated. At the end of a 4-month incubation, viscosity had increased over sixfold in one asphalt sample. The resin components of asphalt appear to be susceptible to degradation by microbes. These resins are probably converted into high molecular weight microbial polysaccharides which increase the viscosity.

Iron bacteria grow upon crude tars with asphaltic properties and at times are known to clog heat exchangers with fibrous precipitates. These precipitates consist primarily of iron and manganese salts (Garnett, 1960).

BIOGEOCHEMICAL PROSPECTING

This field offers some potential but it presents more problems than solutions. It would appear that geobotanical prospecting should receive the greater attention primarily because the seed plants (spermatophytes) are easier to identify and samples are readily obtainable. Lichens growing upon the surface of rocks may offer good source material for analysis in some instances, but lichens do not have extensive root systems and thus their effective spatial range of activity for concentrating minerals is quite limited.

Geomicrobiological prospecting can be used in two instances: first, where the chemical environment is such that it acts as a natural selective enrichment for one or a particular ecological group of microbes; second, where the microbes in question selectively concentrate a mineral and also act as sensitive indicators of the mineral. In both instances the microbe has to be isolated and identified, and quantitative microbial counts have to be made. In the second instance the microbe and the mineral level will have to be determined. Both analyses are expensive.

Since the environment is being altered chemically by the microbe, it may be just as easy to analyze directly for the mineral in question rather than for a microbe which serves only as an indicator of the mineral. Also some microbial environments are microecological, a fact which may give anomalous results. Environments can always be enlarged and extended by extensive ground water movement. For the ground water to serve as an enrichment medium to aid in the selection of one or a group of microbes it must have an unique chemical composition. Again it may be more useful to analyze the ground water for a chemical than to isolate an indicator microbe. Specific microbes found in high numbers in a locality are a result of a specialized ecological habitat with a definite chemical composition. The greater the chemical anomaly, the greater the chance of isolating microbes of biogeochemical significance. This anomalous condition can be local or widespread depending on the size of a particular deposit and the degree of movement of ground waters carrying the mineral or chemical in solution.

Progress in biogeochemical prospecting will be slow because of the limitations described. The microbes which do have some potential in biogeochemical prospecting include lichens, fungi (Basidiomycetes), algae, and bacteria which form blooms in lakes and streams.

Lichens

There are several thousand recognized species of lichens. They consist of a green algae and a fungus living symbiotically. The fungus grows next to the rock obtaining minerals and water from its lithospheric environment, while the algae grows as a thin layer of cells on the surface of the fungus (Fig. 18.1). The thickness of both fungus and algae together rarely exceeds 0.5 mm. Lichens grow slowly and some colonies have been estimated to be 2000 years old (Lamb, 1959). They do not have a root system and growth is normally restricted to the surface of the bedrock. *Parmelia conspersa* is a typical lichen. Some lichens are stalkformers like *Cladonia bellidiflora* and utilize some type of decaying organic matter.

(a)

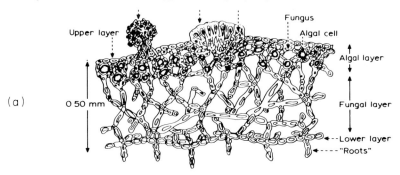

Spectrographic analyses (PPM) of Lichen.

Turkey Creek, Jefferson Co., Colorado.

(b)

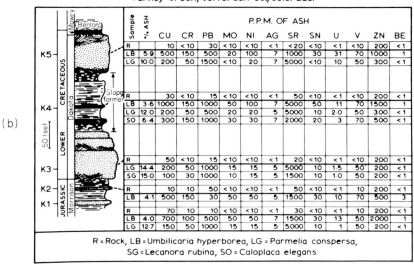

Sample	% ASH	P.P.M. OF ASH											
		CU	CR	PB	MO	NI	AG	SR	SN	U	V	ZN	BE
R		10	<10	30	<10	<10	<1	<20	<10	<1	<10	200	<1
LB	5.9	500	150	500	20	100	7	1000	30	31	70	1000	1
LG	10.0	200	50	1500	<10	20	7	5000	<10	10	50	300	<1
R		30	<10	15	<10	<10	<1	50	<10	<1	<10	200	<1
LB	3.6	1000	150	1000	50	100	7	5000	50	11	70	1500	1
LG	12.0	200	50	500	20	20	5	5000	10	2.0	50	300	<1
SO	6.4	300	150	1000	30	30	7	2000	20	3	70	500	<1
R		50	<10	15	<10	<10	<1	20	<10	<1	<10	200	<1
LG	14.4	200	50	1000	15	15	5	5000	10	1.5	50	200	<1
SG	15.0	100	30	1000	10	15	5	1500	10	1.0	50	200	<1
R		10	10	50	<10	<10	<1	50	<10	<1	10	200	<1
LB	4.1	500	150	30	50	50	5	1500	10	10	70	500	3
R		70	10	10	<10	<10	<1	30	<10	<1	10	200	<1
LB	4.0	700	100	500	50	50	7	1500	30	13	50	2000	1
LG	12.7	150	50	1000	15	15	5	5000	10	1	50	200	<1

R = Rock, LB = Umbilicaria hyperborea, LG = Parmelia conspersa, SG = Lecanora rubina, SO = Caloplaca elegans.

FIG. 18.1. (a) Schematic cross section of a lichen; (b) summary of trace elements in a lichen. [(b) From LeRoy and Koksoy, 1962.]

LeRoy and Koksoy (1962) spectrographically analyzed lichens obtained from bedrock outcroppings of Jurassic and Lower Cretaceous periods in Colorado. The formations involved were Morrison, Dakota, and Mowry (Fig. 18.1). Over 500 ppm Cu^{2+} was found in the ash of *Umbilicaria hyperborea* (L.B.) and up to 1000 ppm of lead in *U. hyperborea, Parmelia conspersa, Caloplaca elegans*, and *Lecanora rubina*. Ash of *Parmelia conspersa* contained 5000 ppm of strontium. All these forms contained 200 ppm of zinc. Other elements were concentrated in lesser amounts.

BIOELECTRICITY

The relevancy of bioelectricity to biogeochemistry might be questioned, but its occurrence in nature is so common that it must be considered a potent mechanism for mobilizing cations and anions.

A fuel cell is defined by the National Electrical Association as an electrochemical cell which can continuously change the chemical energy of a fuel with an oxidant to electrical energy by an isothermal process involving an essentially invariant electrode electrolyte system. Contained within this definition is the fact that the efficiency of energy conversion (from chemical to electrical energy) is not limited by the Carnot cycle. This is attributed to the isothermal nature of a fuel cell.

History of Fuel Cells

Fuel cells are quite old. In 1802 Sir Humphrey Davy used 20 cells in series, each containing carbon in contact with nitric acid, to decompose water. About 40 years later William R. Grove invented and tested a battery of fuel cells as an energy source. The fuel cell however still remained a curiosity until near the end of the 19th century (1894) when a German physical chemist Wilhelm Ostwald realized the potential of the fuel cell as an efficient, simple, and inexpensive means of obtaining electrical energy from chemical energy.

Biochemical fuel cells simply represent an extension of classic fuel cell technology. In 1912 Potter observed that a potential difference was set up between a sterile solution of nutrient salts and a similar solution in which microorganisms were suspended and actively growing. Potter was able to obtain a current of 1.25 mA from 6 cells containing glucose and inoculated with yeast versus a pure glucose solution. In 1931 Cohen built a bacterial battery capable of an output of 2 mA at a potential difference of

0.35 V. The concept of converting biological energy to electrical energy by direct electrochemical means is quite simple. The disintegration of organic compounds by microorganisms is accompanied by liberation of electrical energy. The electrical effects are an expression of the activity of the microorganism and are influenced by the temperature, the concentration of the nutrient medium, and the number of active organisms present. The bacterial culture during the process of energy conversion is, in a sense, a primary electrical half-cell and as such should conceivably be able to perform work.

Uses of Biochemical Fuel Cells

Biological electrical energy production, because of its potential as a cheap source of power, has aroused much interest. The fuels for biochemical cells are normally all natural products, such as food, oil, waste products, and sewage as well as minerals. This approach could serve as a way of disposing of wastes, i.e., obtaining electrical power from wastes while at the same time degrading and eliminating the waste. Bioelectrochemical power could be a boom to countries whose fossil fuel supplies are small. Such systems require cheap, simple structures and controls, and they must be reliable. This involves minimizing the possibility of contamination of the system by undesirable competitive microorganisms. Selective microbial control could be accomplished by choosing a microorganism which operates near the extremities of the pH scale (Cohn, 1963).

At present, in industrialized countries the inexpensive availability of fossil fuels, hydroelectric power, or atomic fuel relegates relatively expensive biochemical energy to speciality uses. These include silent, high efficiency, high power density (small volume, light weight) power sources for military use. Biochemical fuel cells are used for power signaling devices which must function for long periods without refueling, in remote, relatively inaccessible areas. This would include long-life beacons at sea and, in the future, space devices. Biocells could be used as power sources for the military, campers, archeologists, geologists, naturalists, and boating enthusiasts. Potentially the biochemical fuel cell could be powered in remote areas by almost any natural product (food, vegetation, and waste) and should provide a highly flexible energy source.

Types of Biochemical Fuel Cells

Bioelectrochemical conversion devices may be arbitrarily classified according to the methods of generation of the electrode reaction. These methods are either direct or indirect (Del Duca et al., 1962).

In an indirect fuel cell, biochemical agents are used to generate a particular intermediate substance which is more electrochemically active than the starting material. The intermediate reacts with the electrodes as in a conventional fuel cell. Examples of the indirect method are the formation of hydrogen, ammonia, or methanol from higher molecular weight materials such as sugars, starch, or urea. Hydrogen, a by-product of the biochemical reaction of the organism *Clostridium butyricum* with glucose, for example, may then be used in a classic hydrogen-oxygen fuel cell.

Some possible indirect fuel-air primary cells are summarized in Table 18.II.

The operational features of a typical indirect biochemical fuel cell are shown in Fig. 18.2.

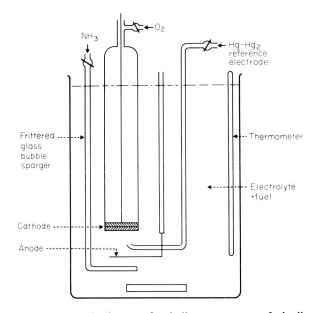

FIG. 18.2. Schematic diagram of an indirect urea-oxygen fuel cell.

There are many examples of the indirect process. *Bacillus pasteurii* enzymatically hydrolyzes urea to ammonia and CO_2 (Table 18.II):

$$NH_2—CO—NH_2 + H_2O \rightarrow 2\,NH_3 + CO_2$$

The electrode reaction is:

$$2\,NH_3 + 6\,OH^- \rightarrow N_2 + 6\,H_2O + 6\,e^-$$

TABLE 18.II

INDIRECT FUEL-AIR PRIMARY CELLS[a]

Natural fuel	Natural source	Catalyst	Electrochemical reactant	Electrode half-cell potential *vs* S.H.E.[b] (v)
NH$_3$ generators				
Urea	Urine	*Bacillus pasteurii* (Microorganism)	NH$_3$	−0.4
Urea	Urine	Urease (Crystalline enzyme)	NH$_3$	−0.4
Amino acid	Protein digest	Amino acid oxidase (Enzyme)	NH$_3$	−0.4
Hydrogen generators				
Formic acid	—	Formic dehydrogenase (enzyme)	H$_2$	−0.4
Glucose	Molasses	*Clostridium butyricum* (Microorganism)	H$_2$	−0.4
Methane generators				
Propionic acid	Fats Oils	*Methanobacterium proprionicum* (Microorganism)	CH$_4$	−0.4

[a] After Del Duca *et al.* (1962).
[b] S.H.E. = standard hydrogen electrode.

Algae which produce oxygen during photosynthesis are active in converting light into electrical energy.

$$6 \, CO_2 + 6 \, H_2O \rightleftharpoons C_6H_{12}O_6 + 6 \, O_2$$

At the electrode oxygen readily picks up electrons.

$$O_2 + 2 \, H_2O + 4 \, e^- \rightarrow 4 \, OH^-$$

The second type of biochemical cell is the direct type. In the direct fuel cell microoorganisms grow and multiply by extracting energy from substrates or foods which they oxidize. The oxidation is catalyzed by enzymes. In biochemical fuel cells of the direct type the enzyme itself reacts with the electrode. Since enzymes are large molecules with molecular weights often exceeding 100,000, diffusion is slow, and reactions with the electrodes could be rate limiting if the reactant had to diffuse to the electrode. If enzymes were to orient themselves on the surface of the electrode it would conceivably be possible for them to continuously accept and donate electrons. A definite orientation of an electron transfer system, on the cytochromes, does occur within cells. Where contact of the enzyme is essential, agitation should improve current flow. Many of these reactions involve oxidation and reduction in which electrons are transferred from one compound to another. During this process instead of transferring electrons to oxygen the inert electrode is capable of accepting the electrons. It is not known exactly what does occur, but two possible mechanisms are postulated. In the first a net positive ion (hydrogen ion) is hypothesized to be transported from anode to cathode where the ions are then bound. In the second mechanism a net negative ion (hydroxyl ion) is transported in the opposite direction. It is believed that pH and ion transport rates partly determine which of the two mechanisms prevails.

In general the dehydrogenation process may be formulated as follows

$$AH_2 + B \rightarrow BH_2 + A$$

A specific example of this is where A is any substrate capable of undergoing biochemical oxidation while B is oxygen

$$AH_2 + \tfrac{1}{2} O_2 \rightarrow A + H_2O$$

AH_2 is the reduced form of the substrate while A is the oxidized form of the substrate (a chemical species different from AH_2).

Figure 18.3 represents a schematic diagram of the process which occurs in a direct biochemical fuel cell.

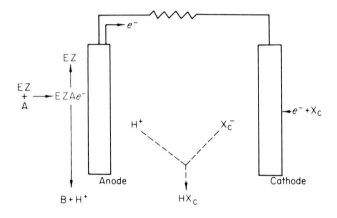

EZ   —   Enzyme
A   —   Substrate
EZAe⁻ —   Enzyme substrate (molecule) complex
B   —   Product
X⁻   —   Oxidant

FIG. 18.3. Schematic diagram of a direct biochemical conversion cell.

TABLE 18.III

TABLE OF A POSSIBLE DIRECT BIOCHEMICAL FUEL CELL

System	Catalyst	Volts vs S.H.E.[a]
Xanthine	Xanthine oxidase (Enzyme)	−0.58
Acetaldehyde/ethyl alcohol	Alcohol dehydrogenase (Enzyme)	−0.20
Pyruvate/lactate	Lactate dehydrogenase (Enzyme)	−0.18
Methane/oxidized forms	*Pseudomonas methanica* (Microorganism)	−0.50
Sulfides/sulfates	*Thiobacillus denitrificans* (Microorganism)	?

[a] S.H.E. = standard hydrogen electrode.

Examples of direct biochemical fuel cells are presented in Table 18.III, while a schematic diagram of a typical indirect fuel cell is shown in Fig. 18.2. A simple and excellent laboratory design of a biofuel cell has been described by Fischer *et al.* (1965). A bacterial methane fuel cell has been utilized by van Hees (1965) and is shown in Fig. 18.4.

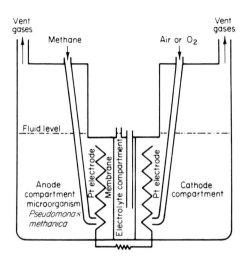

FIG. 18.4. Schematic diagram of a direct methane-oxygen biochemical fuel cell.

There is also a third type of microbial electrical energy production which is intermediate between the direct and indirect systems (Brake *et al.*, 1965). While the mechanism is not well defined it involves an effect of the bacteria on the electrode process which leads to increased current densities and power output through depolarization of the electrodes, especially if the electrodes are mild steel.

An example of this type of process is the effect of *Desulfovibrio desulfuricans* on the hydrogen evolution process. In general the organism reduces sulfates to sulfides with hydrogen by the reaction

$$SO_4^{2-} + 4\,H_2 \rightarrow S^{2-} + 4\,H_2O$$

By attaching themselves and providing a protective coating on the mild steel electrodes the organisms remove absorbed hydrogen reducing the hydrogen over-potential. In removing the absorbed hydrogen, the rate-determining step, namely desorption, is accelerated.

Possible Types of Biochemical Systems

There are three main biological systems which could be utilized in a fuel cell. These are bacteria, fungi, and algae; the use of pure enzymes falls into a special category. Of particular interest is the algae which utilize radiant energy from the sun to convert carbon dioxide and water into carbohydrates or sugar. Such carbohydrates could serve as a substrate for a variety of microorganisms capable of operating in a fuel cell. This process is especially attractive because the energy produced from the fuel cell stems ultimately from sunlight, one of the few truly free sources of energy. The physical and chemical requirements of a bio-fuel cell are simple; electrolytes, a membrane, electrodes, microbes, substrates, and equipment for measuring the amperage (volts) produced.

Types of Membranes

Aqueous membranes. These include buffered liquid electrolytes for use in a biochemical fuel cell. At present, however, the type of fuel cell utilizing this type of electrolyte seems to have little commercial possibility. Immobilized electrolytes, in which acid, alkaline, or salt electrolytes are held in gels such as finely divided silica, asbestos fibers, or agar, are plausible.

Solid Membranes. Included in this category are inorganic and organic membranes, both cationic and anionic. In general cationic membranes are favored over anionic membranes because of their higher conductivities.

Types of Electrodes

The types of electrodes used in biochemical fuel cells are many and vary in design and materials. In popular use at present are carbon, platinum foil, gauze wire, nickel wire mesh, platinized nickel mesh, and mild steel electrodes.

Fuel-Cell Thermodynamics

Carnot Cycle Efficiency

In the Carnot cycle heat flows from a higher temperature to a lower temperature with part of the heat being converted to work. Since heat is a random quantity and work is a directed process a price must be paid for transforming heat to work and this price is a limitation in the Carnot cycle. In a fuel cell the chemical energy produced within the cell is not randomized as is heat. Thermodynamically this may be expressed by

$$\Delta E = T \Delta S - (P \Delta V + W) \tag{1}$$

$$H = PV + E \tag{2}$$

$$G = H - TS \tag{3}$$

Substituting Eq. (2) into Eq. (3) and differentiating

$$\Delta G = \Delta E + P \Delta V + V \Delta P - T \Delta S - S \Delta T \tag{4}$$

Substituting Eq. (1) into Eq. (4)

$$\Delta G = T \Delta S - P \Delta V - W + P \Delta V + V \Delta P - T \Delta S - S \Delta T$$

$$\Delta G = W + V \Delta P - S \Delta T \tag{5}$$

At constant temperature (isothermal) and constant pressure

$$\Delta G = - W$$

Also

$$G = n\zeta\varepsilon$$

Therefore

$$G = W = -n\zeta\varepsilon \tag{6}$$

Defining terms, T = temperature (°K); P = pressure; V = volume; G = Gibb's free energy; H = heat content; ζ = Faraday constant; ε = open circuit voltage; W = electrical energy; E = energy of the system; n — number of electrons in the oxidation reduction of one molecule or one atom.

From Eq. (6) it can be seen that all the useful realizable work (ΔG) of the system may be utilized electrically if the system is isothermal and at constant pressure.

GENERAL THERMODYNAMICS OF A BIOCHEMICAL FUEL CELL

A biological system can only cause a change of electrode potential in a medium, as compared to a sterile medium, by changing the electro-chemical activity of the medium (Young et al., 1966). This happens upon the addition or removal of an electrochemically active component or both.

Thus there are three basic types of biochemical fuel cells within the direct and indirect classifications.

1. The depolarization cells where an electrochemical product is removed by the biological system is a type of direct fuel cell.
2. The product cell where the biological system produces an electro-chemically active product which is utilized in a conventional fuel cell is a type of indirect fuel cell.
3. Another type of indirect fuel cell is the Redox cell where a biological system converts electrochemical products back to reactants. An example is biological reduction of ferricyanide to ferrocyanide which then undergoes electrochemical oxidation to ferricyanide. This type of cell is simply a cyclic regenerating system.

In general the equation for the oxidation-reduction potential of a system consisting of a mixture of oxidized and reduced forms of a substrate is

$$\varepsilon = \varepsilon \cdot + \frac{RT}{n\zeta} \ln[C_{ox}/C_{red}] \tag{7}$$

where $\varepsilon \cdot$ = potential of system when activities of oxidized and reduced states are present in equal amounts; R = molar gas constant; ζ = Faraday's constant = 96,500 coulombs or 23,000 cal; C_{ox} = concentration (activity) of the oxidized form of the reagent; C_{red} = concentration (activity) of the reduced form of the reagent.

When $T = 25°C = 298°K$ and $R = 1.987$ cal/mole/°K, $\zeta = 23,000$ cal

$$\varepsilon = \varepsilon + \frac{0.060}{n} \log[C_{ox}/C_{red}] \tag{8}$$

By Eq. (6) the theoretical emf is related to G (Gibb's free energy) as follows:

$$\varepsilon = -\Delta G/n\zeta \tag{9}$$

And current is defined by

$$I = n\zeta/t \text{ (coulombs/second)}$$
$$= 26.7 \, N \text{ amp} \tag{10}$$

where N is equal to the number of electroactive equivalents reacting in one hour.

In general the kinetics of a biochemical reaction are similar to those of an enzyme-catalyzed reaction

$$v = -\frac{dC}{dt} = (V_{max} + C)/(K_m + C) \tag{11}$$

where v = reaction rate mole/liter/hour; V_{max} = maximum reaction rate; C = concentration of reactant mole/liter; K_m = Michaelis constant, numerically equal to the concentration of the reactant when $v = \frac{1}{2}V_{max}$.

Depolarization Cell

In the depolarization cell the biological system depolarizes one half-cell by removing (for example) oxygen formed at the anode while air or oxygen is supplied to the other half-cell (Young et al., 1966). The emf of such a cell results from a difference in electrode potentials due to the difference in activities of the two half-cells. The resulting emf can be derived from Eq. (8)

$$\varepsilon = (0.060/n) \log[C_{cath}/C_{anode}] \text{ (volts)} \tag{12}$$

where C_{anode} = concentration, activity, or partial pressure of the reactant at the anode and C_{cath} = concentration, activity, or partial pressure of the reactant at the cathode. From Eq. (12) it can be seen that for a high emf C_{anode} must be very small. C_{anode} in an oxygen concentration cell is maintained at low levels by oxygen utilization by aerobic microorganisms. Power output of such a cell is εI.

From Eqs. (9), (10), and (12)

$$\text{Power} = (0.06/n)\log[C_{cath}/C_{anode}]n\zeta(V_{max} C_{anode})/(K_m + C_{anode})$$
$$= 1.6V_{max}/(1 + K_m/C_{anode})\log[C_{cath}/C_{anode}] \tag{13}$$

From Eq. (13) it can be seen that in an oxygen concentration cell, for high power outputs, microorganisms with a large oxygen uptake (low C_{anode}) and low values of K_m are required.

Product Cell

The product cell operating at high substrate concentration and producing an electrochemically active substance at a maximal rate will give higher power densities and efficiencies than the concentration cell.

Of primary consideration in the design of a product cell is the physical state of any electrochemically active products produced. If the products are liquid, or soluble in the medium, then either the product plus medium or the separated product must be fed to a conventional fuel cell. The low concentrations of electrochemically active products in a medium makes a fuel cell, which has to run on these low concentrations, impractical, especially when the physical size of a practical unit is considered. A better method is to separate the product from the medium and then utilize it to operate a fuel cell. This is limited however by the difficulty of product separation. For this reason the majority of indirect cells employ microbiological systems which produce gaseous products such as methane, ammonia, hydrogen sulfide, or hydrogen. The biological production of electrochemically active substances is of such low efficiency that, for economic reasons, the systems employed would have to be of such a nature that the electrochemically active reagent is not the major product but instead a by-product. This would include hydrogen and methane produced by fermentation processes. Although methane and ammonia can be produced relatively cheaply by biological systems they can only be used with some difficulty, or at a high cost, in conventional fuel cells.

Redox Cell

In a Redox cell a chemically inert electrode is immersed in a solution containing both oxidized and reduced forms of an electrochemically active substance. The electrode potential depends on the ratio of oxidized to reduced forms, and if it is connected to another half-cell at a different potential the two half-cells will come to equilibrium at a common potential. A current will flow in the external circuit as the potential of one electrode becomes more positive, as electrons are given up during oxidation, and the potential of the other electrode becomes more negative, as it takes up electrons.

The equations which apply to the Redox cell are as follows.

$$\varepsilon = \varepsilon^\circ + (0.060/n)\log[C_{ox}/C_{red}]$$
$$\varepsilon = \varepsilon^\circ + (0.06/2.30n)\ln(C_{ox}/Ct - C_{ox}) \tag{14}$$

where Ct = total concentration of oxidized and reduced forms. The rate of reaction may be found by differentiating

$$\frac{\partial \varepsilon}{\partial t} = (0.06/2.30n)[(1/C) + \langle 1/(Ct - C)\rangle]\, \partial C/\partial t$$
$$= (0.06/2.30n)[Ct/C(Ct - C)](dC/dt)$$

and

$$\frac{dC}{dt} = (d\varepsilon/dt)(n \cdot Ct)[(2.30/0.06)(C/Ct)[(1 - C)/Ct]] \tag{15}$$

where C = concentration of oxidized form.

As current is drawn off the Redox cell, it is necessary to regenerate the redox medium by a chemical reaction so that the ratio of oxidized to reduced forms of the reactant is kept constant in order to keep the electrode potential constant. This is accomplished biologically.

CHELATION: MINERAL SOLUBILIZATION AND DEPOSITION

There are four well-known mechanisms which affect the solubilization and deposition of minerals. Some of these operate in a marine environment, and they all operate in the lithosphere.
  1. The oxidative state of the element. Metals with several valence states usually possess different degrees of solubility in aqueous solution in each state. In some instances the most oxidized condition of a metal is the least soluble, e.g., Fe^{3+} and Mn^{2+}, whereas, the most oxidized conditions for some polyvalent metals are the most soluble, e.g., U^{6+} and V^{5+}. Solubility of a particular ionic species also varies with the pH, where the pH affects the rate and degree of hydration.
  2. The base exchange (cation) reactions. In terrestrial systems where the solid phase is greatly increased over the aqueous phase many solids behave as if they contained insoluble polyvalent anions. These systems provide binding sites for cations, and an equilibrium is established between the cations present in ground water and those bound to the solid phase. Both clays and insoluble organic matter act in this capacity. As carbon dioxide is produced by microbes, hydrogen ions are generated and bound cations are released for movement in ground water.
  3. The anion exchange reactions. Clays such as halloysite and bentonite (Schell and Jordan, 1959) possess a lattice structure which has anion exchange properties. Anions tend to be more weakly bound in the lithosphere and are readily leached.
       Iron, manganese, vanadium, chromium, nickel, gallium, and zinc are transported in water as colloidal hydroxides or are adsorbed by different colloids. Copper, calcium, strontium, and barium migrate as solutions. Copper migrates in the form of copper sulfate whereas the latter three metals will be found in water high in carbonate. Factors affecting distribution of the elements in any ecological environment are pH, E_h, CO_2, O_2, and the concentration of organic matter.

4. The formation of coordination compounds or metal chelates. Coordination compounds synthesized by microbes can be located either inter- or extracellularly. If extracellular the metal may be immobilized by the formation of a coordination complex with an insoluble "organic complex," or it can be rapidly leached out of a particular formation because of the high solubility of the metal chelate in aqueous solutions.

Ion movement through terrestrial formations also depends on the specific cations and anions involved. For example, where Ca^{2+} and SO_4^{2-} ions exist in the same solution gypsum is usually deposited. Ion interaction and interference is a complex field and to add to this complexity microbes influence both solubilization and deposition of ions.

Mortensen (1963) has reviewed metals complexed by soil organic matter and Perlman (1965) gives an integrated presentation of the literature on the microbial synthesis of metallo-organic compounds.

MICROBIAL CHELATES

Chelation involves the combination of an organic molecule, or ligand, with a metal ion. The concept of chelation is an extension of Werner's work on inorganic coordination compounds. It can be described functionally in situations where an electron pair of a suitable donor group, of the type encountered in the nitrogen atom of ammonia, is made available on a conditional basis to a metal atom such as platinum. Under such an arrangement the platinum is able to fill an electron shell and to form a coordinate covalent bond. Many nitrogen-containing microbial chelators are linked by carbon to give a ring arrangement with the metal occupying a central position. A typical example would be the iron chelated in cytochrome. Metals chelated in this manner are often shielded or protected in a three-dimensional type of organic molecule. The stability of the chelate depends on many factors. The metal-combining capacity of the organic molecule which is usually higher for polyvalent ions than for di- or monovalent ions affects stability. The size of the ring system of the chelate is also important, with maximal stability encountered in five- and six- atom rings which form a three-dimensional organic ligand with the metal. The degree of chelation is pH dependent because hydrogen ions compete with the metal for the binding atom. In microbial systems the chelater can be either inter- or extracellular. If located within the cell, movement of the chelate may be restricted until released by cell lysis. Removal of the metal will then depend upon physical and chemical conditions, as well as a gradual degradation of the chelate by other microbes.

Another type of chelation is the binding of alkaline earth metals by polyhydroxy and polyuronic acids. There is a tendency to relate this primarily to pectins where calcium is chelated, however most microbes form capsular polysaccharides with similar binding capacities. The many metals found associated with marine algae are bound in this manner.

METAL-CONTAINING METABOLITES

Microbes synthesize numerous metal-binding compounds. These have been grouped as (Table 18.IV):

1. Amino acid chelates.
2. Polypeptide or protein chelates.
3. Metallophenolic complexes.
4. Metallophosphate complexes.
5. Organic acid chelates.
6. Polyhydroxy compounds.
7. Metalloporphyrins.

Glycine, histidine, and glutamic acid readily bind cobalt, while copper is bound by alanine and aspartic acids. Zinc cysteine has been prepared. Zinc should be weakly bound by most sulfur-containing amino acids. Of the more complex polypeptides or proteins, bacitracin chelates zinc, cobalt, manganese, and possibly nickel. Vallee (1956) has found zinc present in several alcohol dehydrogenases. Copper is complexed with diphenol oxidase, galactose oxidase, tyrosinase, and protein present in *Bordetella pertussis* (Amaral *et al.*, 1963; Kerterz and Zito, 1957; Hammersten *et al.*, 1959). Iron is bound in the enzyme pyrocatechase and the protein ferredoxin (Suda *et al.*, 1951; Mortenson *et al.*, 1962).

Iron is complexed with many phenolic compounds. It reacts readily with salicylic acid and pyrocatechin. Aluminum complexes with alizarin and many metals are chelated by pyrogallol.

Adenosine mono-, di-, and triphosphates not only contain phosphates but they bind such metals as manganese, zinc, and magnesium. Phytic acid which also contains phosphate chelates a host of bivalent metals (Vohra *et al.*, 1965). Organic acids chelate an impressive list of metals.

TABLE 18.IV

CHELATES SYNTHESIZED BY MICROBES

Type of chelate	Metal chelated	Pertinent references
Amino acid–metal chelates		
Triglycino-cobalt	Co^{2+}	—
Cobalt histidine	Co^{2+}	—
β-Alanine-Cu	Cu^{2+}	—
Cobaltous glutamate	Co^{2+}	—
Copper aspartic acid	Cu^{2+}	—
Zinc cysteine	Zn^{2+}	—
Polypeptide or protein metal chelates		
Bacitracin (antibiotic)	$Zn^{2+}, Co^{2+}, Mn^{2+}$	—
Alcohol dehydrogenase	Zn^{2+}	Vallee (1956)
Diphenol oxidase	Cu^{2+}	—
Galactose oxidase	Cu^{2+}	Amaral et al. (1963)
Pyrocatechase	Fe^{2+}	Suda et al. (1951)
Tyrosinase	Cu^{2+}	Kerterz and Zito (1957)
Ferredoxin	Fe	Mortenson et al. (1962)
Rubredoxin	Fe	Lovenberg and Sobel (1965)
Azurin (*Bordetella pertussis*)	Cu^{2+}	Hammersten et al. (1959); Sutherland and Wilkinson (1963)
Metallophenolic complexes		
Phenolics	Fe	—
Salicylate	Fe	—
Pyrocatechin	Fe	—
Alizarin	Al	—
Pyrogallol	Metals	—
Metallophosphate complexes		
Adenosine monophosphate	PO_4, metal	—
Adenosine diphosphate	PO_4, metal	—
Adenosine triphosphate	PO_4	—
Phytic acid	$PO_4, Na^+, Cu^{2+}, Zn^{2+}, Ni^{2+}, Co^{2+}, Mn^{2+}, Fe^{3+}, Ca^{2+}$	Vohra et al. (1965)
Aliphatic acid	Ca^{2+}	—
Ketosuccinate derivatives	Ca^{2+}	—

(Continued)

TABLE 18.IV—*(Continued)*

Type of chelate	Metal chelated	Pertinent references
2-Ketogluconic acid	Ca^{2+}	—
Gluconate ion	Fe^{3+}	—
Citrate	Metals	—
Tartrate	Metals	—
Ascorbate	Metals	—
Polyhydroxy compounds		
Polyalcohol, ammonia	Cu	—
Polyalcohols	B, Ca	—
Metalloporphyrins		
Chlorophyll a	Mg^{2+}	—
Cytochrome *c*	Fe	—
Phthalocyanine complex	Metals	—
Porphino complex	Cu	—

2-Keto gluconic acid chelates Ca^{2+}, whereas gluconic acid reacts with Fe^{3+}. Citrate, tartrate, ascorbate, and some long-chain aliphatic acids bind a number of bivalent metals.

Calcium was liberated from the following insoluble salts and minerals and chelated to the 2-ketogluconate produced by microbes: calcium carbonate, calcium citrate, calcium oxalate, calcium di- and triphosphate, calcium succinate, calcium magnesium phytate, wollastonite ($CaSiO_3$), chabazite ($CaAl_2Si_4O_{12} \cdot 6 H_2O$), tobernite ($Ca_4Si_5O_{14} \cdot n H_2O$), xonotlite ($Ca_6Si_6O_{18} \cdot H_2O$), hydroxyapatite [$Ca_{10}(OH)_2(PO_4)_6$], and calcium saturated montmorillonite (Duff and Webley, 1959). With the microbes tested calcium sulfate, labradorite ($Ca_2NaAl_5Si_7O_{24}$), and heulandite ($CaAl_2Si_6O_{16} \cdot 5 H_2O$) were not affected.

Polyalcohols chelate copper, calcium, and boron. Marine algae, such as *Fucus vesiculosus* and *Laminaria saccharina* contain 15.5 and 8.9 mg of boron/gm dry weight, respectively. The ability of algae to grow in the presence of large quantities of boron has been associated with their high content of polyhydric alcohols which binds boron. The polyalcohol mannitol has been reported in quantities of 2.3–25.7%. Boron made up to 26.6–42.4% of the ash of marine algae of *Alaria tenuifolia, Egregia menziesii, Laminaria cunefolia, Postelsia* sp., and *Rhodymenia* (Igelsrud *et al.*, 1938). The concentration of boron increases in marine shells and coral, making up 173 mg/kg of ashed sample of *Hydrocorallina*. Boron

TABLE 18.V

METAL-CONTAINING METABOLITES SYNTHESIZED BY MICROBES

Microbe	Chelate or organometallic compound	Reference
I. Metabolites		
Scopulariopsis brevicaulis	Trimethylarsine	Gause (1955)
	Methyldiethylarsine	Bickel et al. (1960b)
	Dimethyl-n-propylarsine	Bickel et al. (1960b)
	Dimethylallylarsine	Bickel et al. (1960b)
	Methylethyl-n-propylarsine	Tsukiura et al. (1964)
	Dimethyl telluride	Bird and Challenger (1939)
Saccharomyces cerevisiae	Selenomethionine	
Microbes	Cyanocobalamin (Co)	Blau (1961)
	Vitamin B$_{12}$ (Co)	
	α-(5,6 Dimethyl benzimidazole) Cobamide-Co-cyanide	
Streptomyces verticillus	Phleomycin (Cu)	Rickes et al. (1948)
		Maeda et al. (1956)
II. Antibiotics		
Actinomyces subtropicus	Albomycin (4.2% Fe)	Gause (1955)
Streptomyces albaduncus	Danomycin (3.1% Fe)	Tsukiura et al. (1964)
Streptomyces grisioflavus	Ferrimycin a (5.3% Fe)	Bickel et al. (1960a)
Streptomyces grisioflavus	Ferrimycin a$_2$ (5.3% Fe)	Bickel et al. (1960a)
Streptomyces griseus	Grisein (5.1% Fe)	Reynolds et al. (1947)
Streptomyces spp.	Lepetit (~1.7% Fe)	Sensi and Timbal (1959)
Streptomyces olivochromogenses	Succinimycin (~4.5% Fe)	Haskell et al. (1963)

III. Growth factors

Sarcina lutea and other microbes	Coprogen (6.6% Fe)	Hesseltine et al. (1952)
Aspergillus niger	Ferrichrome (7.4% Fe)	Neilands (1952)
Aspergillus melleus	Ferrichrysin (7.0% Fe)	Zähner et al. (1962)
Aspergillus fumigatus	Ferricrocin (7.4% Fe)	Zähner et al. (1962)
Microbes (primarily Streptomycetes)	Ferrioxamine a	Bickel et al. (1960a)
	Ferrioxamme b	Bickel et al. (1960a)
	Ferrioxamine c	Bickel et al. (1960a)
	Ferrioxamine d_1	Bickel et al. (1960a)
	Ferrioxamine d_2	Bickel et al. (1960a)
	Ferrioxamine e	Zähner et al. (1962)
	Ferrioxamine g	Keller-Schierlein and Prelog (1962)
Aspergillus nidulans	Ferrirhodin (6.2% Fe)	Zähner et al. (1962)
Paecilomyces varioti	Ferrirubin (5.3% Fe)	Zähner et al. (1962)
Mycobacteria	Mycobactins (6.0% Fe)	van der Walt (1952)

IV. Pigments

Ustilago sphaerogena	Ferrichrome a (5.3% Fe)	Garibaldi and Neilands (1955)
Streptomyces sp.	Ferroverdin (9.3% Fe)	Chain et al. (1955)
Candida pulcherima	Pulcherrimin (12.4% Fe)	Kluyver et al. (1953)

increases in calcareous and sedimentary structures of oceanic origin. Boron in the Pacific Ocean varies from 0.125–0.485 mg/kg of water; the amount present in seawater far exceeds that in soils used in agriculture.

Ammonia formed by deamination of amino acids, nitrate reduction, etc., actively binds copper and possibly nickel and zinc.

Metallo-porphyrins are synthesized by almost all microbes. Chlorophyll a found in algae contains magnesium, whereas cytochromes b, c, and a in bacteria and fungi bind iron. The cyanin component of phthalocyanine makes it an active chelater of iron and similar metals. The porphino-complex itself also chelates copper.

Scopulariopsis brevicaulis synthesizes several arsines (Table 18.V): trimethylarsine (Gause, 1955), methyldiethylarsine, dimethyl-*n*-propylarsine, dimethylallylarsine (Bickel *et al.*, 1960b), and methylethyl-*n*-propylarsine (Tsukiura *et al.*, 1964). Dimethyl telluride is also synthesized by this same microbe (Bird and Challenger, 1939). *Saccharomyces cerevisiae* forms selenomethionine (Blau, 1961).

Numerous microbes produce varying quantities of vitamin B_{12} which contains cobalt. The antibiotics which have been purified, and especially those produced by *Streptomyces*, form complexes with iron. Terrimycin produced by *Streptomyces grisioflavus* contains 5.3% Fe (Bickel *et al.*, 1960a,b). Copper is complexed by phleomycin a product synthesized by *Streptomyces verticillus*. Other iron-containing antibiotics are albomycin, danomycin, grisein, lepitit, and succinomycin.

Many of the growth factors isolated from microbes also contain metals. Coprogen from *Sarcina lutea* has 6.6% Fe (Hesseltine *et al.*, 1952). Ferrichrome, ferrichrysin, and ferricrocin produced by different species of *Aspergillus* all contain over 7.0% iron (Neilands, 1952; Zähner *et al.*, 1963). An impressive list of ferrioxamines, e.g., ferroxamine a, b, c, d_1, d_2, e, f, and g are synthesized by cultures of streptomycetes (Bickel *et al.*, 1960a; Zähner *et al.*, 1962; Keller-Schierlein and Prelog, 1962). Ferrirhodin synthesized by *Aspergillus nidulans* contains 6.2% Fe, while the ferrirubrin of *Paecilomyces varioti* contains 5.3% Fe (Zähner *et al.*, 1963). The mycobactins also contain iron (van der Walt, 1952).

Microbial pigments probably chelate numerous metals. Ferrochrome a of *Ustilago sphaerogena*, ferroverdin of *Streptomyces* sp., and pulcherrimin of *Candida pulcherima* all contain iron (Garibaldi and Neilands, 1955; Chain *et al.*, 1955; Kluyver *et al.*, 1953). Pulcherrimin complexes 12.4% iron.

The chemist has also synthesized many chelates. An excellent compilation of synthetic organometallic compounds has been prepared by

Harwood (1963). The important organometallic compounds used industrially are listed in Table 18.VI. Ethylenediaminetetraacetic acid is used frequently by the biochemist. A compound binding a large number of metals is acetylacetonate. Many of these compounds are synthesized by microbes and in time will probably be identified as microbial products.

Chelation will prove of great value and will receive much investigation in biogeochemistry; and it should be helpful in explaining movement and deposition of minerals.

TABLE 18.VI

ORGANOMETALLIC COMPOUNDS OF INDUSTRIAL IMPORTANCE[a]

Synthetic chelates	Other metals chelated
Lithium octylacetoacetate	Na, K
Tetrasodium ethylenediaminetetraacetic acid	Fe, Ca, Pb, Cd, Th, Ir
Copper acetylacetonate	
Copper ethylenebis(aminoacetylacetone)	
Silver complexes of polyvinyl alcohol	
Beryllium acetylacetonate	Ti, Mg, Zn, Hg, Al, Tl, In, Pt, Os, Ir, Zr, Th, Ge, Va, Cr, Mo, U, Se, Mn, Re, Fe, Co, Ni, Rh
Calcium or strontium octylacetoacetate	
Dicalcium salt of ethylenedinitrioloacetic acid	
Zinc-hydroxyindanones	F
Cadmium alkylene polyaminoacetic acid	Zn
Zinc-N,N'-ethylenebis-(α-Co-hydroxyphenyl) Aminoacetic acid	
Mercury complexes-benzoin	Co, Ni
Mercury complex-propronylacetone	Co, Ni
Copper-8-hydroxyquinolinate	Sn, Fe, Al, Mn
Tin-tartaric acid	
Polyols: mannitol, catechol, pyrogallol	As, B
Chromous octylacetoacetate	
Molybdenum decylxanthate	
Cupferron	U
Manganese terephthaloyl diacetone	
Cobalt salicylaldehydroethylenediamine	
Pallidium benzidine	

[a] Compiled from Harwood (1963).

REFERENCES

Amaral, D., Bernstein, L., Morse, D., and Horecker, B. L. (1963). Galactose oxidase of *Polyporus circinatus;* a copper enzyme. *J. Biol. Chem.* **238**, 2281–2284.

Bickel, H., Bosshordt, R., Gaumann, E., Reusser, P., Vischer, E., Voser, W., Wettstein, A., and Zähner, H. (1960a). Metabolic products of Actinomycetaceae. XXVI. Isolation and properties of ferrioxamines A to F representing new sideramine compounds. *Helv. Chim. Acta* **43**, 2118–2128.

Bickel, H., Gaumann, E., Nussberger, N., Reusser, P., Vischer, E., Voser, W., Wettstein, A., and Zähner, H. (1960b). Metabolic products of actinomycetes. XXV. Isolation and characterization of ferrimycin A and A2, new antibiotics of the sideromycin group. *Helv. Chim. Acta* **43**, 2105–2118.

Bird, M. L., and Challenger, F. (1939). Formation of organometalloidal and similar compounds by microorganisms. *J. Chem. Soc.* pp. 163–168.

Blau, M. (1961). Biosynthesis of selenomethionine and selenocystine labeled with selenium-75. *Biochim. Biophys. Acta* **49**, 389–390.

Brake, J., Townsend, R., and Silverman, H. (1965). Electrical energy from microorganisms. *Chem. Eng. Progr.* **61**, 65–68.

Chain, E. B., Tonolo, A., and Carilli, A. (1955). Ferroverdin, a green pigment containing iron produced by a streptomycin. *Nature* **176**, 645.

Cohen, B. (1931). The bacterial culture as an electrical half-cell. *J. Bacteriol.* **21**, 18–19.

Cohn, E. (1962). Perspectives on biochemical electricity. *Develop. Ind. Microbiol.* **4**, 53–58.

Del Duca, M. G., Fuscoe, J. M., and Zurella, R. W. (1962). Direct and indirect bio-electrochemical energy conversion systems. *Develop. Ind. Microbiol.* **4**, 81–91.

Duff, R. B., and Webley, D. M. (1959). 2-Ketogluconic acid as a natural chelator produced by soil bacteria. *Chem. & Ind. (London)* pp. 1376–1377.

Fischer, D. J., Herner, A. E., Landes, A., Batlin, A., and Barger, J. W. (1965). Electrochemical observations in microbiological processes. *Biotechnol. Bioeng.* **7**, 471–490.

Garibaldi, J. A., and Neilands, J. B. (1955). Isolation and properties of ferrichrome. *J. Am. Chem. Soc.* **77**, 2429–2430.

Garnett, P. H. (1960). Growth of iron bacteria in crude tar. *Nature* **185**, 942–943.

Gause, G. F. (1955). Recent studies on albomycin, a new antibiotic. *Brit. Med. J.* **II**, 1177–1179.

Hammersten, E., Palmatierna, H., and Meyer, E. (1959). Separation of acid polyelectrolytes and proteins in cellular extracts. *J. Biochem. Microbiol. Technol. Eng.* **1**, 273–288.

Harwood, J. H. (1963). "Industrial Applications of the Organometallic Compounds." Reinhold, New York.

Haskell, T. H., Bunge, R. H., French, J. C., and Bartz, Q. R. (1963). Succinimycin, a new iron-containing antibiotic. *J. Antibiotics (Tokyo)* **A16**, 67–75.

Hesseltine, C. W., Pidacks, C., Whitehill, A. R., Bohonos, N., Hutchings, B. L., and Williams, J. H. (1952). Coprogen, a new growth factor for coprophilic fungi. *J. Am. Chem. Soc.* **74**, 1362.

Igelsrud, I., Thompson, T. G., and Zwicker, B. M. G. (1938). The boron content of sea water and of marine organisms. *Am. J. Sci.* **35**, 47–63.

Keller-Schierlein, W., and Prelog, V. (1962). Metabolic products from actinomycetes. XXIX. Structure of ferrioxamine D. *Helv. Chim. Acta* **44**, 709–713.

Kertesz, D., and Zito, R. (1957). Polyphenoloxidase (tyrosinase): Purification and molecular properties. *Nature* **179**, 1017–1018.

Kluyver, A. J., van der Walt, J. P., and Triet, A. J. (1953). Pulcherrimin, the pigment of *Candida pulcherrima*. *Proc. Natl. Acad. Sci. U.S.* **39**, 583–593.

Lamb, I. M. (1959). Lichens. *Sci. Am.* **201**, 144–156.

Leathen, W. W., Braley, S. A., and McIntrye, L. D. (1953). Role of bacteria in the formation of acid from certain sulfuritic constituents associated with bituminous coal. I. *Thiobacillus thiooxidans*. *Appl. Microbiol.* **1**, 61–64.

LeRoy, L. W., and Koksoy, M. (1962). The lichen—a possible plant medium for mineral exploration. *Econ. Geol.* **57**, 107–113.

Lovenberg, W., and Sobel, B. E. (1965). Rubredoxin: a new electron-transfer protein from *Clostridium pasteurianum*. *Proc. Natl. Acad. Sci. U.S.* **54** 193–199.

Lovenberg, W., Buchanan, B. B., and Rabinowitz, J. C. (1963). Studies on the chemical nature of Clostridial ferrodoxin. *J. Biol. Chem.* **238**, 3899–3913.

Maeda, K., Kosaka, H., Yagishita, K., Umezawa, H. (1956). A new antibiotic, phleomycin. *J. Antibiotics (Tokyo)* **A9**, 82–85.

Mortensen, J. L. (1963). Complexing of metals by soil organic matter. *Soil Sci. Soc. Am. Proc.* **27**, 179–186.

Mortenson, L. E., Valentine, R. C., and Carnahan, J. E. (1962). An electron transport factor from *Clostridium pasteurianum*. *Biochem. Biophys. Res. Commun.* **7**, 448–452.

Neilands, J. B. (1952). A crystalline organo-iron pigment from rust fungus (*Ustilago sphaerogena*). *J. Am. Chem. Soc.* **74**, 4846–4847.

Perlman, D. (1965). Microbial production of metal-organic compounds and complexes. *Advan. Appl. Microbiol.* **7**, 103–133.

Potter, M. C. (1912). Electrical effects accompanying the decomposition of organic compounds. *Proc. Roy. Soc.* **84**, 260–276.

Reynolds, D. M., Schatz, A., and Waksman, S. A. (1947). Grisein, a new antibiotic produced by a strain of *Streptomyces griseus*. *Proc. Soc. Exptl. Biol. Med.* **64**, 50–54.

Rickes, E. I., Brink, N. G., Koniuszy, F. R., Wood, T. R., and Folkers, K. (1948). Vitamin B_{12}, a cobalt complex. *Science* **108**, 134.

Schell, W. R., and Jordan, J. V. (1959). Anion exchange studies of pure clays. *Plant Soil* **10**, 303–318.

Sensi, P., and Timbal, M. T. (1959). Isolation of two antibiotics of the grisein and albomycin group. *Antibiot. Chemotherapy* **9**, 160–166.

Suda, M., Hashimoto, K., Matsuoka, H., and Kamahora, T. (1951). Further studies on pyrocatechase. *J. Biochem. (Tokyo)* **38**, 289–296.

Sutherland, I. W., and Wilkinson, J. F. (1963). Two cytochromes isolated from the bacterial genus *Bordetella*. *Biochem. J.* **30**, 105–112.

Traxler, R. W., Proteau, P. R., and Traxler, R. N. (1965). Action of microorganisms on bituminous materials. *Appl. Microbiol.* **13**, 838–841.

Tsukiura, H., Okhanishi, M., Ohmori, T., Koshiyama, H., Miyaki, T., Katazima, H., and Kawaguchi, H. (1964). Danomycin a new antibiotic. *J. Antibiotics (Tokyo)* **A17**, 39–47.

Vallee, B. L. (1956). Zinc and metalloenzymes. *Natl. Acad. Sci.—Natl. Res. Council, Publ.* **514**, 9–14.

van der Walt, J. P. (1952). On the yeast *Candida pulcherrima* and its pigment ('5 Graven-hage: Excelsior foto-offset).

van Hees, W. (1965). A bacterial methane fuel cell. *J. Electrochem. Sci.* **112**, 258–262.

Vohra, F., Gray, G. A., and Kratzer, F. H. (1965). Phytic and metal complexes. *Proc. Soc. Exptl. Biol. Med.* **120**, 447–449.

Young, T. G. Hadjipetrou, L., and Lilly, M. D. (1966). The theoretical aspects of bio-chemical fuel cells. *Biotechnol. Bioeng.* **8**, 581–593.

Zähner, H., Bachmann, E., Hütter, R., and Neusch, J. (1962). Sideramine, an iron-containing growth factor of microorganisms. *Pathol. Microbiol.* **25**, 708–726.

Zähner, H., Keller-Schierlein, W., Hütter, R., Hees-Leisinger, K., and Deer, A. (1963). Stoffwechselproducte von Mikroorganismen Sideramine aus Aspergillaceen. *Arch. Mikrobiol.* **45**, 119–135.

19

BIOGEOCHEMICAL OBSERVATIONS ON OTHER ELEMENTS

In the elemental world, only a little information on biochemistry is available. Even the most important metals, such as iron, aluminum, etc., have received superficial study. In the future, books should be required to cover the important biogeochemistry of many elements discussed in this section. Because of the difficulties in using biochemical grouping of the elements, the elements are discussed in relation to their occurrence in the periodic table.

RUBIDIUM AND CESIUM (FAMILY I)

Rubidium

An analysis of the field mushrooms *Tricholoma columbetta*, *Tricholoma acerbum*, *Tricholoma allum*, and *Cortinarius ciolaceus* showed that they all contained between 1.15–2.80 gm of rubidium/kg of dry sample (Bertrand and Bertrand, 1949a).

Fungi contain an average of 150 mg of rubidium/kg dry tissue (Bertrand and Bertrand, 1947). Phanerogams averaged 50 mg of rubidium and this increased to average of 120 mg in cryptograms. Species of *Fomes* and *Lenzites* contained less than 20 mg of rubidium/kg of dry matter. Fungi living on wood contain less rubidium than those found in soil high in organic matter. Basidiomycetes species belonging to *Cortinarius* and *Tricholoma* obtain most of their nutrients from soil organic matter, and some will concentrate rubidium up to 2.80 gm/kg of dry tissue (Bertrand and Bertrand, 1948). Species of *Rhodopaxillus* also concentrate large amounts of rubidium.

Cesium

Cesium was discovered in water from a Bavarian mineral spring by Bunsen and Kirchoff in 1860. It is the rarest of the alkali metals and is also considered a rare element. Cesium forms strong bases, and its salts are generally water soluble. It is the most electropositive of all metals and oxidizes readily. The distinct physical and chemical similarities of cesium and potassium appear to extend to certain physiological processes also. Because of this similarity, cesium is often reported in terms of the ratio of cesium to potassium.

Cesium is widely disseminated in the earth's crust; commercial-grade ores are found in Africa and the United States. Its most important mineral is pollucite which is a hydrated silicate of aluminum containing up to 36% cesium oxide. Granites average 1.0 ppm cesium, sedimentary rocks 4.0 ppm (Horstman, 1957).

Some coals contain 13 ppm cesium (Smales and Salmon, 1955). In igneous rocks, the potassium to cesium ratio is 1 : 7000, while soils average 0.3–26.0 ppm cesium (Bertrand and Bertrand, 1949b). The amount of cesium in seawater is estimated in parts per billion.

There are 21 known isotopes of cesium ranging from cesium-123 to cesium-144, skipping cesium-124. Cesium-134 has a half-life of 2.2 years. The stable isotope of cesium is the naturally occurring cesium-133. In nuclear fallout, cesium-137 and strontium-90 are the two most important radionuclides, with much more cesium-137 than strontium-90 produced. Both have high fission yields, long half-lives, relatively high energies of emitted radiation, and biochemical similarities to either potassium or calcium in physiological processes.

Unialgal cultures of Euglena and *Chlorella* grown in 1–10 μC of cesium-137/liter concentrated this metal in direct proportion to the amount added (Williams, 1960). The uptake of cesium ceases to be linear when the external concentration exceeds 0.04 meq. Potassium and rubidium ions interfere with cesium-137 uptake by *Chlorella*. Morgan and Meyers (1953) found the order of preference for algae for alkali metals was potassium, rubidium, cesium, and sodium. A similar preference is observed in yeast (Rothstein, 1955). *Chlorella* growing in a solution containing 1 meq. of potassium/liter accumulated 100 times the concentration of cesium-137 found in the medium (Williams and Swanson, 1958). Uptake of cesium is fast and a steady state is reached in 15 hours (Morgan and Meyers, 1953). Dead algae cells concentrate cesium from fresh water. After 8 days the "concentration" of cesium over the external medium in dead cells of *Chlorella* was 420. Potassium does not interfere with cesium uptake in dead cells.

The respiration of the marine algae *Porphyra perforata* washed with potassium-free water was stimulated by the addition of cesium, potassium, or rubidium chlorides. Cesium was the least effective growth stimulant (Eppley, 1960). A similar relationship exists in *Nannochloris* (Chipman, 1954). Cesium-137 uptake is increased in *Fucus, Laminaria,* and *Ulva* by the addition of phosphate and nitrate. Light is essential for cesium uptake by marine algae. This has been observed in *Nannochloris* (Chipman, 1954) and *Rhodymenia polmata* (Scott, 1954). The absence of carbon dioxide also stopped cesium uptake. Cesium absorbed by marine algae is firmly bound and virtually none is lost over a period of weeks when the culture is transferred to natural seawater.

TABLE 19.I

CONCENTRATION OF CESIUM-137 FROM SEAWATER

Cesium-137	Concentration factor over medium
Bacillariaceae	
Nitzschia closterium	1.2
Amphora sp.	1.5
Chlorophyceae	
Chlamydomonas sp.	1.3
Carteria sp.	1.3
Chlorella sp.	2.4
Pyraminonas sp.	2.6
Nannochloris atomus	3.1
Rhodophyceae	
Porphyridium cruentum	1.3

[a] After Boroughs *et al.* (1957).

Cesium cannot be substituted for potassium in metabolism and may even be toxic in the absence of potassium.

Phytoplanktonic algae accumulate many elements from seawater and are important in the removal of radionuclides. Radionuclides are accumulated in colloidal, particulate, and ionic states.

Boroughs *et al.* (1957) examined a wide variety of algae in the Bacillariaceae, Chlorophyceae, and Rhodophyceae to find that Cesium-137 was never concentrated over 3.1 times that present in the external medium (Table 19.I).

Cerium

Cerium-144 is concentrated 4500 times over the cerium content of the medium in algae, *Amphidinium klebsi* (Rice and Williams, 1959). Other algal cultures also concentrate high amounts of cerium (Table 19.II).

TABLE 19.II

CONCENTRATION OF CERIUM-144 FROM SEAWATER[a]

Algae	Concentration factor over medium
Carteria sp.	2400
Platymonas sp.	2100
Nitzschia closterium	2000
Thalassiosira sp.	3300
Amphidinium klebsi	4500
Porphyridium cruentum	3300

[a] After Rice and Williams (1959).

MAGNESIUM, BARIUM, AND MERCURY (FAMILY II)

Magnesium

This metal holds an important position in the biological world. Thus, many biological mechanisms exist for interconverting the magnesium found in the lithosphere and biosphere. Chlorophyll contains up to 0.2% magnesium, while magnesium is found in proteins, plasma, and in basic colloids in higher animals. It is a catalyst in phosphate transfer and is needed for phosphatase activity. The enzyme enolase is a magnesium-protein complex. Magnesium ions link thiamine phosphate with the protein component in yeast carboxylase. It activates isocitric dehydrogenase and is needed in many fungal fermentations, of which penicillin production is a good example. Magnesium induces spore formation and even fat synthesis in fungi. Its role in hexokinase reactions is well known, and magnesium salts accelerate pytalin, pancreatin, trypsin, erepsin, pancreatic lipase, and rennin activity. These are but a few of the biochemical reactions and compounds in which magnesium is involved or present. Many of these functions are vital to life itself. Without magnesium, chlorophyll would not function, and the continuous supply of oxygen to the atmosphere synthesized by plants would become limiting.

In many biological systems there is a delicate relationship between magnesium and calcium. An inbalance in this equilibrium can result in an antagonism of one of these ions for the other. Osterhaut (1909) observed antagonism between magnesium and sodium salts on spore formation in *Botritis cinerea*. Examples of antagonism in higher plants and animals are far more common. For a comprehensive review of the biological role of magnesium, see Pleskchitser (1955).

Lactobacillus lactis requires 100–200 mg of $MgSO_4$/liter of medium for growth (Norakowska-Waszczuk, 1965). Bivalent cations such as Mg^{2+} are required for oxidative phosphorylation by *Alcaligenes faecalis*; a specific polynucleotide was also necessary (Shibko and Pinchot, 1961).

MacLeod and Onofrey (1957) isolated marine species of *Flavobacterium* and pseudomonads, requiring halides, Mg^{2+}, Ca^{2+}, and Fe^{3+}. Bromide could replace the chloride requirement, but iodide inhibited growth. Magnesium was not required for all isolates, and Sr^{2+} could be added to reduce the Mg^{2+} requirements. In an examination of the refractive bodies in the fluid core of *Amoeba proteus*, Mg^{2+} and Ca^{2+} were found to be present in high concentrations (Heller and Kopac, 1956). Iron, copper, and zinc were absent.

Panels of tin and aluminum were prepared and placed in water from the Black Sea. Reeflike colonies formed. On panels of brass or bronze cratering was common and peripheral deposits of calcium and magnesium carbonate formed. Carbonate formation was attributed to *Leptothrix* and *Bacillus mycoides* (Kalinenko, 1959).

Barium

Barium is in the alkaline earth series of elements along with calcium, strontium, and radium. Radioactive barium is an abundant fission product. Eight isotopes of barium are known but most have a half-life of seconds or minutes. Barium-140 has a half-life of 12.8 days. Because of the short half-life of these isotopes and the relative unimportance of barium in biochemical reactions, the element has not received serious biological study. Its distribution in biological systems is comparable to that of strontium (Bauer, 1960). Yeast preferentially absorbs lanthanum-140 (Bower and Robinson, 1951), a daughter product resulting from the decay of barium-140.

Mercury

Mercury is one of the most toxic metals. It is a common constituent of fungicides and reacts nonspecifically with —SH groups. *Escherichia coli* contain about 10^8 SH groups/cell (Steel, 1960). In a comparison of gram-negative and gram-positive bacteria, the former are almost twice as resistant to mercury and copper penetration as the latter (Tröger, 1959).

YTTRIUM AND THE FAMILY III ELEMENTS

There is almost no biogeochemical information for most of the elements in this group. Spooner (1949) observed the uptake of yttrium-90 by sessile algae, and his experiments indicated that the algae act like ion-exchange resins. Boron is an important element in this group. It is found in many microbes and has been discussed in the section on chelation. No well-developed biochemical data are available on scandium, lanthanum, gallium, indium, thallium, and actinium.

HAFNIUM, ZIRCONIUM, AND LEAD (FAMILY IV)

Hafnium and Zirconium

Microbiological methods of concentrating and separating hafnium and zirconium are being investigated in several laboratories. Partial success has been achieved with some cultures of *Aerobacter aerogenes*.

The first alkyl derivatives of these metals were the cyclopentadienyls. Alkyl zirconiums are synthesized by reacting alkyl aluminums (halides) with zirconium tetrahalide. Organometallic compounds of hafnium or thorium have not been reported. Aryl zirconium compounds form with anthracene, stilbene, and phenanthrene. Thorium forms similar cyclopentadienyl compounds to those of zirconium $[(C_5H_5)_2ThBr_2]$ (Reynolds and Wilkinson, 1956).

In experiments with *Porphyra* sp., Foreman and Templeton (1955) have reported concentration factors between 200 and 470 for zirconium-95 over the surrounding medium. Fifty percent of this activity was lost in 6 days.

Lead

Arthrobacter was isolated by Ehrlich (1963) from the surface of galena, sphalerite, pyrrhotite, and realger, and *Hyphomicrobium* was detected in a sample of realger. Iron-oxidizing bacteria found in these preparations did not penetrate mineral sulfides but resided on the surface. Lead is normally quite toxic to microbes and most research has been restricted to toxicity studies. Although some titanium is present in microbes, reports on titanium, germanium, and tin are lacking.

PHOSPHORUS AND ARSENIC (FAMILY V)

Metallic phosphorus is too reactive and toxic in aqueous systems to be concentrated by microbes. Once oxidized to phosphate, it enters into many biological reactions.

Phosphates

Phosphate is a vital component in all life processes. Nucleic acids contain a high percentage of phosphate, and adenosine triphosphate is an energy carrier in all biological systems. It is also associated with phospholipids and a variety of carbohydrates. Some microbes form polyphosphates (Lohmann, 1958; Kuhl, 1960). In higher animals polyphosphates have been detected in liver nuclei (Penniall and Griffin, 1964) and mitochondria (Lynn and Brown, 1963). Polyphosphates may play a role in the transport of glucose into yeast cells (van Steveninck and Booij, 1964). Polyphosphates as associated or integral parts of the plasmic membrane probably take part in the enzymatic phosphorylating process preceding the glucose uptake in many cells. Ni^{2+} and UO_2^{2+} bind these polyphosphates and inhibit glucose transport.

Inorganic phosphate (metaphosphate) was identified in yeast by Liebermann as early as 1888. Wiame (1947) isolated a basophilic substance from yeast and identified it as inorganic polyphosphate.

Algae also form polyphosphates (Vagabov and Serenkov, 1963). *Scenedesmus obliquus* synthesizes low molecular weight polyphosphates in early stages of growth and high molecular weight polymers in later stages. *Dunaliella viridis* forms some polyphosphates early in growth, but these later disappear. Inorganic polyphosphate and metaphosphate accumulate in sulfur-starved cells of *Escherichia coli* and *Aerobacter aerogenes*.

The biochemistry of pyrophosphate has been reviewed by Hoffman-Ostenhof (1962) and of inorganic polyphosphate by Kuhl (1960), and, more recently, by Harold (1966). The chemistry of condensed inorganic phosphates is not simple. The term polyphosphates refers to all pentavalent phosphorus compounds in which various numbers of tetrahedral PO_4 are linked by oxygen bridges. Three classes are common:

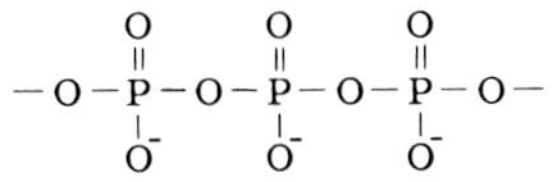

Tripolyphosphate

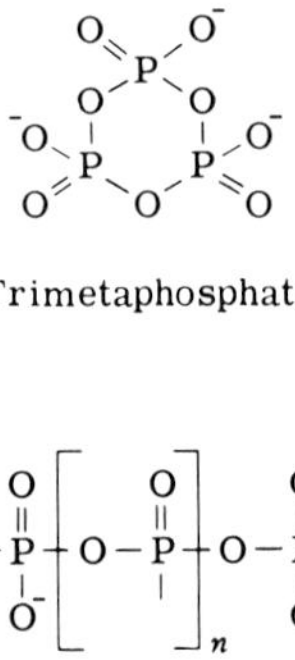

Trimetaphosphate

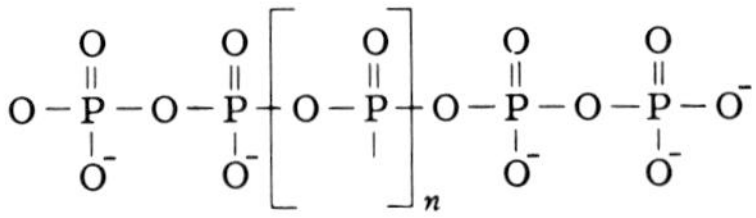

Polyphosphate of chain length $n + 4$

The cyclic condensed phosphates conform to an elementary formula of $M_nP_nO_{3n}$ and are designated metaphosphates. Only tri-, tetra-, and metaphosphates are well known. Unbranched linear condensed phosphates structures have a basic composition of $M_{n+2}P_nO_{3n+1}$ and are called polyphosphates. Cross-linked condensed phosphates (ultraphosphates) occur under conditions where the phosphate at the branching points share three oxygen atoms with an adjacent phosphate group. Branching points are readily hydrolyzed in water.

The chain length of polyphosphates varies with the microbe and its conditions of growth. In yeast, a chain length of $n = 300$ (Liss and Langen, 1960) and, in *A. aerogenes*, $n = 500$ (Harold, 1963) are not uncommon. Long-chain polyphosphates appear to be biosynthesized by a polyphosphate kinase (Yoshida and Yamataka, 1953; Kornberg *et al.*, 1956):

$$\text{Adenosine triphosphate} + P_{i(n)} \xrightleftharpoons[(Mg^{2+})]{} \text{Adenosine diphosphate} + P_{i(n+1)}$$

Polyphosphate fructokinase and polyphosphatase are known (O. Szymona and Szumilo, 1966; Mattenheimer, 1956). Both enzymes function to provide phosphate for reutilization in cellular metabolism.

Extracellular phosphate esters are hydrolyzed by alkaline phosphatase (Fig. 19.1) and provide inorganic phosphate for transport through cellular membranes (Menkina, 1950). Tricalcium phosphate as well as the inorganic phosphates are dissolved by *Bacillus subtilis, Bacillus magaterium, Aspergillus niger, Penicillium ardum, Sclerotium rolfsii*, and other common soil microbes (Kurylowciz and Kermanowa, 1957; Das, 1963; Kobus, 1962). The Soviets are reported to have inoculated soils with phosphobacterin which contains *Bacillus megaterium* var. *phosphaticum*. This bacterium acts to solubilize soil phosphate (Cooper, 1959; Sundara-Rao and Sinha, 1963). The inorganic phosphate may move in the ground water to areas where it may be deposited again as a calcium, magnesium, ammonium, or sodium phosphate. This precipitation is controlled by pH. Such deposits are comparable to the guano deposits of the Pacific

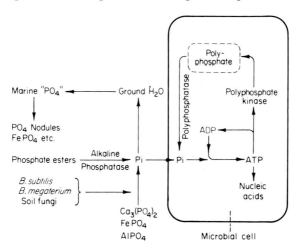

FIG. 19.1. Polyphosphate formation in microbes. (After Harold and Harold, 1965).

Islands, which consist chiefly of bird and animal excrement. Generally bone-beds, shells, teeth, etc., of animals are of minor importance as a source of sedimentary phosphate. In marine environments, phosphate is precipitated from colloidal suspension to form phosphate nodules. Large phosphorite deposits are found in England, Russia, Tunisia, and Algeria. Most phosphate as well as fluorine compounds are found in silicate rocks as the complex fluorophosphate apatite $Ca_{10}(PO_4)F_2$. Calcium phosphate or hydroxy apatite has an affinity for the fluoride ion. Phosphate is also found with apatite-iron ores in the Kiruna district in Northern Sweden and in sedimentary deposits of Jurassic oolite iron. Inorganic phosphate can also be concentrated by microbes in the form of polyphosphate.

Polyphosphates occur in bacteria, fungi, and algae. *Mucor racemosus* accumulates 5–6% phosphorus on a dry weight basis upon cultivation in excess KH_2PO_4; 58% of the phosphorus was in the form of inorganic phosphate and 17% in the nucleic acid fraction (James and Casida, 1964). Some yeasts may accumulate up to 20% of their dry weight as polyphosphate (Liss and Langen, 1962).

In some microbes inorganic polyphosphates, rather than adenosine triphosphate, are involved in phosphorylation (M. Szymona and Ostrowski, 1964; Dirheimer and Ebel, 1962). Hydrolysis of polyphosphates of chain lengths of $n = 8$, 13, 20, 120, and 200 was studied by van Steveninck (1966). Nickelous ions inhibited the hydrolysis of polyphosphates, while uranyl salts acted quite differently and increased the hydrolysis of polyphosphates but not pyrophosphate.

Much of the phosphate in lakes is removed from solution by microbes and falls out as sediment in lake muds, while some precipitates in the form of Fe (PO_4). If anaerobic conditions develop and sulfate is present, the iron becomes reduced and reacts with H_2S produced through sulfate reduction to yield FeS_2 (Sugawara *et al.*, 1957). Phosphates may be released by this process or precipitated as apatite if adequate calcium is present.

Arsenic

This element is widely distributed in nature and occurs with pyrite as FeAsS and other metal sulfides. It is recovered as a by-product from the smelting of copper and lead ores. Arsenic is highly concentrated in coals, and its concentration in atmospheric dust over cities ranges from 50 to 400 ppm. Its concentration ranges from 1–50 μg in seawater (Smales and Pate, 1952; Gorgy *et al.*, 1948; Ishibashi, 1954) though it does not change with depth. Seaweeds along the cost of Nova Scotia concentrate arsenic 200–600 times over the 0.002 ppm contained in seawater. Eleven species of algae contain from 5–94 ppm arsenic, while Phaeophyceae range from 30–70 ppm (Young and Langille, 1958).

Although arsenic is definitely poisonous at high levels, it is often used in medicine, e.g., in the treatment of nutritional disturbances, arthritis, asthma, skin diseases, etc. It is a good pesticide and is used for killing insects, bacteria, rodents, and plant life. Sodium arsenite is a most effective herbicide for submerged weeds. Estimated use in the United States exceeds 600,000 lb of arsenic annually.

The trivalent state of arsenic is generally the most toxic. Some arsenic compounds such as arsenilic acid are used as food additives for such domestic animals as poultry, sheep, and swine (Ishibashi, 1954; Libby and Schaible, 1955; Hanson et al., 1956). Since arsenic is in the same group of elements as phosphorus, its concentration by plants and animals and its wide range of biological effects should not be regarded as unusual.

Fifteen strains of bacteria which oxidize arsenite were present in arsenical cattle-dipping fluids used to control biting insects, etc. Species of *Pseudomonas*, *Xanthomonas*, and *Arthrobacter* were identified (Turner, 1954). A maximal arsenite oxidizing capacity was observed in *Pseudomonas arsenoxydans*, and a cell-free arsenite dehydrogenase was isolated (Legge and Turner, 1954).

A culture of *Pseudomonas*, obtained from cattle dip containing arsenic, catalyzed the oxidation of arsenite to arsenate (Turner, 1949). An adaptive dehydrogenase linked with cytochrome system is proposed.

$$H_3AsO_3 + H_2O \rightarrow H_3AsO_4 + 2 H^+ + 2 \; e^-$$

No evidence was shown that the energy produced by this reaction was utilized for growth.

The other elements in Group V are antimony, bismuth, niobium, and tantalum, but very little biochemical data are available on these metals.

SELENIUM, MOLYBDENUM, CHROMIUM, AND TUNGSTEN (FAMILY VI)

These Group VI metals are involved in several biochemical reactions. Selenium has properties similar to sulfur, while molybdenum increases nitrogen fixation. Both chromium and tungsten are concentrated by certain microbes.

Selenium

Selenium is required for growth of certain *Astragalus* species (Trelease *et al.*, 1960). In *Astragalus bisulcatus*, most of the selenium exists as methyl-selenocysteine. These plants accumulate rather large amounts of selenium. Lipman and Waksman (1923) reported the presence of an autotrophic bacterium which oxidized selenium to selenic acid which is comparable to the oxidation of sulfur to sulfuric acid by *Thiobacillus thiooxidans*. Selenium is similar to sulfur, which also both occurs in Family VI of the periodic table.

This element may not be as toxic as it was formerly believed to be. It was recently observed that cells of *Chlorella vulgaris* concentrate selenium both from selenate and selenomethionine (Shrift, 1961). Initially cell division is inhibited but adaptation is rapid. The enzyme ATP sulfurylase obtained from yeast which catalyzes the formation of adenosine-5'-phosphosulfate (APS) and pyrophosphate from ATP and sulfate will also function when selenate is substituted for sulfate. The adenosine-5'-phosphoselenate was less stable than APS. Selenium has also been substituted in other enzymatic reactions involving sulfur.

Arsenic, selenium, and tellurium have similar chemical properties which should give rise to biochemical similarities in the metabolic processing of these elements.

Molybdenum

Small amounts of Mo^{2+} (0.002 mg/100 ml medium) are added to increase nitrogen fixation in *Azotobacter*. The ammonium moiety of ammonium molybdate also undergoes nitrification. At concentration of 10^{-9} M nitride utilization is increased elevenfold in *Nitrobacter*. Calculations show that 2000 atoms of Mo^{2+} are required for the synthesis of one cell of *Nitrobacter* (Finstein and Delwiche, 1965).

Bryner and Anderson (1957) has shown that *Thiobacillus ferrooxidans* oxidizes molybdenite.

$$2 \, MoS_2 + 9 \, O_2 + 6 \, H_2O \rightarrow 2 \, H_2MoO_4 + 4 \, H_2SO_4$$

Leaching in the presence of microbes was sevenfold over that obtained with sterile water, with molybdic and sulfuric acids being formed.

The uptake of molybdenum in plants is greater than the uptake of copper and tungsten. Thus, the analysis of vegetation for Mo^{2+} may have value in biogeochemical prospecting. Analyses on needles and second-year stems of alpine fir, lodgepole pine, and creosote brush showed 0.1–0.5 ppm of Mo^{2+} in areas where no molybdenum mineralization was observed. Where a modest amount of mineralization occurs, this increased to 2–5 ppm and in mining areas, to 40–60 ppm (Warren et al., 1953).

Chromium

Much chromium occurs in the mineral chromite. Its formula has been given as $FeO \cdot Cr_2O_3$, but the ferrous oxide can be replaced by magnesium oxide and the chromic oxide by aluminum oxide or ferric oxide to give the formula (Mg,Fe) $(Cr,Al,Fe)_2O_4$. Although chromium was first discovered in 1797 by a French chemist, L. N. Vauquelin, in the mineral crocoite, a supply was not discovered in America until 1827. Major chromium-producing countries are Turkey, Africa, U.S.S.R., the Philippines, Rhodesia, Nyasaland, and the U.S.A. (Stillwater complex in Montana). Chromite occurs in such ultrabasic intrusive igneous rocks as periodotite, and in some serpentines.

Municipal water supplies contain about 0.04 ppm chromium and seawater contains much less (5×10^{-5} ppm), with the primary form being CrO_4^{2+}.

Chromium metal is bluish-white and forms cubic crystals. Its properties are similar to those of molybdenum, tungsten, and uranium. In nature it is found in insoluble oxides. It has valence states of Cr^{2+}, Cr^{3+}, Cr^{5+}, and Cr^{6+} which should make it amenable to oxidation by microbes. Hexavalent chromium is easily reduced to the trivalent form by heat, organic matter, or reducing agents. Chloride, nitrate, and sulfate salts of chromium are soluble in water, but the hydroxide and carbonates are not. The major use of chromium is in ferrous alloys to increase resistance to corrosion.

In biological systems the trivalent state of chromium is regarded as less toxic than the hexavalent state. In a study completed by Davis et al. (1958), on organisms of the Columbia river, algae (Table 19.III) were the highest in chromium, concentrating this ion 4000 times over the river water. Minnows and crayfish were lowest, concentrating chromium 200 times.

In all biological systems examined, chromium is found in trace quantities, and apparently has no essential physiological role. Chromium is toxic when encountered in high concentrations to both plants and animals. The hexavalent form is regarded to be at least 100 times more toxic than the trivalent form. Chromium-51 is the only isotope with special environmental importance because it is an effluent waste from the Hanford reactor and other reactors which are being actively installed throughout the world.

TABLE 19.III

RELATIVE AMOUNTS OF CHROMIUM-51 IN COLUMBIA RIVER ORGANISMS[a]

Biological sample	Concentration of chromium-51 $(10^{-4} (\mu C/gm))$	Concentration factor
River water (filtered)	0.02	—
Algae		
Stigeoclonium sp.	80	4000
Sponge		
Spongilla sp.	50	2000
Insect larvae		
Hydropsyche sp.	60	3000
Snail (shell excluded)		
Stagnicola sp.	50	2000
Minnow		
Richardsonius sp.	4	200
Crayfish		
Astacus sp.	5	200

[a] After Davis *et al.* (1958).

Tungsten

This element was discovered by C. W. Scheele in 1781 and isolated by J. J. de Elhurjar and F. de Elhurjar in 1783. Tungsten occurs in minerals of scheelite ($CaWO_4$), wolframite [$(Fe,Mn)WO_4$], and stolzite ($PbWO_4$). Deposits are found in China, Malaya, Mexico, Alaska, South America, and Portugal, and ores contain from 0.4–1.0% tungsten trioxide. It has a melting point of 3370°C which is higher than any known metal; at normal temperatures, it is hard and brittle. Important soluble salts are sodium and potassium tungstates.

Soils contain 10–2500 ppm tungsten. For growth of pasture plants, tungsten can replace molybdenum. There appears to be a relationship between tungsten and molybdenum in bacterial nitrogen fixation. Tungsten competitively inhibits the normal function of molybdenum in *Azotobacter* (Keeler and Varner, 1957; Takahashi and Nason, 1957) and may be incorporated into tungstoproteins by this microorganism. Tungsten inhibition of nitrification is greatly influenced by pH. Thus, at a pH of 7.7 tungsten slightly inhibited nitrification but stimulated activity at 8.8. The reverse effect was observed for sodium molybdate at these pH levels. With a mixture of tungsten and molybdenum, nitrification was stimulated at pH 8.1 (Zavarzin, 1958). There is also an antagonism of molybdenum uptake in *Aspergillus niger* by tungsten (Higgins, 1956).

Metabolic studies are easily completed with tungsten-185 which has a half-life of 73.2 days.

FAMILY VII

Iodine is accumulated by algae and is required in hormone synthesis by higher animals. Biologically and medically important, iodine has been used in the treatment of goiter for over one hundred years. Fluorine has protective value in prohibiting tooth decay, but its function in microbial metabolism is unknown as is the function of bromine and rhenium.

Iodine

Iodine is one of 64 trace elements which comprise 0.4% of the earth's mass. In the lithosphere it is present in 0.3 ppm, ranking with silver, platinum, iridium, and gold. It was first isolated from the ash of seaweed by Courtois (1813). Iodine is ubiquitous; however, the oceans contain far more iodine than the land. The iodine in the oceans originated from erosion of land masses. It is probably present in an ionic state although its form in ocean waters is still conjectural. Seawater averages 50 μg of iodine/liter (Vinogradov, 1953). Marine microbes concentrate iodine in amounts exceeding $10^{-3}\%$ based on dry weight. Iodine is found in air, probably as an evaporate from the oceans.

Some of the brown algae *Phaeophyceae*, especially species of *Laminaria*, contain up to 0.2% iodine in the living plant. The brown algae have been used as a commercial source of iodine. In 1930, Dixit reported 0.09% iodine in an air-dried sample of *Asparagopsis* from Indian waters. Red algae are not good sources of iodine.

The availability of iodine-131 has made it possible to follow the microbial metabolism of this element. In *Laminaria flexicaulis* and *Laminaria saccharina*, over 80% of the accumulated iodine-131 remained as inorganic iodide. The biosynthesis of mono- and iodotyrosine has been observed (Roche and Yagi, 1952). The red algae *Asparagopsis taxisformis* concentrated iodine-131 over 18,000 times that found in the surrounding water (Palumbo, 1955).

FAMILY VIII AND THE RARE EARTHS

Plutonium

Plutonium became the first of the man-made elements in 1942. Because of the superiority of plutonium-239 as nuclear fuel, it is destined to have an even greater role in human affairs.

Chemically it belongs to the actinide series of rare earth elements. It forms insoluble fluorides, hydroxides, and oxides. It exists in tri- and tetravalent cations or in the hexavalent form (PuO_2^{2+}). Soluble organic complexes are formed with citrate and with thenoyltrifluoracetone. Because of its ability to react with organic materials and its radioactive properties, it can be an extremely hazardous material. The danger from external decay is not as great as from internal sources since it decays by emission of alpha particles which have a very short range.

Ruthenium and Rhodium

Ruthenium and rhodium have received very little study in biological systems and appear to be absent in biological tissue analyzed by conventional techniques. Both ruthenium and rhodium occur in large amounts in nuclear explosions. Ruthenium is eliminated from biological systems rapidly, and this may be correlated with its many valence states, e.g., Ru^{2+}, Ru^{3+}, Ru^{4+}, Ru^{6+}, and Ru^{8+} which may facilitate removal.

Radioactive rhodium does not occur in high amounts in nature, and it is exceptional among the Group VIII metals of the periodic table in that it has a plus 3 valence state only.

Ruthenium-103, ruthenium-106, and rhodium-106 present a problem in effluent wastes from fuel reprocessing plants. Ruthenium is not as easily removed from wastes as strontium, cesium, and the rare earths, and, in a nuclear war its fallout would be a prime contributor to both the local and worldwide gamma irradiation problem.

Ruthenium tetroxide (RuO_4) starts to volatilize at 45°C which adds to its mobility in air.

Rare Earths

Rare earths include those elements in the periodic table of atomic number 57 through 71 (lanthanum to lutetium). Because of their similarities in chemical behavior, scandium and yttrium are also often included. In aqueous solutions the rare earths are normally encountered in the trivalent state.

Trace amounts of the rare earths are concentrated in the marine algae *Lithothamnium*. Elements observed were prasiodymium, neodymium, samarium, cerium, lanthanum, and yttrium (Vinogradov, 1953). Yttrium has been found in radiolarians, globigerina salts, and gorgonaceans. Samarium has been identified in two species of corals.

Of the various species of soil microbes, *Azotobacter* and *Pseudomonas* accumulate detectable amounts of radium, uranium, and thorium. Mycobacteria are less effective (Krasilnikov *et al.*, 1958). These microbes could be used as indicators of the state of radioactivity of soils.

Of the rare earths, strontium has been investigated most thoroughly.

Strontium Biogeochemistry

Strontium was identified by L. Dieulafait in 1877 in fossil brachiopods, an indication that strontium was present in substantial amounts in ancient oceans. Strontium and its nearest chemical relative, calcium, are both leached from the lithosphere and carried by rivers to the ocean in the form of bicarbonates.

The ionic radii of strontium and calcium are 1.27 and 1.06 Å, respectively. During the cooling of magmas, the calcium ions are more strongly bound than the larger strontium ions because of the dependence of electrostatic forces upon interionic distances. Potassium is close to strontium with a radius of 1.33 Å; however, it is univalent and there is a preferential capture of the elements with higher valency during early crystallization. Slow crystallization permits escape of strontium from magmas as compared to minerals which have undergone rapid cooling as in some volcanic rocks.

Potash minerals which crystallize early contain strontium as do calcium minerals which crystallize late. Polyhalite and anhydrite may contain 1000–7000 ppm of SrO. This is explained by the diadochic replacement principle. In biochemical deposits of calcium more strontium is found in aragonite than in calcite (Odum, 1957).

Both strontium and calcium are concentrated by biogenic activity in sediments and occur in the lithosphere through orogenic uplifts.

The two major minerals of strontium are strontianite ($SrCO_3$) and celestite ($SrSO_4$). Large deposits of these minerals are not common, and most strontium occurs as a minor constituent of other minerals. Some strontium is formed from the radiogenic decay of radioactive rubidium. The amounts produced in nature this way are small.

Biochemistry of Strontium

Odum (1957) found the concentration of strontium in organisms to be quite variable and highly dependent on amounts present in the external environment in his examination of hundreds of skeletons. Marine algae possessed a high Sr/Ca ratio of 3.96–10.8 reported as Sr/Ca/1000 atoms. This ratio decreased to 0.52 in freshwater algae. The protozoan *Acantharian radiolaria* concentrates celestite in its skeleton, and Sr/Ca ratios of 14,300 are not uncommon. The biochemistry of this deposition is not known. In comparing brown, green, and red algae (Table 19.IV), the brown algae possess a Sr/Ca ratio several times that of the green and red algae.

Rice (1956) has found a strontium uptake in *Carteria* proportional to the rate of cell division. In *Chlorella pyrenoidosa* strontium has been found at a concentration 10 times greater than calcium (Knauss and Porter, 1954). Webb (1937) found 4–5 times the Sr/Ca ratio in brown algae

TABLE 19.IV

STRONTIUM-CALCIUM RATIO IN MARINE ALGAE[a]

Algae	Sr/Ca (atoms/1000 atoms)
Brown algae	
Laminaria agardii	40
Ascophyllum nodosum	30
Sargassum (pelagic)	43
Fucus (Baltic)	26
Green algae	
Ulva lactuca	7
Red algae	
Chondrus crispus	9
Ceramium sp.	4

[a] After Odum (1957).

over that present in the medium. The ash of certain *Macrocystis* species may contain 2.0% strontium (Wilson and Fields, 1942).

Open-sea phytoplankton of *Gymnodinium simplex* and *Katodinium rotundata* were cultivated on media containing 5–100 μC of either strontium-90 or yttrium-90. All cells became radioactive, with the uptake of strontium proportional to the amount added. The cells concentrated radioactive strontium 18.6–382 times over that in the medium. Concentration factors for radioactive strontium of marine algae was 18.6 for *Gymnodinium simplex*, and 382 for *Katodinium rotundata*.

Cell growth was not inhibited in media containing 10 times the strontium present in seawater (Corcoran and Kimball, 1963; Rice, 1956). Strontium does not fulfill the requirement for calcium in algae.

The Sr/Ca ratio does not vary greatly between deposits precipitated directly and deposits, such as aragonite, where bacteria and algae are involved (Odum, 1957).

Deposits	Sr/Ca (atoms/1000/atoms)
Aragonite reef corals	9–12
Aragonite green algae	10–13
Unconsolidated and consolidated oolite	9
Drewite	9.3
Nonforaminifera rock fragments	9.5
Oolite (Great Salt Lake)	4.2

The low Sr/Ca ratio in oolite from the Great Salt Lake is similar to the levels observed in oolite from freshwater lakes. The higher levels of strontium in all the other deposits may be attributed to its concentration by calareous algae although this is a moot question. In ancient oolite deposits, Sr/Ca ratio of 9 atoms/1000 atoms suggest a paleoecological activity involving algal photosynthesis near the site of deposition.

Ground waters are normally high in strontium if the salt content is high. Strontium sulfate is more insoluble than calcium sulfate, and it actually will replace some gypsum. Celestite deposition requires only moderate levels of soluble or insoluble sulfate. Celestite forms in limestones along water courses and in glades, and is found associated with some lead and silver ores, as well as in the cap rocks of salt domes, and with sulfur in Texas (Brown, 1931).

Strontium may become enriched in percolating waters as it is depleted from carbonate sediments. During the recrystallization of aragonite to calcite, some strontium is released. Since the solubility of strontium carbonate in salt water is approximately 100 times less than strontium sulfate, strontianite should replace celestite and celestite should replace gypsum. Strontium becomes enriched in marine but not freshwater sediments (Katchenkov, 1952). Fertile soils vary from 0.013–0.266% strontium (Vinogradov et al., 1945). Soils in Florida which contain higher amounts of phosphates are lower in strontium, containing 0.001–0.1% (Rogers et al., 1939). This is associated with sedimentary replacement processes. The strontium content increases in phosphate mixed with coral rock.

The Sr/Ca ratio of sediments depends first on the concentration of these elements in the water and then on the degree of enrichment through algal and skeletal sediments. Marine sediments almost always have a higher Sr/Ca ration than open freshwaters. Highest Sr/Ca values are found for coral reefs or areas suspected of chemical precipitation. Littoral deposits vary greatly in their strontium content.

REFERENCES

Bauer, G. C. H. (1960). Radioisotopes in the skeleton. *Symp. Radioisotopes Biosphere,* *Univ. Minn.,* 1959 pp. 354–365.

Bertrand, G., and Bertrand, D. (1947). Studies on rubidium in cryptograms. *Ann. Agron.* **17**, 323–328.

Bertrand, G., and Bertrand, D. (1948). The relatively high rubidium content of certain fungi. *Compt. Rend.* **227**, 1128–1130.

Bertrand, G., and Bertrand, D. (1949a). High rubidium content of certain fungi. *Ann. Inst. Pasteur* **76**, 199–202.

Bertrand, G., and Bertrand, D. (1949b). Sur la presence et la teneur en cesium des terres arables. *Compt. Rend.* **229**, 533–535.

Boroughs, H., Chipman, W. A., and Rice, T. R. (1957). Laboratory experiments on the uptake, accumulation and loss of radionuclides by marine organisms. *Natl. Acad. Sci.—Natl. Acad. Res. Council, Publ.* **551**, 80–87.

Bowen, V. T., and Robinson, A. C. (1951). Uptake of Lanthanum by yeast. *Nature* **167**, 1032.

Brown, L. S. (1931). Cap-rock petrograph, *Bull. Am. Assoc. Petrol. Geologists.* **15**, 509–529.

Bryner, L. C., and Anderson, R. (1957). Microorganisms in leaching sulfide minerals. *Ind. Eng. Chem.* **49**, 1721–1724.

Chipman, W. A. (1954). Accumulation and loss of Ru-106, Sr-89 and Cs-137 by marine fish and shellfish. *U.S. Fish Wildlife Serv., Progr. Rept.* pp. 2–12.

Cooper, R. (1959). Bacterial fertilizers in the Soviet Union. *Soils Fertilizers* **22**, 327–333.

Corcoran, E. F., and Kimball, J. F., Jr. (1963). The uptake, accumulation and exchange of strontium-90 by open sea phytoplankton. *Proc. 1st Nat. Symp. Radioecol. Fort Collins, Col., 1961* pp. 187–191. Reinhold, New York.

Courtois, B. (1813). Découverte d'une substance nouvelle (iode) dans vareck. *Ann. Chim. (Phys.)* [1] **88**, 304–310.

Das, A. C. (1963). Utilization of insoluble phosphates by soil fungi. *J. Indian Soc. Soil Sci.* **11**, 203–207.

Davis, J. J., Perkins, R. W., Palmer, R. F., Hanson, W. C., and Cline, J. F. (1958). Radioactive materials in aquatic and terrestrial organisms exposed to reactor effluent water. *Proc. 2nd U.N. Intern. Conf. Peaceful Uses At. Energy, Geneva, 1958* Vol. 18, pp. 423–428. United Nations, New York.

Dieulafait, L. (1877). La strontiane, sa diffusion dans la nature minerale et dans la nature vivante, a l'epoque actuelle et dans la serie des temps geologiques. Consequences relatives aux eaux minerales saliferes. *Compt. Rend.* **84**, 1303.

Dirheimer, G. and Ebel, J. P. (1962). Mise en évidence d'une polyphosphate-glucose-phosphotransferase dans *Corynebacterium xerosis. Compt. Rend.* **254**, 2850–2852.

Dixit, S. C. (1930). A note on the percentage of iodine in certain algae. *J. Indian Chem. Soc.* **7**, 959.

Ehrlich, H. (1963). Microbial association with some mineral sulfides. *Biochem. Sulfur Isotopes., Proc. Symp., Yale Univ., 1963*, pp. 153–168. Yale Univ. Press. New Haven, Connecticut.

Eppley, R. W. (1960). Respiratory responses to cations in a red algae and their relations to ion transport. *Plant Physiol.* **35**, 637–644.

Finstein, M. S., and Delwiche, C. C. (1965). Molybdenum as a micronutrient for *Nitrobacter. J. Bacteriol.* **89**, 123–128.

Foreman, E. E., and Templeton, W. L. (1955). The uptake of Zr^{95} and Nb^{95} by *Porphyra* sp. *U.K. At. Energy Authority, Risley, IGR and DB* (**w**)–**TN–187**, 1–6.

Gorgy, S., Bakestraw, N. W., and Fox, D. L. (1948). Arsenic in the sea. *J. Marine Res.* **7**, 22–32.

Hanson, L. E., Hill, E. G., and Ferrin, E. F. (1956). The comparitive value of antibiotics and arsenic acids for growing pigs. *J. Animal Sci.* **15**, 280–287.

Harold, F. M. (1963). Inorganic polyphosphate of high molecular weight from *Aerobacter aerogenes. J. Bacteriol.* **86**, 885–887.

Harold, F. M. (1966). Inorganic polyphosphates in biology: structure, metabolism and functions. *Bacteriol. Rev.* **30**, 772–794.

Harold, F. M., and Harold, R. I. (1965). Degradation of inorganic polyphosphate in mutants of *Aerobacter aerogenes. J. Bacteriol.* **89**, 1262–1270.

Heller, I. M., and Kopac, M. J. (1956). Cytochemical reactions of normal and starving *Amoeba proteus*. II. Localization of minerals. *Exptl. Cell Res.* **11**, 206–248.

Higgins, E. S. (1956). Tungstate antagonism of molybdate in *Aspergillus niger. Proc. Soc. Exptl. Biol. Med.* **92**, 509–511.

Hoffman-Ostenhof, O. (1962). Some biological functions of the polyphosphates. *Colloq. Intern. Centre Natl. Rech. Sci. (Paris)* **106**, 640–650.

Horstman, E. L. (1957). The distribution of lithium, rubidium and cesium in igneous and sedimentary rocks. *Geochim. Cosmochim. Acta* **12**, 1–28.

Ishibashi, M. (1954). "Minute Elements in Seawater." Univ of Kyota, Japan.

James, A. W., and Casida, L. E., Jr. (1964). Accumulation of phosphorus compounds by *Mucor racemosus. J. Bacteriol.* **87**, 150–155.

Kalinenko, V. O. (1959). Bacterial colonies on metal panels in seawater. *Mikrobiologiya* **28**, 750–756.

Katchenkov, S. M. (1952). Characteristics of the lower Permian sediments with chemical distribution, determined by spectral analyses. *Compt. Rend. Acad. Sci. URSS* **82**, 961–994.

Keeler, R. F., and Varner, J. E. (1957). Tungstate as an antagonist of molybdate in *Azotobacter vinelandii. Arch. Biochem. Biophys.* **70**, 585–590.

Knauss, H. J., and Porter, J. W. (1954). The absorption of inorganic ions by *Chlorella pyrenoidosa. Plant Physiol.* **29**, 229–234.

Kobus, J. (1962). The distribution of microorganisms mobilizing phosphorus in different soils. *Acta Microbiol. Polon.* **11**, 255–264.

Kornberg, A., Kornberg, S. R., and Simms, E. S. (1956). Metaphosphate synthesis by an enzyme from *Escherichia coli. Biochim. Biophys. Acta* **26**, 215–227.

Krasilnikov, N. A., Drobkov, A. A., and Shirkov, O. G. (1958). Accumulation of naturally radioactive elements by soil microbes. *Dokl. Akad. Nauk. SSSR* **120**, 1136–1137.

Kuhl, A. (1960). Die Biologie der Bondensterten anorganischen Phosphate. *Ergeb. Biol.* **23**, 144–185.

Kurylowicz, B., and Kermanowa, J. (1957). Preliminary investigations on tests for soil bacteria making available to plants phosphorus from insoluble phosphates. *Zeszyty Nauk. Szkoly Glownej Gospodarst. Wiejskiego Warszawie, Rolnictwo* **1**, 75–84.

Legge, J. W., and Turner, A. W. (1954). Bacterial oxidation of arsenite. III. Cell-free arsenite dehydrogenase. *Australian J. Biol. Sci.* **7**, 496–503.

Libby, D. A., and Schaible, P. J. (1955). Observations on the growth responses to antibiotics and arsenic acids in poultry feeds. *Science* **121**, 733–734.

Lipman, J. G., and Waksman, S. A. (1923). The oxidation of selenium by a new group of autotrophic microorganisms. *Science* **57**, 60.

Liss, E., and Langen, P. (1960). Über ein hochmolekulares Polyphosphat der hefe. *Biochem. Z.* **333**, 193–201.

Liss, E., and Langen, P. (1962). Versuche zur Polyphosphat-Überkompensation in Hefezeellen nach phosphatverarmung. *Arch. Mikrobiol.* **41**, 383–392.

Lohmann, K. (1958). *In* " Kondensierte: Phosphate in Lebensmitteln," pp. 29–44. Springer, Berlin.

Lynn, W. S., and Brown, R. H. (1963). Synthesis of polyphosphate by rat liver mitochondria. *Biochem. Biophys. Res. Commun.* **11**, 367–371.

MacLeod, R. A., and Onofrey, E. (1957). Nutrition and metabolism of marine bacteria. VI. Quantitative requirements for halides, magnesium, calcium and iron. *Can. J. Microbiol.* **3**, 753–759.

Mattenheimer, J. (1956). Die Substratspezifität anorganischer Poly- und Metaphosphatasen. I. Optamale Werkungsbedingungen für den enzymateschen Abbau von Poly- und Metaphorphaten. *Z. Physiol. Chem.* **303**, 107–114.

Menkina, R. A. (1950). Bacteria which mineralize organic phosphorus compounds. *Mikrobiologiya* **10**, 308–316.

Morgan, L. O., and Meyers, J. (1953). " Biological Accumulation of Inorganic Materials by Algae," AEC Rept. Lab. Algae Physiol., Univ. Texas, Austin, Texas.

Norakowska-Waszczuk, A. (1965). The effect of certain cations on the growth of *Lactobacillus lactis* and *Bactobacillus delbrueckii.* I. The effect of magnesium. *Acta Microbiol. Polon.* **14**, 73–85.

Odum, H. T. (1957). Biogeochemical deposition of strontium. *Publ. Inst. Marine Sci.* **4**, 38–114.

Osterhout, W. (1909), The protective action of sodium for plants. *Jahrbu. Wiss. Botan.* **46**, 121–136.

Palumbo, R. F. (1955). Uptake of I^{131} by the red algae *Asparagopsis taxiformis*. *U.S. At. Energy Comm., Rept.* UWFL–**44**, 1–11.

Penniall, R., and Griffen, J. B. (1964). Studies of phosphorus metabolism by isolated nuclei. IV. Formation of polyphosphate. *Biochim. Biophys. Acta* **90**, 429–431.

Pleskchitser, A. I. (1955). Biologicheskaia Rol 'Magniia. *Usp. Sovreme. Biol.* **40**(1), 52–67.

Reynolds, L. T., and Wilkinson, G. (1956). π-Cyclopentadienyl compounds of uranium-IV and thorium-IV. *J. Inorg. Nucl. Chem.* **2**, 246–253.

Rice, T. R. (1956). The accumulation and exchange of strontium by marine planktonic algae. *Limnol. Oceanog.* **1**, 123–138.

Rice, T. R., and Williams, V. M. (1959). Uptake, accumulation and loss of radioactive cerium-144 by marine planktonic algae. *Limmol. Oceanog.* **4**, 277–290.

Roche, J., and Yagi, Y. (1952). Fixation of radioactive iodine by (marine) algae; iodinated constituents of *Laminaria*. *Compt. Rend. Soc. Biol.* **146**, 642–645.

Rogers, L. H., Gall, O. E., Gaddum, L. W., and Barnette, R. M. (1939). Distribution of macro and micro elements in some soils of peninsular Florida. *Florida, Univ., Agr. Expt. Sta. (Gainsville), Bull.* **341**, 1–31.

Rothstein, A. (1955). Relationship of the cell surface to electrolyte metabolism in yeast. *Symp. Marine Biol. Lab. Woods Hole, Massachussetts, 1954*, pp. 65–100.

Scott, R. (1954). A study of cesium by marine algae. *Proc. 2nd Radioisotope Conf., Oxford., Engl., 1954* Vol. 1, pp. 373–380. Butterworth, London and Washington, D.C.

Shibko, S., and Pinchot, G. B. (1961). Effects of magnesium and polyanions on oxidative phosphorylation in bacteria. *Arch. Biochem. Biophys.* **93**, 140–146.

Shrift, A. (1961). Biochemical interrelations between selenium and sulfur plants and microorganisms. *Federation Proc.* **20**, 695–702.

Smales, A. A., and Pate, B. D. (1952). The determination of submicrogram quantities of arsenic by radioactivation. Part III. The determination of arsenic in biological material *Analyst* **77**, 196–202.

Smales, A. A., and Salmon, L. (1955). Determination by radioactivation of small amounts of rubidium and cesium in seawater and related materials of geochemical interest. *Analyst* **80**, 37–50.

Spooner, G. M. (1949). Observations on the absorption of radioactive strontium and yttrium by marine algae. *J. Marine Biol. Assoc. U.K.* **28**, 587–625.

Steel, K. J. (1960). Reaction of mercuric chloride with bacterial SH groups. *J. Pharm. Pharmacol.* **12**, 59–61.

Sugawara, K., Koyama, T., and Kamata, E. (1957). Recovery of precipitated phosphate from lake muds as related to sulfate reduction. *J. Earth Sci., Nagoya Univ.* **5**, 60–67.

Sundara-Rao, W. V. B., and Sinha, M. K. (1963). Phosphate dissolving microorganisms in the soil and rhizosphere. *Indian J. Agr. Sci.* **33**, 272–278.

Szymona, M., and Ostrowski, W. (1964). Inorganic polyphosphate glucokinase of *Mycobacterium phlei*. *Biochim. Biophys. Acta* **85**, 283–295.

Szymona, O., and Szumilo, T. (1966). Adenosine triphosphate and inorganic polyphosphate fructokinases of *Mycobacterium phlei*. *Acta Biochim. Polon.* **17**, 129–144.

Takahashi, H., and Nason, A. (1957). Tungstate as a competitive inhibitor of molybdate in nitrate assimilation and in nitrogen fixation by *Azotobacter*. *Biochim. Biophys. Acta* **23**, 433–434.

Trelease, S. F., DiSomma, A. A., and Jacobs, A. L. (1960). Seleno-amino acids found in *Astragalus bisulcatus*. *Science* **132**, 618.

Tröger, R. (1959). Detection of metals in bacteria treated with mercury or copper. *Arch. Mikrobiol.* **33**, 186–190.

Turner, A. W. (1949). Bacterial oxidation of arsenite. *Nature* **164**, 76–77.

Turner, A. W. (1954). Bacterial oxidation of arsenite. I. Description of bacteria isolated from arsenical cattle-dipping fluids. *Australian J. Biol. Sci.* **7**, 452–478.

Vagabov, V. M., and Serenkov, G. P. (1963). Polyphosphates in two species of algae. *Vestn. Mosk. Univ., Ser. VI: Biol., Pochvoved.* **18**, 38–47.

van Steveninck, J. (1966). The influence of metal ions on the hydrolysis of polyphosphates. *Biochemistry* **5**, 1998–2002.

van Steveninck, J., and Booij, H. J. (1964). The role of polyphosphates in the transport mechanisms of glucose in yeast cells. *J. Gen. Physiol.* **48**, 43–60.

Vinogradov, A. P. (1953). "The Elementary Chemical Composition of Marine Organisms." Sears Found. Marine Res. Memoir II. Yale Univ., Press. New Haven, Connecticut.

Vinogradov, A. P., and Borovik-Romanova, T. F. (1945). On the geochemistry of strontium. *Compt. Rend. Acad. Sci. URSS* **46**, 193–196.

Warren, H. V., Delavault, R. E., and Routley, D. G. (1953). Preliminary studies of the biogeochemistry of molybdenum. *Trans. Roy. Soc. Can., Sect. IV* [3] **47**, 71–75.

Webb, D. A. (1937). Studies on the ultimate composition of biological material. Pt. II. Spectrographic analyses of marine invertebrates with special reference to the chemical composition of their environment. *Sci. Proc. Roy. Dublin Soc.* **21**, 505–539.

Wiame, J. M. (1947). The metachromatic reaction of hexametaphosphate. *J. Am. Chem. Soc.* **69**, 3146–3147.

Williams, L. G. (1960). Uptake of cesium-137 by cells and ditritus of *Euglena* and *Chlorella*. *Limnol. Oceanog.* **5**, 301–311.

Williams, L. G., and Swanson, H. D. (1958). Concentration of cesium-137 by algae. *Science* **127**, 187–188.

Wilson, S. H., and Fields, M. M. (1942). Studies in spectrographic analysis. II. Minor elements in a seaweed (*Macrocystic pyrifera*). *New Zealand J. Sci. Technol.* **D23**, 47–48.

Yoshida, A., and Yamataka, A. (1953). On the metaphosphate of the yeast. *J. Biochem.* **40**, 85–94.

Young, E. G., and Langille, W. M. (1958). The occurrence of inorganic elements in marine algae of the Atlantic provinces of Canada. *Can. J. Botany* **36**, 301–310.

Zavarzin, G. A. (1958). The excitation of the second stage of nitrification. II. The effect of heavy metals on nitrification. *Mikrobiologiya* **27**, 542–546.

AUTHOR INDEX

Numbers in italics refer to the pages on which the complete references are listed.

SUBJECT INDEX